Selected Titles in This Series

(*Continued in the back of this publication*)

Operator Theory for Complex and Hypercomplex Analysis

CONTEMPORARY
MATHEMATICS

212

Operator Theory for Complex and Hypercomplex Analysis

Operator Theory for Complex
and Hypercomplex Analysis
December 12–17, 1994
Mexico City, Mexico

E. Ramírez de Arellano
N. Salinas
M. V. Shapiro
N. L. Vasilevski
Editors

American Mathematical Society
Providence, Rhode Island

This volume contains the proceedings of an international conference that took place in Mexico City, December 12–17, 1994. The subject of the conference was chosen to highlight the interplay between operator theory and complex and hypercomplex analysis as well as to study some classes of operators appearing in them.

Support was provided by the Center for Research and Advanced Study (CINVESTAV-IPN), the School of Physics and Mathematics of the National Polytechnic Institute (ESFM-IPN), the Consejo Nacional de Ciencia y Tecnología (CONACYT-México), and the National Science Foundation (NSF-USA).

1991 *Mathematics Subject Classification.* Primary 47–02;
Secondary 47D30, 47G10, 30G35, 32–02.

Library of Congress Cataloging-in-Publication Data
Operator theory for complex and hypercomplex analysis : a conference on operator theory for complex and hypercomplex analysis, December 12–17, 1994, Mexico City, Mexico / E. Ramírez de Arellano... [et al.], editors.
    p. cm. — (Contemporary mathematics, ISSN 0271-4132 ; 212)
    Includes bibliographical references.
    ISBN 0-8218-0677-7 (softcover : alk. paper)
    1. Operator theory—Congresses.   2. Functions of complex variables—Congresses.   3. Mathematical analysis—Congresses.   I. Ramírez de Arellano, E. (Enrique)   II. Series: Contemporary mathematics (American Mathematical Society) ; v. 212.
QA329.O6365   1997
515'.724—dc21   97-28599
                                                                                       CIP

CONTENTS

PREFACE

Integral representations for different classes of functions in analysis motivate the introduction and study of a series of important operators: singular integral, Toeplitz, Bergman, convolution operators on Lie groups, some classes of pseudo-differential operators, etc. Investigation of these operators develops and enriches "pure" operator theory and exerts a stimulating influence on important areas of analysis.

This volume contains the proceedings of an international conference that took place in Mexico City, December 12–17, 1994. The subject of the conference was chosen to highlight the interplay between operator theory and complex and hyper-complex analysis as well as to study some classes of operators appearing in them.

The editors wish to thank the contributors to this volume and also the referees for their many helpful comments to the authors. They wish to apologize for the unfortunate delay of its publication but it was not easy to interact through the pitfalls of communication among different countries, computer systems, and various incompatible codes. The task of the editors was made not only possible but infinitely much easier by the cooperation and technical assistance of Mrs. Larisa Martin, who retyped some of the manuscripts in the appropriate form, formatted all the manuscripts according to the Contemporary Math. Journal specifications, and put together the final version of the volume for publication. The editors are indebted to her.

The conference was made possible thanks to the generous support of the Center for Research and Advanced Study (CINVESTAV-IPN) and of the School of Physics and Mathematics of the National Polytechnic Institute (ESFM-IPN). Special thanks is due to the Consejo Nacional de Ciencia y Tecnología (CONACYT-México) and to the National Science Foundation (NSF-USA) for their financial support.

The Editors

Contemporary Mathematics
Volume **212**, 1998

The Bergman Projection on Sectorial Domains

David E. Barrett

ABSTRACT. This paper provides a synthesis formula for the Bergman projection on domains with sector fibers over a base manifold, and shows how this synthesis formula can be combined with residue techniques to provide asymptotic expansions in some cases. An application is provided to the study of the $\bar{\partial}$-Neumann problem on pseudoconvex domains with complex curves in the boundary.

1. Introduction

In this paper we study the Bergman projection on domains Ω of the form

$$\{(z,w) \in X \times \mathbb{C} : \alpha(z) < \arg w < \beta(z)\},$$

where X is a complex manifold and α and β are, respectively, upper- and lower-semicontinuous functions on X with $\alpha < \beta \le \alpha + 2\pi$. (We will call Ω the *sectorial domain with base X and angle functions α, β.*)

We will show in Section 3 that the Bergman projection on a sectorial domain can be synthesized from weighted projections on the base. In some cases methods of the residue calculus can be applied to the synthesis formula to obtain asymptotic expansions near $w = 0$. Section 2 begins the discussion of a couple of families of sectorial domains for which this is possible. The asymptotic expansions themselves are developed in Sections 4, 5, 6, and 7.

Section 8 applies the results of Sections 5 and 6 to the study of the $\bar{\partial}$-Neumann problem on pseudoconvex domains with complex curves in the boundary.

2. Examples

EXAMPLE 2.1. Consider a domain $D \subset \mathbb{C}^{n+1}$ defined by an inequality of the form

$$\operatorname{Re} \zeta_{n+1} > \phi(\zeta_1, \ldots, \zeta_n),$$

where ϕ is a non-negative upper-semicontinuous function satisfying a homogeneity relation

$$(2.1) \qquad \phi(\lambda^{a_1}\zeta_1, \ldots, \lambda^{a_n}\zeta_n) = \lambda\phi(\zeta_1, \ldots, \zeta_n) \qquad \text{for } \lambda \in \mathbb{R}^+$$

1991 *Mathematics Subject Classification.* Primary 32H10; Secondary 32M99.
Supported in part by a grant from the National Science Foundation.

with $a_1, \ldots, a_n \in \mathbb{R}^+$.

(See [Sta] for information on how to recognize local biholomorphic copies of such domains, assuming real-analyticity.)

We transform D by means of the biholomorphic map

$$D \to \Omega$$

(2.2)
$$(\zeta_1, \ldots, \zeta_n, \zeta_{n+1})(z_1, \ldots, z_n, w) \overset{\text{def}}{=} (\zeta_{n+1}^{-a_1} \zeta_1, \ldots, \zeta_{n+1}^{-a_n} \zeta_n, \zeta_{n+1})$$

(In effect we have performed a "weighted blow-up" of the origin.)

The domain Ω is invariant under dilations $(z, w) \mapsto (z, \lambda w)$ in the w variable and hence its fibers over $\mathbb{C}^n$ are unions of sectors.

If, for example,

(2.3) ϕ is a function of $(|\zeta_1|, \ldots, |\zeta_n|)$

then Ω is in fact a sectorial domain with base

(2.4) $X = \{z : \phi(z) < 1\}$

and angle functions

(2.5) $\alpha(z) = -\arccos(\phi(z)), \quad \beta(z) = \arccos(\phi(z)).$

More generally, define an equivalence relation on Ω by setting $(z, w) \sim (\tilde{z}, \tilde{w})$ if and only if w and $\tilde{w}$ lie in the same sector of the fiber over $z = \tilde{z}$. Let X be the quotient space. Then it is easy to see that the natural projection $X \to \mathbb{C}^n$ is a local homeomorphism.

Assume now that Ω is pseudoconvex. Then $-\log(\text{dist}\,(\cdot, b\Omega))$ satisfies a maximum principle along the fibers so that fibers can not split as z varies. Thus X is a Hausdorff complex manifold spread over $\mathbb{C}^n$ and Ω is a sectorial domain over the base manifold X. (The angle functions α, β are single-valued since they satisfy $-(\pi/2) \leq \alpha < \beta \leq \pi/2$.)

Assume further that

(2.6) $\phi > 0$ on $\mathbb{C}^n \setminus \{0\}$.

Then X is bounded and $b\Omega$ consists of points $(z_1, \ldots, z_n, |w|e^{i\theta})$ such that $|\theta| \leq \pi/2$ and

$$\phi(e^{ia_1\theta} z_1, \ldots, e^{ia_n\theta} z_n) \geq \cos\theta \geq \lim_{s \to (e^{ia_1\theta} z_1, \ldots, e^{ia_n\theta} z_n)} \inf \phi(s).$$

If $\cos\theta < 1$ then clearly z is an interior point of X. Thus $\alpha(z), \beta(z) \to 0$ as $z \to bX$, or, as we recast this for later use,

(2.7)   for every $\epsilon > 0$ there is $K \subset X$ compact such that $|\alpha|, |\beta| < \epsilon$ on $X \setminus K$.

EXAMPLE 2.2. Consider a sectorial domain Ω whose fibers are half-planes; that is,

(2.8) $\beta = \alpha + \pi.$

Using the map $(z, w) \mapsto (z, -w)$ we see that the Levi-form at a point (z, w) in the boundary of Ω in $X \times \hat{\mathbb{C}}$ has the opposite sign of the Levi-form at $(z, -w)$; thus this portion of the boundary is pseudoconvex if and only if it is Levi-flat. In this situation, α and β are pluriharmonic and the leaves of the Levi-foliation are locally graph hypersurfaces of the form $w = Ce^{if(z)}$, where f is a holomorphic function with $\operatorname{Re} f = \alpha$ and $C \in \mathbb{R} \cup \{\infty\}$. (Here $\hat{\mathbb{C}}$ denotes the Riemann sphere $\mathbb{C} \cup \{\infty\}$.)

If

(2.9)
$$\overline{X} = X \cap bX \text{ is a compact bordered Riemann surface}$$
$$\text{in an ambient Riemann surface } Y$$

and

$$\alpha \text{ is smooth on } \overline{X} \text{ and harmonic on } X$$

then the boundary $b\Omega$ of Ω in $Y \times \hat{\mathbb{C}}$ consists of the portion discussed above together with a second Levi-flat piece

$$\{(z, w) \in bX \times \hat{\mathbb{C}} : \alpha(z) \le \arg w \le \beta(z)\}.$$

The two pieces intersect in a totally real edge, and $b\Omega$ is locally biholomorphic to the boundary of the bidisk Δ^2. Moreover, Ω is covered by Δ^2 via a map of the form

$$\Delta \times \Delta \cong \widetilde{X} \times \{w : \operatorname{Re} w > 0\} \to \Omega(\tilde{z}, w) \mapsto (z, e^{i\alpha(z) - \tilde{\alpha}(\tilde{z})} w),$$

where $\tilde{\alpha}$ denotes a harmonic conjugate for α.

3. Synthesis formula

Let Ω be a sectorial domain of dimension $n + 1$.

Let $L^2_{(n+1,0)}(\Omega)$ denote the Hilbert space of square-integrable $(n+1, 0)$-forms

$$F = f(z, w) \, dz_1 \wedge \cdots \wedge dz_n \wedge dw$$

on Ω with norm

$$\|F\|^2 = \frac{i^{(n+1)^2}}{2^{n+1}} \int_\Omega F \wedge \overline{F}.$$

Let

$$B(\Omega) = \{F \in L^2_{(n+1,0)}(\Omega) : F \text{ is holomorphic on } \Omega\},$$

$$B^\sharp(\Omega) = \{F \in L^2_{(n+1,0)}(\Omega) : f(z, w) \, dw \text{ is holomorphic on } \Omega_z \text{ for a.e. } z\},$$

$$C^\sharp(\Omega) = \{F \in L^2_{(n+1,0)}(\Omega) : f(z, w) \, dw \text{ extends continuously to } \overline{\Omega_z} \text{ for a.e. } z\}$$

and

$$A^\sharp(\Omega) = B^\sharp(\Omega) \cap C^\sharp(\Omega);$$

here Ω_z denotes the fiber over $z \in X$ and $\overline{\Omega_z}$ denotes the closure of Ω_z in $\hat{\mathbb{C}}$.

Note that $A^\sharp(\Omega)$ is dense in $B^\sharp(\Omega)$.

The Bergman projection P_Ω is the orthogonal projection from $L^2_{(n+1,0)}(\Omega)$ onto $B(\Omega)$.

Our main tool for the study of P_Ω will be a partial Mellin transform. For a form $F \in A^\sharp(\Omega)$ we define a section $\mathcal{M}F$ of the pullback $T^*_{(n,0)}(X \times \mathbb{R})$ of $T^*_{(n,0)}(X)$ to $X \times \mathbb{R}$ by

(3.1)
$$\mathcal{M}F = \mathcal{M}f(z, \xi) \, dz_1 \wedge \cdots \wedge dz_n,$$
$$\mathcal{M}f(z, \xi) = \int_0^\infty f(z, w) w^{-i\xi} \, dw.$$

(The integral is taken over any ray joining 0 to ∞ in Ω_z.)

REMARK. The transform $\mathcal{M}$ defined here differs from the standard Mellin transform by premultiplication by w. This choice simplifies the formulae below.

The Plancherel formula

$$(3.2) \qquad \|F\|^2 = \frac{i^{n^2}}{2^{n+1}\pi} \int\limits_{X \times \mathbb{R}} \gamma_\xi(z)\mathcal{M}F \wedge \overline{\mathcal{M}F} \wedge d\xi,$$

$$\gamma_\xi(z) \overset{\text{def}}{=} \begin{cases} \dfrac{e^{-2\xi\alpha(z)} - e^{-2\xi\beta(z)}}{2\xi}, & \xi \neq 0 \\ \beta(z) - \alpha(z), & \xi = 0 \end{cases}$$

allows us to extend $\mathcal{M}$ so that $\sqrt{2\pi}\mathcal{M}$ maps $B^\sharp(\Omega)$ isometrically onto the weighted L^2 space $L^2_{(n,0)}(X \times \mathbb{R}, \gamma_\xi(z)\,d\xi)$ with inversion formula

$$(3.3) \qquad f(z,w) = \frac{1}{2\pi} \int\limits_{\mathbb{R}} \mathcal{M}f(z,\xi)w^{i\xi-1}\,d\xi$$

valid and absolutely convergent for all $F \in B^\sharp(\Omega)$. These assertions follow from standard facts about the Fourier transform; details are in Appendix A.

We note for future reference that

$$(3.4) \qquad \frac{w^{i\xi}}{\sqrt{\gamma_\xi(z)}} \quad \text{decays exponentially as } \xi \to \pm\infty$$

for fixed $w \in \Omega_z$.

For $\xi \in \mathbb{R}$ let $Q_\xi : L^2_{(n,0)}(X,\gamma_\xi) \to B(X,\gamma_\xi)$ denote the Bergman projection corresponding to the positive weight γ_ξ; thus

$$\int\limits_X \gamma_\xi F \wedge \overline{H} = \int\limits_X \gamma_\xi\, Q_\xi F \wedge \overline{H}$$

for $H \in B(X,\gamma_\xi)$.

The Q_ξ piece together to form an operator

$$Q : L^2_{(n,0)}(X \times \mathbb{R}, \gamma_\xi(z)\,d\xi) \to L^2_{(n,0)}(X \times \mathbb{R}, \gamma_\xi(z)\,d\xi)$$

which is again the orthogonal projection onto the space of forms holomorphic along X.

THEOREM 3.1. (Synthesis formula).

$$P_\Omega = \mathcal{M}^{-1}Q\mathcal{M}P^\sharp,$$

where $P^\sharp : L^2_{(n+1,0)}(\Omega) \to B^\sharp(\Omega)$ is the fiberwise Bergman projection.

PROOF. $\mathcal{M}^{-1}Q\mathcal{M}P^\sharp$ maps $L^2_{(n+1,0)}(\Omega)$ to $B(\Omega)$, so it suffices to check that $P_\Omega F$ and $\mathcal{M}^{-1}Q\mathcal{M}P^\sharp F$ match when paired with $H \in B(\Omega)$.

But in fact

$$\int_\Omega \mathcal{M}^{-1} Q \mathcal{M} P^\sharp F \wedge \overline{H} = \frac{1}{2\pi} \int_{X \times \mathbb{R}} \gamma_\xi Q \mathcal{M} P^\sharp F \wedge \overline{\mathcal{M} H} \wedge d\xi \quad \text{(by (3.2), polarized)}$$

$$= \frac{1}{2\pi} \int_{X \times \mathbb{R}} \gamma_\xi \mathcal{M} P^\sharp F \wedge \overline{\mathcal{M} H} \wedge d\xi$$

$$= \int_\Omega P^\sharp F \wedge \overline{H} \quad \text{(by (3.2) again)}$$

$$= \int_\Omega F \wedge \overline{H}$$

$$= \int_\Omega P_\Omega F \wedge \overline{H}$$

as required.

4. Complex weights

By Theorem 3.1 and (3.3), for

$$F = f(z, w) \, dz_1 \wedge \cdots \wedge dz_n \wedge dw \in L^2_{(n+1,0)}(\Omega)$$

we may represent

$$P_\Omega F = (P_\Omega f)(z, w) \, dz_1 \wedge \cdots \wedge dz_n \wedge dw$$

by the integral

$$(4.1) \qquad (P_\Omega f)(z, w) = \frac{1}{2\pi} \int_{\mathbb{R}} (Q_\xi \mathcal{M} P^\sharp f)(z, \xi) w^{i\xi - 1} \, d\xi.$$

Our goal in the next few sections will be to use methods of contour integration to obtain asymptotic expansions for (4.1) as $w \to 0$.

To obtain such expansions we need

(1) a meromorphic continuation of the integrand to $\xi \in \mathbb{C}$; (2) decay estimates for the extended integrand as $\operatorname{Re} \xi \to \pm\infty$.

In this section we find conditions sufficient to guarantee that the operator-valued function $\xi \mapsto Q_\xi$ extends meromorphically from $\mathbb{R}$ to $\mathbb{C}$.

We will use the generic notation M_η to represent the operation of multiplication by η.

PROPOSITION 4.1. *Let ψ, η be a pair of positive weight functions on a complex manifold X. Assume that $\log \psi \in L^\infty_{loc}(X)$ and $\log \eta \in L^\infty(X)$ so that in particular*

$$L^2_{(n,0)}(X, \psi) = L^2_{(n,0)}(X, \eta\psi).$$

Let $P_\psi, P_{\eta\psi}$ be the associated weighted Bergman projections. Then

$$(4.2) \qquad P_{\eta\psi} M_{\eta^{-1}} P_\psi M_\eta = P_{\eta\psi}.$$

PROOF. Both $P_{\eta\psi}M_{\eta^{-1}}P_\psi M_\eta$ and $P_{\eta\psi}$ map $L^2_{(n,0)}(\Omega,\psi)$ to $B(\Omega,\psi)$, so it suffices to check that $P_{\eta\psi}M_{\eta^{-1}}P_\psi M_\eta F$ and $P_{\eta\psi}F$ agree when paired with $H \in L^2_{(n,0)}(\Omega,\eta\psi)$. But

$$\int_X \eta\psi P_{\eta\psi}M_{\eta^{-1}}P_\psi M_\eta F \wedge \overline{H} = \int_X \psi P_\psi M_\eta F \wedge \overline{H}$$
$$= \int_X \eta\psi F \wedge \overline{H}$$
$$= \int_X \eta\psi P_{\eta\psi}F \wedge \overline{H}$$

as required.

For application to the study of complex weights it will be useful to recast this result as follows.

COROLLARY 4.2. *Let X,ψ,η be as in Proposition 4.1. Then the Bergman-Toeplitz operator*

$$P_\psi M_\eta|_{B(X,\psi)}$$

is invertible in $\mathcal{L}(B(X,\psi))$; moreover,

(4.3) $$P_{\eta\psi} = \left(P_\psi M_\eta|_{B(X,\psi)}\right)^{-1} P_\psi M_\eta.$$

PROOF. Proposition 4.1 shows that

$$P_{\eta\psi}M_{\eta^{-1}}|_{B(X,\psi)}$$

is both left- and right-inverse to

$$P_\psi M_\eta|_{B(X,\psi)}.$$

THEOREM 4.3. *Let X be a relatively compact pseudoconvex domain in some ambient Stein manifold. Assume that the L^2-minimizing solution operator*

$$S_X : \{G \in L^2_{(n,1)}(X) : \overline{\partial}G = 0\} \to L^2_{(n,0)}(X)$$

for $\overline{\partial}$ is compact, and let α and β be C^1 functions on $\overline{X}$ such that $\delta \overset{\text{def}}{=} \beta - \alpha > 0$ is constant. Then $\xi \mapsto Q_\xi$ extends to a meromorphic map from $\mathbb{C}$ to the space $\mathcal{L}(L^2_{(n,0)}(X))$ of bounded operators on $L^2_{(n,0)}(X)$. This map has no poles in the strip $\{\xi : |\mathrm{Im}\,\xi| \leq \pi/4\|\alpha\|_{L^\infty(X)}\}$.

PROOF. We have

(4.4) $$\gamma_\xi(z) = \frac{e^{-2\xi\alpha(z)}e^{-\delta\xi}\sinh(\delta\xi)}{\xi};$$

since constant factors in the weight function do not affect the projection operator this yields

$$Q_\xi = P_{\gamma_\xi} = P_{e^{-2\xi\alpha}} \qquad \text{for } \xi \in \mathbb{R}.$$

So we try to extend $P_{e^{-2\xi\alpha}}$ to $\xi \in \mathbb{C}$.

Using the formula $P_X = I - S_X\overline{\partial}$ (see for example [Kra, 8.1.3]) we have

(4.5)
$$\begin{aligned} P_X M_{e^{-2\xi\alpha}}|_{B(X)} &= (I - S_X\overline{\partial})M_{e^{-2\xi\alpha}}|_{B(X)} \\ &= \left(M_{e^{-2\xi\alpha}} - S_X M_{\overline{\partial}e^{-2\xi\alpha}}\right)|_{B(X)}; \end{aligned}$$

but $M_{e^{-2\xi\alpha}}$ is invertible and $S_X M_{\overline{\partial}e^{-2\xi\alpha}}$ is compact, so

$$\mathbb{C} \to \mathcal{L}(L^2_{(n,0)}(\Omega))$$
$$\xi \mapsto M_{e^{-2\xi\alpha}} - S_X M_{\overline{\partial}e^{-2\xi\alpha}}$$

is an entire mapping into the space of Fredholm operators of index zero. Since the range of this map includes some invertible operators (in particular $0 \mapsto I$), the "analytic Fredholm theorem" [RS, VI.14] furnishes a meromorphic inverse map

$$\mathbb{C} \setminus \{\text{poles}\} \to \mathcal{L}(L^2_{(n,0)}(\Omega))$$
$$\xi \mapsto (M_{e^{-2\xi\alpha}} - S_X M_{\overline{\partial}e^{-2\xi\alpha}})^{-1}.$$

Combining (4.3) with (4.5) we now have

$$Q_\xi = (M_{e^{-2\xi\alpha}} - S_X M_{\overline{\partial}e^{-2\xi\alpha}})^{-1} P_X M_{e^{-2\xi\alpha}},$$

which clearly extends meromorphically to $\xi \in \mathbb{C}$.

To prove the final statement of the Theorem note that

$$P_X M_{e^{-2\xi\alpha}} H = 0$$

implies

$$\text{Re} \int_X e^{-2\xi\alpha} H \wedge \overline{H} = \text{Re} \int_X (P_X M_{e^{-2\xi\alpha}} H) \wedge \overline{H} = 0$$

forcing $H \equiv 0$ or $|\text{Im}\, 2\xi\alpha(z)| > \pi/2$ at some $z \in X$.

If X satisfies (2.9) then S_X maps the Sobolev space $W^s_{(2,1)}(X)$ to $W^{s+1}_{(2,0)}(X)$ for each s [Bel, p. 61] [Rau, Thm. 5.9.3]. Thus S_X is compact not only from $L^2_{(2,1)}(X)$ to $L^2_{(2,0)}(X)$ but from $W^s_{(2,1)}(X)$ to $W^s_{(2,0)}(X)$ for each s, yielding the following Addendum.

ADDENDUM 4.4. *If Ω satisfies the hypotheses (2.8), (2.9), (2.10) then $\xi \mapsto Q_\xi$ is meromorphic from $\mathbb{C}$ to $\mathcal{L}(W^s_{(2,0)}(X))$ for all s. In particular, the residues of $\xi \mapsto Q_\xi$ lie in $\mathcal{L}(W^s_{(2,0)}(X))$ for all s.*

Similar remarks apply to smoothly bounded domains in $\mathbb{C}^n$ which are strictly pseudoconvex [FK] or finite type [Cat].

Theorem 4.3 does not apply to the sectorial domains of Example 2.1 since the boundary vanishing condition (2.7) contradicts the constancy of δ. To handle these domains we have instead the following result.

THEOREM 4.5. *Let X be a complex manifold and let α and β be, respectively, upper- and lower-semicontinuous functions on X satisfying $\alpha < \beta \leq \alpha + 2\pi$ and the boundary vanishing condition (2.7). Then $\xi \mapsto Q_\xi$ extends to a meromorphic map from $\mathbb{C}$ to the space $\mathcal{L}(L^2_{(n,0)}(X, \beta - \alpha))$ of bounded operators on $L^2_{(n,0)}(X, \beta - \alpha)$. This map has no poles in the strip $\{\xi : |\text{Im}\,\xi| \leq \pi/4 \max\{\|\alpha\|_{L^\infty(X)}, \|\beta\|_{L^\infty(X)}\}\}$.*

8 DAVID E. BARRETT

PROOF. By (4.3) we have

$$(4.6) \qquad Q_\xi = (P_{\beta-\alpha} M_{\gamma_\xi/(\beta-\alpha)}|B(X,\beta-\alpha))^{-1} P_{\beta-\alpha} M_{\gamma_\xi/(\beta-\alpha)}$$

for $\xi \in \mathbb{R}$.

By (2.7), for each $\xi \in \mathbb{C}, \epsilon \in \mathbb{P}^+$ there is a compact set $K \in \mathbb{C}$ such that

$$\frac{\gamma_\xi}{\beta-\alpha} = \frac{e^{-2\xi\alpha} - e^{-2\xi\beta}}{2\xi(\beta-\alpha)}$$

satisfies

$$\left|1 - \frac{\gamma_\xi}{\beta-\alpha}\right| < \epsilon \quad \text{on } X \setminus K.$$

It follows then from Cauchy-type estimates [Hör1, Thm. 1.2.4] and Montel's Theorem that

$$M_{1-\gamma_\xi/(\beta-\alpha)}|B(X,\beta-\alpha)$$

is a compact operator, so that

$$P_{\beta-\alpha} M_{\gamma_\xi/(\beta-\alpha)}|B(X,\beta-\alpha) = (I - P_{\beta-\alpha} M_{1-\gamma_\xi/(\beta-\alpha)})|B(X,\beta-\alpha)$$

is Fredholm index 0 for each $\xi \in \mathbb{C}$. As in the proof of Theorem 4.3 this implies that

$$(P_{\beta-\alpha} M_{\gamma_\xi/(\beta-\alpha)}|B(X,\beta-\alpha))^{-1}$$

is meromorphic in ξ so that (4.6) furnishes the desired meromorphic extension of $\xi \mapsto Q_\xi$.

The final statement of the Theorem follows as in the proof of Theorem 4.3.

PROPOSITION 4.6. *If $\psi : X \to \mathbb{R}^+$ satisfies $\log \psi \in L^\infty_{loc}$ and $h : X \to \mathbb{C} \setminus \{0\}$ is holomorphic with $h, h^{-1} \in L^\infty(\Omega)$ then*

$$(4.7) \qquad M_{h^{-1}} P_\psi M_h = P_{\psi|h|^2}.$$

PROOF. As usual, it suffices to check that

$$\int_X \psi|h|^2 M_{h^{-1}} P_\psi M_h F \wedge \overline{H} = \int_X \psi P_\psi M_h F \wedge \overline{hH}$$
$$= \int_X \psi|h|^2 F \wedge \overline{H}$$
$$= \int_X \psi|h|^2 P_{\psi|h|^2} F \wedge \overline{H}$$

for $F \in L^2_{(n,0)}(X,\psi), H \in B(X,\psi)$.

COROLLARY 4.7. *In the situation of Theorem 4.3, if α is the real part of a holomorphic function h then the meromorphic map $\xi \mapsto Q_\xi$ has no poles.*

PROOF. By (4.7) and analytic continuation in ξ we have

$$Q_\xi = P_{e^{-\xi h} e^{-\xi \overline{h}}}$$
$$= M_{e^{\xi h}} Q_0 M_{e^{-\xi h}}.$$

5. Function spaces

THEOREM 5.1. *If $F \in C^\sharp(\Omega)$ then*

(5.1)
$$(\mathcal{M}P^\sharp f)(z,\xi) = \frac{i}{2\gamma_\xi(z)} \int_{\Omega_z} f(z,w)\overline{w}^{-i\xi-1} dw \wedge d\overline{w}$$

on a.e. fiber Ω_z.

PROOF. We may restrict attention to the case $n = 0$, in which case it will suffice to check that

(5.2)
$$\int_{\mathbb{R}} \gamma_\xi \left(\frac{i}{2\gamma_\xi} \int_\Omega f(w)\overline{w}^{-i\xi-1} dw \wedge d\overline{w} \right) \overline{\mathcal{M}h(\xi)} \, d\xi = \int_{\mathbb{R}} \gamma_\xi \mathcal{M}P^\sharp f(\xi) \overline{\mathcal{M}h(\xi)} \, d\xi$$

when $h(w)\,dw \in B(\Omega)$.

But the right-hand side of (5.2) is

$$\pi i \int_\Omega P^\sharp f(w)\,dw \wedge \overline{h(w)\,dw} = \pi i \int_\Omega f(w)\,dw \wedge \overline{h(w)\,dw}$$

by (3.2), while application of Fubini's Theorem (recalling (3.2) and (3.4)) and the inversion formula (3.3) to H reduces the left-hand side of (5.2) to

$$\pi i \int_\Omega f(w)\,dw \wedge \overline{h(w)\,dw}$$

as required.

THEOREM 5.2.

$$\mathcal{M}P^\sharp \left(C^\infty_{0,(n+1,0)}(\Omega) \right) = \{ \Upsilon = \upsilon(z,\xi)\,dz_1 \wedge \cdots \wedge dz_n \in L^2_{(n,0)}\left(X \times \mathbb{R}, \gamma_\xi(z)\,d\xi\right) :$$

the projection of supp Υ *on X is compact,*

$\gamma_\xi \Upsilon$ *extends to a form in $C^\infty_{(n,0)}(X \times \mathbb{C})$*

which is independent of $d\xi$ and holomorphic in ξ,

and there are $B \in \mathbb{R}^+$ and $\tilde{\alpha}, \tilde{\beta} \in C^\infty(X)$ with

$\alpha < \tilde{\alpha} < \tilde{\beta} < \beta$ such that for every linear differential

operator L on X we have

$|L(\gamma_\xi(z)\upsilon(z,\xi))| \leq A_L(e^{-(\operatorname{Re}\xi)\tilde{\alpha}(z)} + e^{-(\operatorname{Re}\xi)\tilde{\beta}(z)})e^{B|\operatorname{Im}\xi|}$

for some constant A_L}.

Here $C^\infty_{0,(n+1,0)}(\Omega)$ denotes the space of smooth compactly supported $(n+1,0)$-forms on Ω.

PROOF. $\subset$: Pick $\tilde{\alpha}, \tilde{\beta} \in C^\infty(X)$ such that $\alpha < \tilde{\alpha} < \tilde{\beta} < \beta$ and $\tilde{\alpha}(z) < \arg w < \tilde{\beta}(z)$ when $(z,w) \in$ supp Υ. Then

$$|\overline{w}^{-i\xi}| \leq (e^{-(\operatorname{Re}\xi)\tilde{\alpha}(z)} + e^{-(\operatorname{Re}\xi)\tilde{\beta}(z)})e^{B|\operatorname{Im}\xi|}$$

on supp Υ, where

$$B \overset{\text{def}}{=} \sup_{\text{supp}\,\Upsilon} \big|\log|w|\big|.$$

The desired conclusions now follow easily from differentiation of (5.1).

$\supset$: Use of a partition of unity on X allows us to reduce to consideration of the case where X is a domain in $\mathbb{C}^n$.

The estimates on $\gamma_\xi(z)v(z,\xi)$ guarantee that the Laplace transform

$$\Phi(z,\zeta) = \int_0^\infty \gamma_\xi(z)v(z,\xi)e^{-\zeta\xi}\,d\xi$$

defines a smooth function on

$$(X \times \mathbb{C}) \setminus \{(z,\zeta) : -\tilde{\beta}(z) \le \operatorname{Re}\zeta \le -\tilde{\alpha}(z), |\operatorname{Im}\zeta| \le B\}$$

which is holomorphic in ζ (see [Boa, 5.3]); here the integral defining $\Phi(z,\zeta)$ may be taken along any ray $\arg\xi = \theta$ satisfying

$$\operatorname{Re}e^{i\theta}(\sigma - \zeta) < 0 \text{ on } \{\sigma : -\tilde{\beta}(z) \le \operatorname{Re}\sigma \le -\tilde{\alpha}(z), |\operatorname{Im}\sigma| \le B\}.$$

By [Boa, 5.3.5] we have

(5.3)
$$\gamma_\xi(z)v(z,\xi) = \frac{1}{2\pi i}\int_{|\zeta|=M} \Phi(z,\zeta)e^{\zeta\xi}\,d\zeta$$

for M large.

Pick a function

$$\mu \in C^\infty(X \times \mathbb{C})$$

such that $\mu \equiv 0$ in a neighborhood of $\{(z,\zeta) : -\tilde{\beta}(z) \le \operatorname{Re}\zeta \le -\tilde{\alpha}(z), |\operatorname{Im}\zeta| \le B\}$ and $\operatorname{supp}(1-\mu) \subset \{(z,\zeta) : -\alpha(z) \le \operatorname{Re}\zeta \le -\beta(z), |\operatorname{Im}\zeta| \le B+1\}$.

Let

$$\rho(z,\zeta) = \begin{cases} (\mu\Phi)(z,\zeta) & \text{if } (z,\zeta) \in \operatorname{supp}\mu, \\ 0 & \text{if } (z,\zeta) \notin \operatorname{supp}\mu. \end{cases}$$

Then

$$\gamma_\xi(z)v(z,\xi) = \frac{1}{2\pi i}\int_{|\zeta|=M} \rho(z,\zeta)e^{\zeta\xi}\,d\zeta$$

$$= \frac{1}{2\pi i}\int_{\zeta\in\mathbb{C}} e^{\zeta\xi}\,\overline{\partial}(\rho(z,\zeta)\,d\zeta) \qquad \text{(by Stokes' Theorem)}$$

$$= \frac{1}{2\pi}\int_{\alpha < \arg w < \beta} \overline{w}^{-i\xi-1}\partial(\rho(z,\log\overline{w})\,d\overline{w}) \qquad \text{(setting } \zeta = -i\log\overline{w}).$$

Hence by Theorem 5.1

$$\Upsilon = \mathcal{M}P^\sharp((\pi i)^{-1}dz_1 \wedge \cdots \wedge dz_n \wedge \partial(\rho(z,\log\overline{w})))$$

as required.

COROLLARY 5.3. *If* $\Upsilon = v(z,\xi)\, dz_1 \wedge \cdots \wedge dz_n \in \mathcal{M}P^\sharp(C^\infty_{0,(n+1,0)}(X))$ *then for* $z \in X$ *fixed,* $v(z,\xi)$ *extends meromorphically to* $\xi \in \mathbb{C}$ *with, at most, simple poles at points in*

$$\frac{i\pi}{(\beta-\alpha)(z)}\mathbb{Z}\setminus\{0\}.$$

PROOF. This follows from

$$\gamma_\xi(z) = e^{-\xi(\alpha+\beta)(z)}\frac{\sinh(\xi(\beta-\alpha)(z))}{\xi}$$

and Theorem 5.2.

COROLLARY 5.4. *Assume that* $\delta = \beta - \alpha$ *is constant. Then for* $\xi_0 \notin (i\pi/\delta)\mathbb{Z}\setminus\{0\}$ *the space*

$$\mathcal{M}P^\sharp(C^\infty_{0,(n+1,0)}(X))(\cdot,\xi_0)$$

is dense in $L^2_{(n,0)}(X)$, *and for* $\xi_0 \in (i\pi/\delta)\mathbb{Z}\setminus\{0\}$ *the space*

$$\mathrm{Res}_{\xi_0}\mathcal{M}P^\sharp(C^\infty_{0,(n+1,0)}(X))$$

is dense in $L^2_{(n,0)}(X)$.

PROOF. If $G \in C^\infty_{0,(n,0)}(X)$ has small enough support then there is $\theta \in \mathbb{R}$ so that $\alpha < \theta < \beta$ on $\mathrm{supp}\, G$. Then Theorem 5.2 implies that

$$e^{-\xi\theta}G \in \gamma_\xi(\cdot)\mathcal{M}P^\sharp(C^\infty_{0,(n+1,0)}(X)).$$

In view of (4.4) this implies that

(5.4) $$e^{\xi_0(2\alpha(\cdot)-\theta)}G$$

is in

$$\mathcal{M}P^\sharp(C^\infty_{0,(n+1,0)}(X))(\cdot,\xi_0)$$

if $\xi \notin (i\pi/\delta)\mathbb{Z}\setminus\{0\}$, and in

$$\mathrm{Res}_{\xi_0}\mathcal{M}P^\sharp(C^\infty_{0,(n+1,0)}(X))$$

if $\xi \in (i\pi/\delta)\mathbb{Z}\setminus\{0\}$. But linear combinations of forms of of the type (5.4) are dense in $L^2_{(n,0)}(X)$.

For $F = f(z,w)\, dz_1 \wedge \cdots \wedge dz_n \wedge dw$ we let $(\partial/\partial w)^j F = (\partial^j f/\partial w^j)(z,w)\, dz_1 \wedge \cdots \wedge dz_n \wedge dw$.

THEOREM 5.5. *If* $\beta - \alpha \equiv \pi$ *and* $(\partial/\partial w)^j F \in B^\sharp(\Omega)$ *for* $j = 0,\ldots,k$ *then*

$$\sqrt{\gamma_\xi(z)\cdot(\xi+ik)\cdots(\xi+i)}\mathcal{M}F$$

extends to a form in $L^2_{(n,0)}(X)\otimes H^1(\Sigma_k)$, *where* $H^1(\Sigma_k)$ *is the Hardy space for the strip*

$$\Sigma_k \overset{\mathrm{def}}{=} \{\xi : -k < \mathrm{Im}\,\xi < 0\}.$$

Thus $\mathcal{M}F$ *extends meromorphically to* $\xi \in \Sigma_k$ *with, at most, simple poles at* $-i,\ldots,-i(k-1)$.

PROOF. If suffices to show that if $(\partial^j f/\partial w^j)(z,w)\,dw \in B(\Omega_z)$ for $j = 0, \ldots, k$ then

$$\sqrt{\gamma_\xi(z) \cdot (\xi + ik) \cdots (\xi + i)} \mathcal{M}f(z,\xi)$$

extends to a function in $H^1(\Sigma_k)$ with norm bounded by a constant times

$$\|f(z,w)\,dw\|_{B(\Omega_z)} + \cdots + \left\|\tfrac{\partial^k f}{\partial w^k}(z,w)\,dw\right\|_{B(\Omega_z)}.$$

We may approximate $f(z,w)\,dw$ by forms with derivatives $(\partial^j f/\partial w^j)(z,w)\,dw$ extending continuously to $\overline{\Omega_z}$ for $j = 0, \ldots, k$, and so it will suffice to work with forms in this latter class.

Pick a cut-off function $\mu \in C_0^\infty(\mathbb{C})$ with $\mu \equiv 1$ in a neighborhood of the origin. Let

$$\mathcal{I}(\xi) = \int\limits_{\Omega_z} f(z,w)\overline{w}^{-i\xi-1}\,dw \wedge d\overline{w}$$

$$= \int\limits_{\Omega_z} (1 - \mu(w))f(z,w)\overline{w}^{-i\xi-1}\,dw \wedge d\overline{w}$$

$$+ \frac{(-1)^k}{(i\xi + 1 - k) \cdots (i\xi)} \int\limits_{\Omega_z} \mu(w)f(z,w) \left(\frac{\partial}{\partial \overline{w}} + e^{2i\alpha(z)}\frac{\partial}{\partial w}\right)^k$$

$$\cdot \overline{w}^{-i\xi-1+k}\,dw \wedge d\overline{w}.$$

Since the vector field $(\partial/\partial\overline{w}) + e^{2i\alpha(z)}(\partial/\partial w)$ is tangential to $b\Omega_z$, we may integrate by parts to obtain

(5.5)

$$\mathcal{I}(\xi) = \int\limits_{\Omega_z} (1 - \mu(w))f(z,w)\overline{w}^{-i\xi-1}\,dw \wedge d\overline{w}$$

$$+ \frac{1}{(i\xi + 1 - k) \cdots (i\xi)} \int\limits_{\Omega_z} \left(\left(\frac{\partial}{\partial \overline{w}} + e^{2i\alpha(z)}\frac{\partial}{\partial w}\right)^k \left(\mu(w)f(z,w)\right)\right)$$

$$\cdot \overline{w}^{-i\xi-1+k}\,dw \wedge d\overline{w}.$$

Direct estimation of the integrals and an induction on k show that $\mathcal{I}(\xi)$ is holomorphic in a neighborhood of the closure $\overline{\Sigma_k}$ of Σ_k in $\mathbb{C}$. Using (5.1), (5.5), and (3.2) we find that

$$\int\limits_{\mathbb{R}} \frac{|\mathcal{I}(\xi)|^2}{\gamma_\xi(z)}\,d\xi = 8\pi\|f(z,w)\,dw\|_{B(\Omega_z)}^2$$

and

$$\int\limits_{\mathbb{R}-ik} \frac{|\mathcal{I}(\xi)|^2}{\gamma_{\mathrm{Re}\,\xi}(z)}\,d\xi \lesssim \left(\|f(z,w)\,dw\|_{B(\Omega_z)} + \cdots + \left\|\tfrac{\partial^k f}{\partial w^k}(z,w)\,dw\right\|_{B(\Omega_z)}\right)^2.$$

Moreover, direct estimation of the integrals in (5.5) reveals that

$$h(\xi) \stackrel{\text{def}}{=} \tfrac{i}{2}\sqrt{\frac{(\xi + ik) \cdots (\xi + i)}{\gamma_\xi(z)}}\mathcal{I}(\xi) = O\left(|\xi|^{(k+1)/2}\right)$$

on $\overline{\Sigma_k}$. This shows that h is integrable with respect to harmonic measure on bΣ_k (since harmonic measure decays exponentially), and the Phragmén-Lindelöf Principle [Ran, Thm. 2.3.5] shows that h coincides with the Poisson integral of its boundary values and thus belongs to $H^1(\Sigma_k)$. But $h(\xi)$ coincides with $\sqrt{\gamma_\xi(z)} \cdot (\xi + ik) \cdots (\xi + i)\mathcal{M}F$ when $\xi \in \mathbb{R}$, so we are done.

6. Asymptotic expansions for Example 2.2

THEOREM 6.1. *Let Ω be a sectorial domain satisfying the hypotheses (2.8), (2.9), (2.10). Then for $F \in C^\infty_{0,(2,0)}$ and $R > 0$ we have*

$$(6.1) \quad (P_\Omega f)(z,w) = \frac{i}{4\pi^2 w} \sum_{-R \le \operatorname{Im} \xi < 0} \left(\operatorname{Res}_\xi (Q_\xi \mathcal{M}P^\sharp f)(z,\xi) \right) w^{i\xi} + \mathrm{O}(|w|^{R-1}).$$

On any compact subset of $\overline{\Omega} \setminus \{(z,w) : w = \infty\}$ the estimate holds uniformly and the sum converges uniformly and absolutely.

Let

$$E = \{\xi : |\operatorname{Re} \xi| < 1, |\operatorname{Im} \xi| < R + 1\}$$

and let τ_j denote the translation $\xi \mapsto \xi - j$.

LEMMA 6.2.

$$T : \xi \mapsto (Q_\xi \mathcal{M}P^\sharp f)(\cdot, \xi)w^{i\xi - 1}|w|^{\operatorname{Im} \xi + 1}$$

defines a meromorphic map from $\mathbb{C}$ to $L^\infty_{(1,0)}(\Omega)$. Moreover, for some $k \in \mathbb{R}^+$ there are constants $\nu > 0, C > 0$ so that the distributions T_j induced by $T \circ \tau_j|_E$ satisfy bounds of the form

$$(6.2) \qquad \|T_j\|_{W^{-k}(E, L^\infty_{(1,0)}(\Omega))} \le Ce^{-\nu|j|}.$$

Here meromorphic maps are interpreted as distributions through the use of Cauchy principal values; see [BG, 3.6,8].

PROOF OF THEOREM 6.1 ASSUMING LEMMA 6.2. The sum is obtained by integrating the distributional derivative $(\partial/\partial\bar\xi)(M_{w|w|^{-\operatorname{Im} \xi - 1}}T)$, so the assertions about the convergence of the sum follow from the exponential decay estimate (6.2).

To establish (6.1), pick a function $\mu \in C^\infty_0(E)$ so that

$$\sum_{j \in \mathbb{Z}} \mu(\xi + j) \equiv 1$$

in a neighborhood of $\{\xi : |\operatorname{Im} \xi| \le R\}$.

Then

$$P_\Omega f(z,w)$$
$$= \frac{1}{2\pi} \int_{\mathbb{R}} (Q_\xi \mathcal{M}P^\sharp f)(z,\xi)w^{i\xi - 1}\, d\xi \qquad \text{(by (4.1))}$$

$$= \lim_{N \to \infty} \frac{1}{2\pi} \int_{\mathbb{R}} \sum_{|j| \le N} \mu(\xi + j)(Q_\xi \mathcal{M} P^\sharp f)(z, \xi) w^{i\xi - 1} \, d\xi$$

$$= \lim_{N \to \infty} \frac{1}{2\pi} \int_{\operatorname{Im} \xi < 0} \frac{\partial}{\partial \bar{\xi}} \left(\sum_{|j| \le N} \mu(\xi + j)(Q_\xi \mathcal{M} P^\sharp f)(z, \xi) \right) w^{i\xi - 1} \, d\xi \wedge d\bar{\xi}$$

(Stokes)

$$= \lim_{N \to \infty} \left\{ \frac{1}{2\pi} \int_{-R \le \operatorname{Im} \xi < 0} \frac{\partial}{\partial \bar{\xi}} \left(\sum_{|j| \le N} \mu(\xi + j)(Q_\xi \mathcal{M} P^\sharp f)(z, \xi) \right) w^{i\xi - 1} \, d\xi \wedge d\bar{\xi} \right.$$

$$\left. + \frac{1}{2\pi} \int_{\operatorname{Im} \xi < -R} \frac{\partial}{\partial \bar{\xi}} \left(\sum_{|j| \le N} \mu(\xi + j)(Q_\xi \mathcal{M} P^\sharp f)(z, \xi) \right) w^{i\xi - 1} \, d\xi \wedge d\bar{\xi} \right\}$$

(Residues arising from poles on the line $\operatorname{Im} \xi = -R$ are thrown into the first term.) By Lemma 6.2 and the Residue Theorem, the first term is of the form

$$\frac{i}{4\pi^2} \sum_{\substack{-R \le \operatorname{Im} \xi < 0 \\ |\operatorname{Re} \xi| \le N-1}} \left(\operatorname{Res}_\xi (Q_\xi \mathcal{M} P^\sharp f)(z, \xi) \right) w^{i\xi - 1} + \mathrm{O}\left(e^{-\nu|N|} |w|^{-1} \right)$$

while the second term is $\mathrm{O}(|w|^{R-1})$ uniformly in N, both estimates holding off of a neighborhood of $\overline{\Omega} \times \{\infty\}$. Thus passing to the limit we obtain (6.1).

PROOF OF LEMMA 6.2. Let $\Gamma_1, \dots, \Gamma_K$ be a basis for $H^1(X, \mathbb{Z})$. Since periods of holomorphic differentials on open Riemann surfaces can be prescribed freely [For, Thm. 28.6], we may choose a vector $\alpha = (\alpha_1, \dots, \alpha_K)$ of functions smooth on $\overline{X}$ and harmonic on X such that

(1) $\int_{\Gamma_k} *d\alpha_\ell = 2\pi \delta_{k\ell}$;

(2) $\alpha = \alpha \cdot \mathbf{v}$ for some $\mathbf{v} \in \mathbb{R}^K$.

For any $\mathbf{n} \in \mathbb{Z}^k$ the harmonic conjugate $\widetilde{\mathbf{n} \cdot \alpha}$ for $\mathbf{n} \cdot \alpha$ has periods in $2\pi \mathbb{Z}$ so that

$$h_{\mathbf{n}} \stackrel{\mathrm{def}}{=} e^{\mathbf{n} \cdot \alpha + i \widetilde{\mathbf{n} \cdot \alpha}}$$

is a well-defined holomorphic function on X satisfying

$$|h_{\mathbf{n}}| = e^{\mathbf{n} \cdot \alpha}.$$

Thus in particular

$$e^{-2(\xi - j)\alpha} = |h_{[j\mathbf{v}]}|^2 e^{-2\xi\alpha + 2(j\mathbf{v} - [j\mathbf{v}]) \cdot \alpha}$$

so that by Proposition 4.6 we have

(6.3) $$Q_{\xi - j} = M_{h_{[j\mathbf{v}]}^{-1}} P_{e^{-2\xi\alpha + 2(j\mathbf{v} - [j\mathbf{v}]) \cdot \alpha}} M_{h_{[j\mathbf{v}]}}.$$

(Here $[\cdot]$ denotes the componentwise greatest integer function $[\mathbf{v} \stackrel{\mathrm{def}}{=} ([x_1], \dots, [x_K])$.) As in Theorem 4.3 and Addendum 4.4, $\xi \mapsto P_{e^{-2\xi\alpha + 2(j\mathbf{v} - [j\mathbf{v}]) \cdot \alpha}}$ extends to a meromorphic map from $\mathbb{C}$ to $\mathcal{L}(W_{(1,0)}^s(X))$; (6.3) extends to $\xi \in \mathbb{C}$ by analytic continuation.

Since the distributions $\xi \mapsto P_{e^{-2\xi\alpha+2\mathbf{u}\cdot\alpha}}$ vary continuously with $\mathbf{u} \in [0,1]^K$, there is an upper bound on the order of poles encountered in the compact set $\overline{E} \subset \mathbb{C}$; thus for suitable $k > 0$

(6.4) the distributions $\xi \mapsto P_{e^{-2\xi\alpha+2\mathbf{u}\cdot\alpha}}$ form a bounded subset of $W^{-k}(E, \mathcal{L}(W_{(1,0)}^s(X)))$ as $\mathbf{u}$ ranges over $[0,1]^K$.

By Theorem 5.2, z-derivatives of $e^{-j\alpha(z)}\gamma_{\xi-j}(z)\mathcal{M}P^\sharp f(z,\xi-j)$ satisfy estimates of the form

$$|L(e^{-j\alpha(z)}\gamma_{\xi-j}(z)\mathcal{M}P^\sharp f(z,\xi-j))| \le p_L(|j|)(e^{j(\pi-\epsilon)} + e^{\epsilon j})$$

on $E \times X$; here and below p_L denotes a polynomial depending on L which will change from line to line.

Since $|h_{[j\mathbf{v}]}(z)| \asymp e^{j\alpha(z)}$ on E, inspection of derivatives of $h_{[j\mathbf{v}]}(z)e^{j\alpha(z)}/\gamma_{\xi-j}(z)$ reveals that they satisfy estimates of the form

$$\left| L\left(\frac{h_{[j\mathbf{v}]}(z)e^{j\alpha(z)}}{\gamma_{\xi-j}(z)} \right) \right| \le \frac{p_L(|j|)}{e^{2j\pi} + 1}$$

on $E \times X$ for $|j| > 1$.

Thus

$$|L(h_{[j\mathbf{v}]}(z)\mathcal{M}P^\sharp f(z,\xi-j))| = \left| L\left(\frac{h_{[j\mathbf{v}]}(z)e^{j\alpha(z)}}{\gamma_{\xi-j}(z)} e^{-j\alpha(z)}\gamma_{\xi-j}(z)\mathcal{M}P^\sharp f(z,\xi-j) \right) \right|$$

$$\le \frac{p_L(|j|)}{e^{(\pi+\epsilon)j} + e^{-\epsilon j}}$$

on $E \times X$ for $|j| > 1$. By (6.4) we therefore have

$$\left\| P_{e^{-2\xi\alpha+2(j\mathbf{v}-[j\mathbf{v}])\cdot\alpha}} h_{[j\mathbf{v}]}(\cdot)\mathcal{M}P^\sharp f(\cdot,\xi-j) \right\|_{W^{-k}(E, W_{(1,0)}^s(X))} \le \frac{q_s(|j|)}{e^{(\pi+\epsilon)j} + e^{-\epsilon j}}$$

for $|j| > 0$, where q_s denotes a polynomial depending on s. Taking $s = 2$ and invoking the Sobolev Embedding Theorem we obtain z-uniform estimates

$$\left\| P_{e^{-2\xi\alpha+2(j\mathbf{v}-[j\mathbf{v}])\cdot\alpha}} h_{[j\mathbf{v}]}(\cdot)\mathcal{M}P^\sharp f(\cdot,\xi-j) \right\|_{W^{-k}(E, C_{(1,0)}(\overline{X}))} \lesssim \frac{q_2(|j|)}{e^{(\pi+\epsilon)j} + e^{-\epsilon j}}.$$

Again using $|h_{[j\mathbf{v}]}(z)| \asymp e^{j\alpha(z)}$ as well as $\alpha(z) < \arg w < \alpha(z) + \pi$ we find that

$$|h_{[j\mathbf{v}]}^{-1}(z)w^{i(\xi-j)-1}| \lesssim (e^{j\pi} + 1)|w|^{-\operatorname{Im}\xi-1}$$

on $E \times \Omega$. Hence

$$\left\| (Q_{\xi-j}\mathcal{M}P^\sharp f)(\cdot,\xi-j)w^{i(\xi-j)-1}|w|^{\operatorname{Im}\xi+1} \right\|_{W^k(E, L_{(1,0)}^\infty(\Omega))}$$

$$= \left\| (M_{h_{[j\mathbf{v}]}^{-1}} P_{e^{-2\xi\alpha+2(j\mathbf{v}-[j\mathbf{v}])\cdot\alpha}} M_{h_{[j\mathbf{v}]}}\mathcal{M}P^\sharp f)(\cdot,\xi-j)w^{i(\xi-j)-1} \right.$$

$$\left. \cdot |w|^{\operatorname{Im}\xi+1} \right\|_{W^k(E, L_{(1,0)}^\infty(\Omega))}$$

$$\le C_1 q_2(|j|)e^{-\epsilon|j|}$$

$$\le C_2 e^{-(\epsilon/2)|j|},$$

for $|j| > 1$.

Thus an estimate of the form (6.2) holds for $|j| > 1$; after adjusting the values of k and C_2 to account for additional poles coming from zeros of γ_ξ along the imaginary ξ-axis, it will hold for all j.

ADDENDUM 6.3. *It is clear from the proof that expansions obtained from (6.1) by differentiation with respect to z are likewise valid.*

7. Asymptotic expansions for Example 2.1

THEOREM 7.1. *Let Ω be a sectorial domain with angle functions α and β satisfying the boundary vanishing condition (2.7). Assume also that X is Stein (or admits a complete Kähler metric) and that α and $-\beta$ are plurisubharmonic and C^2 and satisfy*

$$(7.1) \qquad i\partial\alpha \wedge \overline{\partial}\alpha \leq C\, i\partial\overline{\partial}\alpha, \quad i\partial\beta \wedge \overline{\partial}\beta \leq -C\, i\partial\overline{\partial}\beta$$

for some constant C. Then for $F \in C^\infty_{0,(n,0)}$ and $R > 0$ we have

$$(7.2) \quad (P_\Omega f)(z,w) = \frac{i}{4\pi^2 w} \sum_{-R \leq \operatorname{Im}\xi < 0} \left(\operatorname{Res}_\xi (Q_\xi \mathcal{M} P^\sharp f)(z,\xi) \right) w^{i\xi} + \mathrm{O}(|w|^{R-1}).$$

On any compact subset of $\Omega \cup (X \times \{0\})$ the estimate holds uniformly and the sum converges uniformly and absolutely.

Note that, unlike in the preceding section, poles of $\mathcal{M}P^\sharp f$ arising as in Corollary 5.3 are immediately canceled by the right-most factor in (4.6) so that the only residues appearing in the sum arise from poles of $\xi \mapsto Q_\xi$.

COROLLARY 7.2. *Let ϕ be a C^2 function on $\mathbb{C}^n$ satisfying (2.3), (2.6), and the homogeneity condition (2.1). Define X, α, β by (2.4) and (2.5) and assume that the corresponding sectorial domain Ω is pseudoconvex. Then the asymptotic expansion (7.2) is valid.*

PROOF. By Theorem 7.1 it suffices to show that (7.1) holds.

We claim first that $\log\phi$ is plurisubharmonic. In fact, X is Reinhardt and pseudoconvex (since α is a plurisubharmonic exhaustion function for X) so

$$\log X \stackrel{\text{def}}{=} \{(x_1,\dots,x_n) \in \mathbb{R}^n : (e^{x_1},\dots,e^{x_n}) \in X\}$$

is convex. The homogeneity relation (2.1) implies that

$$\log\phi(e^{x_1},\dots,e^{x_n})$$

is the signed distance from $(x_1,\dots,x_n)$ to bX along the family of lines parallel to $(a_1,\dots,a_n)$ and is thus a convex function of $(x_1,\dots,x_n)$. This implies that $\log\phi$ is plurisubharmonic on $\mathbb{C}^n \setminus \{z : z_1 \cdots z_n = 0\}$, hence on $\mathbb{C}^n$.

It follows now that

$$i\partial\alpha \wedge \overline{\partial}\alpha = \frac{\phi^2}{1-\phi^2} i\partial\log\phi \wedge \overline{\partial}\log\phi$$

$$\leq \frac{1}{2}\left\{\frac{\phi}{\sqrt{1-\phi^2}} + \frac{\phi^3}{(1-\phi^2)^{3/2}}\right\} i\partial\log\phi \wedge \overline{\partial}\log\phi$$

$$+ \frac{\phi}{2\sqrt{1-\phi^2}} i\partial\overline{\partial}\log\phi$$

$$= \frac{1}{2} i\partial\overline{\partial}\alpha.$$

Since $\beta = -\alpha$, (7.1) holds.

REMARK. The *existence* of an asymptotic expansion attached to Example 2.1 can be established more simply by other means [Bar1], but Corollary 7.2 shows that understanding the coefficients of this expansion is tantamount to understanding the residues of Q_ξ. Theorem 7.1 covers domains not treated by [Bar1].

LEMMA 7.3. *There are constants $\nu > 0, C > 0$ such that*

$$(7.3) \qquad \|Q_\xi \mathcal{M} P^\sharp F(\cdot, \xi)\|_{L^2_{(n,0)}(X, \gamma_{\mathrm{Re}\,\xi})} \leq Ce^{-\nu|\mathrm{Re}\,\xi|}$$

when $|\mathrm{Im}\,\xi| \leq R + 1$ and $|\mathrm{Re}\,\xi|$ is large.

In particular, there are only finitely many poles in the strip $|\mathrm{Im}\,\xi| \leq R + 1$.

PROOF OF THEOREM 7.1 ASSUMING LEMMA 7.3. For any relatively compact domain $X_0 \subset X$ the L^2 estimates of Lemma 7.3 give rise via the Solid Mean Value Property to L^∞ estimates of the form

$$\sqrt{\gamma_{\mathrm{Re}\,\xi}(z)} \left|Q_\xi \mathcal{M} P^\sharp f(z, \xi)\right| \lesssim e^{-\nu'|\mathrm{Re}\,\xi|}$$

for $z \in X_0, |\mathrm{Im}\,\xi| \leq R + 1, |\mathrm{Re}\,\xi|$ large.

Using the estimate

$$|w^{i\xi-1}||w|^{\mathrm{Im}\,\xi+1} \lesssim \sqrt{|\mathrm{Re}\,\xi|\gamma_{\mathrm{Re}\,\xi}(z)} \quad \text{for } w \in \Omega_z, \ |\mathrm{Re}\,\xi| \text{ large}$$

and adopting the notations E and T_j from Lemma 6.2 we find that

$$\|T_j\|_{L^\infty(E, L^\infty_{(1,0)}(\{(z,w)\in\Omega : z\in X_0\}))}$$

decays exponentially as $|j| \to \infty$.

The rest of the proof follows that of Theorem 6.1.

LEMMA 7.4. $|\gamma_\xi/\gamma_{\mathrm{Re}\,\xi}| \to 1$ as $|\mathrm{Re}\,\xi| \to \infty$, *uniformly in the strip* $|\mathrm{Im}\,\xi| < R + 1$.

PROOF. For reasons of symmetry it suffices to treat the case $\mathrm{Re}\,\xi \to +\infty$. Writing

$$(7.4) \qquad \frac{\gamma_\xi}{\gamma_{\mathrm{Re}\,\xi}} = \frac{\mathrm{Re}\,\xi}{\xi} e^{-2i\mathrm{Im}\,\xi\,\alpha}\left(1 - \frac{e^{-2i\mathrm{Im}\,\xi(\beta-\alpha)} - 1}{e^{2\mathrm{Re}\,\xi(\beta-\alpha)} - 1}\right),$$

it suffices to note that

$$(7.5) \qquad \frac{e^{-2i\operatorname{Im}\xi(\beta-\alpha)} - 1}{e^{2\operatorname{Re}\xi(\beta-\alpha)} - 1} = \begin{cases} O(\frac{1}{\operatorname{Re}\xi}) & \text{when } \beta - \alpha \geq \frac{\log\operatorname{Re}\xi}{2\operatorname{Re}\xi}, \\ O(\frac{\log\operatorname{Re}\xi}{\operatorname{Re}\xi}) & \text{when } \frac{1}{\operatorname{Re}\xi} \leq \beta - \alpha \leq \frac{\log\operatorname{Re}\xi}{2\operatorname{Re}\xi}, \\ O(\frac{1}{\operatorname{Re}\xi}) & \text{when } \beta - \alpha \leq \frac{1}{\operatorname{Re}\xi}. \end{cases}$$

LEMMA 7.5. $\|Q_\xi\|_{\mathcal{L}(L^2_{(n,0)}(X,\gamma_{\operatorname{Re}\xi}))} \to 1$ *as* $|\operatorname{Re}\xi| \to \infty$, *uniformly in the strip* $|\operatorname{Im}\xi| < R + 1$.

PROOF. $\|Q_\xi\|_{\mathcal{L}(L^2_{(n,0)}(X,\gamma_{\operatorname{Re}\xi}))} \geq 1$ so it suffices to show that

$$\limsup_{|\operatorname{Re}\xi|\to\infty} \|Q_\xi\|_{\mathcal{L}(L^2_{(n,0)}(X,\gamma_{\operatorname{Re}\xi}))} \leq 1.$$

Again it suffices to treat the case $\operatorname{Re}\xi \to +\infty$.

From (4.3) we have

$$\begin{aligned} Q_\xi &= \left(Q_{\operatorname{Re}\xi} M_{\gamma_\xi/\gamma_{\operatorname{Re}\xi}}|_{B(X,\gamma_\xi)}\right)^{-1} Q_{\operatorname{Re}\xi} M_{\gamma_\xi/\gamma_{\operatorname{Re}\xi}} \\ &= \left((I - S_{\operatorname{Re}\xi}\overline{\partial})M_{\gamma_\xi/\gamma_{\operatorname{Re}\xi}}|_{B(X,\gamma_\xi)}\right)^{-1} Q_{\operatorname{Re}\xi} M_{\gamma_\xi/\gamma_{\operatorname{Re}\xi}} \\ &= \left(M_{\gamma_\xi/\gamma_{\operatorname{Re}\xi}}\left(I - M_{\gamma_\xi/\gamma_{\operatorname{Re}\xi}}^{-1} S_{\operatorname{Re}\xi} M_{\overline{\partial}(\gamma_\xi/\gamma_{\operatorname{Re}\xi})}\right)|_{B(X,\gamma_\xi)}\right)^{-1} Q_{\operatorname{Re}\xi} M_{\gamma_\xi/\gamma_{\operatorname{Re}\xi}}, \end{aligned}$$

where $S_{\operatorname{Re}\xi}$ denotes the $L^2_{(n,0)}(X,\gamma_{\operatorname{Re}\xi})$-minimizing solution operator for $\overline{\partial}$.

Now

$$\|Q_{\operatorname{Re}\xi}\|_{\mathcal{L}(L^2_{(n,0)}(X,\gamma_{\operatorname{Re}\xi}))} = 1,$$

and Lemma 7.4 shows that

$$\|M_{\gamma_\xi/\gamma_{\operatorname{Re}\xi}}\|_{\mathcal{L}(L^2_{(n,0)}(X,\gamma_{\operatorname{Re}\xi}))}, \|M_{\gamma_\xi/\gamma_{\operatorname{Re}\xi}}^{-1}\|_{\mathcal{L}(L^2_{(n,0)}(X,\gamma_{\operatorname{Re}\xi}))} \to 1,$$

so it will suffice to show that

$$\|S_{\operatorname{Re}\xi} M_{\overline{\partial}(\gamma_\xi/\gamma_{\operatorname{Re}\xi})}\|_{\mathcal{L}(B(X,\gamma_{\operatorname{Re}\xi}),L^2_{(n,0)}(X,\gamma_{\operatorname{Re}\xi}))} \to 0.$$

But the modified Hörmander estimates of [Dem, Thm. 4.1], [Ohs] show that

$$\|S_{\operatorname{Re}\xi} g\|_{L^2_{(n,0)}(X,\gamma_{\operatorname{Re}\xi})} \leq \|g\|_{L^2_{(n,1)}((X,\omega_{\operatorname{Re}\xi}),\gamma_{\operatorname{Re}\xi})},$$

where $\omega_{\operatorname{Re}\xi}$ is the (possibly degenerate) metric induced by $-\partial\overline{\partial}\log\gamma_{\operatorname{Re}\xi}$. Now

$$(7.6) \qquad -\partial\overline{\partial}\log\gamma_{\operatorname{Re}\xi} = \frac{e^{-2\operatorname{Re}\xi\alpha}\partial\overline{\partial}\alpha - e^{-2\operatorname{Re}\xi\beta}\partial\overline{\partial}\beta}{\gamma_{\operatorname{Re}\xi}} + \frac{e^{-2\operatorname{Re}\xi(\alpha+\beta)}\partial(\beta-\alpha)\wedge\overline{\partial}(\beta-\alpha)}{\gamma_{\operatorname{Re}\xi}^2},$$

and referring to (7.4) and (7.5) we have

$$(7.7) \qquad \overline{\partial}\left(\frac{\gamma_\xi}{\gamma_{\operatorname{Re}\xi}}\right) = O(\overline{\partial}\alpha) + O\left(\frac{e^{-2\operatorname{Re}\xi\beta}}{e^{-2\operatorname{Re}\xi\alpha} - e^{-2\operatorname{Re}\xi\beta}}\overline{\partial}(\beta-\alpha)\right).$$

By (7.6) we have in particular $-i\partial\overline{\partial}\log\gamma_{\operatorname{Re}\xi} \geq 2\operatorname{Re}\xi\, i\partial\overline{\partial}\alpha$ so that the hypothesis (7.1) guarantees that the $\omega_{\operatorname{Re}\xi}$-norm of the first term of (7.7) is $O(1/\sqrt{\operatorname{Re}\xi})$; moreover, examination of the last term of (7.6) reveals that the $\omega_{\operatorname{Re}\xi}$-norm of the last term of (7.7) is $O(e^{2\operatorname{Re}\xi(\alpha-\beta)}/\operatorname{Re}\xi)$, hence also $O(1/\sqrt{\operatorname{Re}\xi})$.

Thus

$$\|M_{\overline{\partial}(\gamma_\xi/\gamma_{\mathrm{Re}\,\xi})}\|_{\mathcal{L}(B(X,\gamma_{\mathrm{Re}\,\xi}),L^2_{(n,1)}((X,\omega_{\mathrm{Re}\,\xi}),\gamma_{\mathrm{Re}\,\xi}))} = \mathrm{O}\left(\frac{1}{\sqrt{\mathrm{Re}\,\xi}}\right)$$

and

$$\|S_{\mathrm{Re}\,\xi}M_{\overline{\partial}(\gamma_\xi/\gamma_{\mathrm{Re}\,\xi})}\|_{\mathcal{L}(B(X,\gamma_\xi),L^2_{(n,0)}(X,\gamma_{\mathrm{Re}\,\xi}))} = \mathrm{O}\left(\frac{1}{\sqrt{\mathrm{Re}\,\xi}}\right),$$

completing the proof.

PROOF OF LEMMA 7.3. Once again it suffices to treat the case $\mathrm{Re}\,\xi \to +\infty$. But Theorem 5.2 guarantees that

(7.8) $$\|\mathcal{M}P^\sharp F\|_{L^2_{(n,0)}(X,\gamma_{\mathrm{Re}\,\xi})} \lesssim e^{-\nu\mathrm{Re}\,\xi}$$

when $|\mathrm{Im}\,\xi| \leq R+1$ and $\mathrm{Re}\,\xi$ is large. Applying Lemma 7.5 we immediately obtain (7.3).

8. $\overline{\partial}$-Neumann on domains with curves in the boundary

Let D be a smoothly bounded pseudoconvex domain in $\mathbb{C}^2$. Suppose that some neighborhood N of $\overline{D}$ contains a one-dimensional submanifold Y such that $Y \cap D = \emptyset$ and $Y \cap bD$ is a bordered Riemann surface $\overline{X} = X \cup b_Y X$. Shrinking N, we may and shall assume that $\overline{Y}$ is a bordered Riemann surface retracting onto $\overline{X}$ and that a neighborhood U of Y is biholomorphic to the product $Y \times \Delta$ of Y with the unit disk via a map $\Psi = (Z,W) : U \to Y \times \Delta$ [Siu, Cor.1] [For, Thm. 30.3].

Let r be a smooth real-valued function on N with non-vanishing gradient on bD such that D is defined by the inequality $r < 0$, and let $\alpha : Y \to \mathbb{R}/2\pi\mathbb{Z}$ be defined by

$$\alpha(z) = \arg\frac{\partial(r \circ \Psi^{-1})}{\partial\overline{w}}(z,0).$$

Then α is harmonic on X (see for example [BF]) and smooth on Y (shrinking further if necessary).

LEMMA 8.1. *After possible modification of the map Ψ, α can be lifted to a single-valued map $Y \to \mathbb{R}$.*

PROOF. Let $p_\alpha : \pi_1(Y) \to 2\pi\mathbb{Z}$ be the period map for α. Let ω be a bounded holomorphic differential on Y with additive periods given by $-ip_\alpha$ [For, Thm. 28.6]. Then replacing Ψ by $\widetilde{\Psi} \stackrel{\text{def}}{=} (Z, R\exp(\int_{z_0}^Z \omega)W)$, R large, and U by $\widetilde{U} = \widetilde{\Psi}^{-1}(Y \times \Delta)$ we find that $\alpha(z)$ is replaced by $\tilde{\alpha}(z) = \alpha(z) + \mathrm{Im}\int_{z_0}^z \omega$ so that $p_{\tilde{\alpha}} = 0$ as required.

We assume henceforth that α is single-valued. Let $\beta = \alpha + \pi$ and define γ_ξ and Q_ξ as in Section 3. Then Theorem 4.3 guarantees that $\xi \mapsto Q_\xi$ extends meromorphically to $\xi \in \mathbb{C}$.

Let $W^k_{(2,0)}(D)$ denote the space of $(2,0)$-forms on D with k with coefficients in the Sobolev space W^k.

THEOREM 8.2. *For D as above, if Q_ξ has a pole at a point ξ_0 with $-k <$ $\operatorname{Im}\xi_0 < 0$ then the Bergman projection P_D for D does not map $W^k_{(2,0)}(D)$ to $W^k_{(2,0)}(D)$.*

If α has a single-valued harmonic conjugate than $\xi \mapsto Q_\xi$ is entire by Corollary 4.7 and, indeed, if bD is strictly pseudoconvex (or finite type) off of $\overline{X}$ then a result of Boas and Straube [BS] guarantees that in this case P_D maps $W^k_{(2,0)}(D)$ to $W^k_{(2,0)}(D)$ for all k.

Theorem 8.2 was proved earlier in [Bar2] in the special case of the Diederich-Fornæss worm domains. Christ has recently shown [Chr] that in fact the Bergman projection operators for these domains fail to map $C^\infty_{(2,0)}(\overline{D})$ to $C^\infty_{(2,0)}(\overline{D})$ (i.e., "Condition R" does not hold for these domains).

In general, Proposition 4.6 shows that for fixed X the poles of $\xi \mapsto Q_\xi$ depend only on the conjugate periods of α. We claim that for a dense set of periods we will in fact encounter poles in the lower half-plane. Indeed, [Bar3, Sec. 4] shows that for a dense set of periods there is ξ_0 on the negative imaginary axis and $\sigma \in L^2_{(1,1)}(X), \sigma \not\equiv 0$ with

$$(8.1) \qquad\qquad \sigma = -2\xi_0 \overline{\partial}\alpha \wedge S_X \sigma.$$

We claim that Q_ξ has a pole at ξ_0, for otherwise (4.2) and analytic continuation yield

$$(8.2) \qquad\qquad Q_{\xi_0}\left(e^{2\xi_0\alpha} P_X S_X \sigma\right) = Q_{\xi_0}\left(e^{2\xi_0\alpha} S_X \sigma\right).$$

But the left-hand side of (8.2) vanishes since $P_X \circ S_X = 0$ while the right-hand side does not vanish since $e^{2\xi_0\alpha} S_X \sigma$ is non-zero and holomorphic (by (8.1)); the contradiction establishes the claim. A general argument [Aup, Thm. 3.4.24] now shows that poles are encountered off of a pluripolar set of periods.

Let $\Omega \subset Y \times \mathbb{C}$ be the sectorial domain with base X and angle functions α, β, and let Ω' be the sectorial domain with base Y and angle functions $\alpha - \pi/2, \beta + \pi/2$.

LEMMA 8.3. *$B(\Omega')|_\Omega$ is dense in $B(\Omega)$.*

PROOF. Letting γ'_ξ denote the weight function associated to Ω', it will suffice by Plancherel's formula (3.2) to show that the space of z-holomorphic functions in $L^2_{(1,0)}(Y \times \mathbb{R}, \gamma'_\xi)$ is dense in the space of z-holomorphic functions in $L^2_{(1,0)}(X \times \mathbb{R}, \gamma_\xi)$. But this follows easily from the density of $B(Y)$ in $B(X)$ [Bel, Cor. 15.4] [HL, Thm. 2.12.3(ii)].

PROOF OF THEOREM 8.2. For $\lambda \in \mathbb{R}^+$ let $\Psi_\lambda = (Z, \lambda W), \Omega_\lambda = \Psi_\lambda(U \cap D)$.

Fix $F \in C^\infty_{0,(2,0)}(Y \times \mathbb{C})$ with supp $F \subset \Omega$. For λ large we have supp $F \subset \Omega_\lambda$ and thus $\Psi^*_\lambda F \in C^\infty_{0,(2,0)}(U \cap D)$. These forms have L^2-norm independent of λ. Hence the forms

$$P_D\left(\chi_{U \cap D} \Psi^*_\lambda F\right) \in B(D)$$

are uniformly bounded in $B(D)$, and the forms

$$\chi_{\Omega_\lambda}(\Psi^{-1}_\lambda)^* P_D\left(\chi_{U \cap D} \Psi^*_\lambda F\right)$$

are uniformly bounded in $L^2_{(2,0)}(Y \times \mathbb{C})$ and holomorphic on Ω_λ.

Since each compact $K \subset \Omega$ is contained in Ω_λ for $\lambda \geq \lambda_K$, we may invoke Cauchy-type estimates and Montel's Theorem to conclude that for some sequence $\lambda_j \to \infty$ the forms

$$\chi_{\Omega_{\lambda_j}} (\Psi_{\lambda_j}^{-1})^* P_D \left(\chi_{U \cap D} \Psi_{\lambda_j}^* F \right)$$

converge weakly in $L_{(2,0)}^2 (Y \times \mathbb{C})$ to a form $\widetilde{P}F$ which is holomorphic in Ω and $\equiv 0$ off of $\overline{\Omega}$.

We claim that

(8.3) $$\widetilde{P}F = P_\Omega F$$

on Ω. By Lemma 8.3 it suffices to show that

(8.4) $$\int_\Omega \widetilde{P}F \wedge \overline{H} = \int_\Omega F \wedge \overline{H}$$

for all $H \in B(\Omega')$. But

$$\int_\Omega \widetilde{P}F \wedge \overline{H} = \lim_{j \to \infty} \int_{Y \times \mathbb{C}} \chi_{\Omega_{\lambda_j}} (\Psi_{\lambda_j}^{-1})^* P_D \left(\chi_{U \cap D} \Psi_{\lambda_j}^* F \right) \wedge \overline{H}$$

$$= \lim_{j \to \infty} \int_{U \cap D} P_D \left(\chi_{U \cap D} \Psi_{\lambda_j}^* F \right) \wedge \overline{\Psi_{\lambda_j}^* H}.$$

Picking a cutoff function $\mu \in C_0^\infty(\Delta)$ with $\mu \equiv 1$ near 0 and noting that the mass of $\Psi_\lambda^*(F)$ concentrates along $W = 0$ as $\lambda \to \infty$ we have

$$\lim_{j \to \infty} \int_{U \cap D} P_D \left(\chi_{U \cap D} \Psi_{\lambda_j}^* F \right) \wedge \overline{\Psi_{\lambda_j}^* H}$$

$$= \lim_{j \to \infty} \int_{U \cap D} \mu(W) P_D \left(\chi_{U \cap D} \Psi_{\lambda_j}^* F \right) \wedge \overline{\Psi_{\lambda_j}^* H}$$

$$= \lim_{j \to \infty} \int_{U \cap D} \mu(W) \Psi_{\lambda_j}^* F \wedge \overline{\Psi_{\lambda_j}^* H}$$

$$\qquad - \lim_{j \to \infty} \int_{U \cap D} \Psi_{\lambda_j}^* F \wedge \overline{(I - P_D)(\chi_{U \cap D} \mu(W) \Psi_{\lambda_j}^* H)}$$

$$= \lim_{j \to \infty} \int_{\Omega_{\lambda_j}} \mu(w/\lambda_j) F \wedge \overline{H}$$

$$\qquad - \lim_{j \to \infty} \int_{U \cap D} \Psi_{\lambda_j}^* F \wedge \overline{S(\overline{\partial}(\chi_{U \cap D} \mu(W) \Psi_{\lambda_j}^* H))},$$

where S is the L^2-minimizing solution operator for $\overline{\partial}$ on D [Hör2, Prop. 4.2.5]. But the first term equals $\int_\Omega F \wedge \overline{H}$ when j is large, and the second term tends to 0 since $\Psi_{\lambda_j}^* H \to 0$ in L^2 on supp $\overline{\partial}(\chi_{U \cap D} \mu(W))$. Thus we have established (8.4) and hence (8.3).

Now for $\lambda > 1$ the W^k-norms of $\chi_{U \cap D} \Psi_\lambda^* F$ are $O(\lambda^k)$. Thus if P_D maps W^k to W^k then the W^k-norms of $P_D (\chi_{U \cap D} \Psi_\lambda^* F)$ are also $O(\lambda^k)$ and so the L^2-norms of

$$\left(\frac{\partial}{\partial w} \right)^k (\Psi_\lambda^{-1})^* P_D (\chi_{U \cap D} \Psi_\lambda^* F)$$

on $\Omega_\lambda \cap \Omega$ are uniformly bounded as $\lambda \to \infty$. Thus $(\partial/\partial w)^k P_\Omega F$ is square-integrable on Ω. By interpolation it follows that $(\partial/\partial w)P_\Omega F, \ldots, (\partial/\partial w)^{k-1} P_\Omega F$ are also square-integrable on Ω. Theorem 5.5 now tells us that $\mathcal{M}P_\Omega F$ is meromorphic on

$\Omega \times \{\xi : -k < \operatorname{Im} \xi < 0\}$ with, at most, simple poles along $\Omega \times i\mathbb{Z}$. Theorems 3.1 and 5.2 now implies that $\mathcal{M}P^\sharp F$ lies in the kernel of the principal part of Q_ξ at ξ_0 if $i\xi_0 \notin \mathbb{Z}$ and $\operatorname{Res}_{\xi_0} \mathcal{M}P^\sharp F$ lies in the kernel of the principal part of Q_ξ at ξ_0 if $i\xi_0 \in \mathbb{Z}$. In view of the density results in Corollary 5.4, this implies that there is no principal part and hence no pole at ξ_0.

Appendix A. Mellin transform formulae

PROOF OF BASIC FACTS ABOUT $\mathcal{M}$.

Setting $dz = dz_1 \wedge \cdots \wedge dz_n$ for brevity, we have

$$\|F\|^2 = \frac{i^{(n+1)^2}}{2^{n+1}} \int_\Omega F \wedge \overline{F}$$

$$= \frac{i^{n^2}}{2^n} \int_{\substack{z \in X \\ u \in \mathbb{R} \\ \alpha(z) < \theta < \beta(z)}} |f(z, e^{u+i\theta})|^2 e^{2u} \, dz \wedge \overline{dz} \wedge du \wedge d\theta$$

$$\text{(setting } w = e^{u+i\theta})$$

$$= \frac{i^{n^2}}{2^{n+1}\pi} \int_{\substack{z \in X \\ \xi \in \mathbb{R} \\ \alpha(z) < \theta < \beta(z)}} \left| \int_{u \in \mathbb{R}} f(z, e^{u+i\theta}) e^{u-i\xi u} \, du \right|^2 dz \wedge \overline{dz} \wedge d\xi \wedge d\theta$$

$$\text{(by Plancherel)}$$

$$= \frac{i^{n^2}}{2^{n+1}\pi} \int_{\substack{z \in X \\ \xi \in \mathbb{R} \\ \alpha(z) < \theta < \beta(z)}} e^{-2\xi\theta} \mathcal{M}F \wedge \overline{\mathcal{M}F} \wedge d\xi \wedge d\theta$$

$$\text{(changing back to } w)$$

$$= \frac{i^{n^2}}{2^{n+1}\pi} \int_{X \times \mathbb{R}} \frac{e^{-2\xi\alpha(z)} - e^{-2\xi\beta(z)}}{2\xi} \mathcal{M}F \wedge \overline{\mathcal{M}F} \wedge d\xi$$

for $F \in A^\sharp(\Omega)$, establishing (3.2) and allowing us to extend $\sqrt{2\pi}\mathcal{M}$ to an isometry of $B^\sharp(\Omega)$ into $L^2_{(n,0)}(X \times \mathbb{R}, \gamma_\xi(z) \, d\xi)$.

Since

$$\int_\mathbb{R} |\mathcal{M}f(z,\xi) w^{i\xi-1}| \, d\xi \leq \left(\int_\mathbb{R} |\mathcal{M}f(z,\xi)|^2 \gamma_\xi(z) \, d\xi \right)^{1/2} \left(\int_\mathbb{R} \frac{|w^{2(i\xi-1)}|}{\gamma_\xi(z)} \, d\xi \right)^{1/2}$$

and

$$\int_\mathbb{R} \frac{|w^{2(i\xi-1)}|}{\gamma_\xi(z)} \, d\xi < \infty$$

for $w \in \Omega_z$ by (3.4), the integral $\int_\mathbb{R} \mathcal{M}f(z,\xi) w^{i\xi-1} \, d\xi$ converges when $w \in \Omega_z$ for a.e. z. Thus to establish (3.3) we may restrict attention to F in the dense subspace

$A^\sharp(\Omega)$. In this case the Fourier inversion formula yields

$$f(z,w) = \frac{1}{2\pi w} \int_{\xi \in \mathbb{R}} \left(\int_{x \in \mathbb{R}} e^x w f(z, e^x w) e^{-ix\xi}\, dx \right) d\xi$$

$$= \frac{1}{2\pi w} \int_{\xi \in \mathbb{R}} \left(\int_0^\infty f(z,\zeta) \left(\frac{\zeta}{w} \right)^{-i\xi} d\zeta \right) d\xi$$

$$= \frac{1}{2\pi} \int_{\mathbb{R}} \mathcal{M}f(z,\xi) w^{i\xi-1}\, d\xi$$

as required.

A similar argument establishes that the reverse inversion formula holds on a dense subspace of $L^2_{(n,0)}(X \times \mathbb{R}, \gamma_\xi(z)\, d\xi)$, showing that $\mathcal{M}$ maps $B^\sharp(\Omega)$ onto $L^2_{(n,0)}(X \times \mathbb{R}, \gamma_\xi(z)\, d\xi)$.

References

[Aup] B. Aupetit, *A primer on spectral theory*, Springer-Verlag, 1991.

[Bar1] D. Barrett, *Regularity of the Bergman projection on domains with transverse symmetries*, Math. Ann. **258** (1982), 441–446.

[Bar2] D. Barrett, *Behavior of the Bergman projection on the Diederich-Fornæss worm*, Acta Math. **168** (1992), 1–10.

[Bar3] D. Barrett, *Duality between A^∞ and $A^{-\infty}$ on domains with nondegenerate corners*, Cont. Math. **185** (1995), 77–87.

[BF] D. Barrett and J. E. Fornæss, *On the smoothness of Levi-foliations*, Pub. Math. UAB (Barcelona) **32** (1988), 171–177.

[Bel] S. Bell, *The Cauchy transform, potential theory, and conformal mapping*, CRC Press, 1992.

[BG] C. Berenstein and R. Gay, *Complex variables*, Springer-Verlag, 1991.

[BS] H. Boas and E. Straube, *de Rham cohomology of manifolds containing the points of infinite type, and Sobolev estimates for the $\overline{\partial}$-Neumann problem*, J. Geom. Anal. **3** (1993), 225–235.

[Boa] R. P. Boas, Jr., *Entire Functions*, Academic Press, 1954.

[Cat] D. Catlin, *Subelliptic estimates for the $\overline{\partial}$-Neumann problem*, Ann, Math. **117** (1987), 147–172.

[Chr] M. Christ, *Global C^∞ irregularity of the $\overline{\partial}$-Neumann problem for worm domains*, J. Amer. Math. Soc. **9** (1996), 1171–1185.

[Dem] J.-P. Demailly, *Estimations L^2 pour l'opérateur $\overline{\partial}$ d'un fibré vectoriel holomorphe semi-positif au-dessus d'une variété Kahlérienne compléte*, Ann. Scient. Ec. Norm. Sup., 4e série **15** (1982), 457–511.

[For] O. Forster, *Lecture on Riemann surfaces*, Springer-Verlag, 1981.

[FK] G. B. Folland and J. J. Kohn, *The Neumann problem for the Cauchy-Riemann complex*, Ann. of Math. Studies No. 75, Princeton Univ. Press (1972).

[HL] G. Henkin and J. Leiterer, *Theory of functions on complex manifolds*, Birkhäuser, 1984.

[Hor1] L. Hörmander, *An introduction to complex analysis in several variables*, 3rd ed., North-Holland, 1990.

[Hor2] L. Hörmander, *Notions of convexity*, Birkhäuser, 1994.

[Kra] S. Krantz, *Partial differential equations and con.,plex analysis*, CRC Press, 1992.

[Ohs] T. Ohsawa, *Vanishing theorem on complete Kähler manifolds*, Publ. RIMS, Kyoto Univ. **20** (1984), 21–28.

[Rau] J. Rauch, *Partial differential equations*, Springer-Verlag, 1991.

[RS] M. Reed and B. Simon, *Methods of modern mathematical physics, Vol I: Functional analysis,* revised and enlarged edition, Academic Press, 1980.

[Siu] Y.-T. Siu, *Every Stein subvariety admits a Stein neighborhood*, Invent. math. **38** (1976), 89–100.

[Sta]   N. Stanton, *Homogeneous real hypersurfaces*, Math. Res. Lett. **2** (1995), 311–319.

DEPTMENT OF MATHEMATICS, UNIVERSITY OF MICHIGAN, ANN ARBOR, MI 48109-1003 U.S.A.

E-mail address: barrett@umich.edu

Contemporary Mathematics
Volume **212**, 1998

Subelliptic Geometry

Richard Beals, Bernard Gaveau and Peter Greiner

ABSTRACT. To any subriemannian geometry, one can associate a subelliptic operator. In the riemannian case, this operator is the usual elliptic Laplace operator and, in this situation, the main singularity of the operator is given by $d(x,y)^{2-n}$ where n is the dimension and $d(x,y)$ is the riemannian distance. The analoguous statement for a subelliptic operator (with $d(x,y)$ the subriemannian distance) is not correct. This means that the subriemannian concepts are irrelevant to invert subelliptic operators. In the step 2 case, one can construct an approximate fundamental solution as an integral over a characteristic variety with parameter τ,

$$\int_{\mathbf{R}^p} \frac{V(x,y,\tau)d\tau}{f(x,y,\tau)^n},$$

where $f(x,y,\tau)$ is a complex distance function depending on τ and $V(x,y,\tau)$ a density function. The subriemannian distance is the critical value with respect to τ of this complex distance function. This complex distance function defines a new kind of subelliptic geometric concept.

1. Introduction

Let $X_1, \ldots, X_p$ denote p vector fields on an open set of $\mathbf{R}^n$ and consider the "Laplacian"

(1)
$$\Delta = \frac{1}{2} \sum_{i=1}^{p} X_i^2.$$

We want to find the Green function $G(x,y)$ of Δ,

$$\Delta G(x,y) = \delta(x-y)$$

and the heat kernel $p_u(x,y)$ of Δ,

$$\frac{\partial p_u(x,y)}{\partial u} = \Delta p_u(x,y), \qquad (u>0);$$
$$\lim_{u \to o} p_u(x,y) = \delta(x-y),$$

1991 *Mathematics Subject Classification*. Primary 35A08, 35H05; Secondary 53B99.

Research partially supported by NSF Grant DMS-9213595, by E. U. Grant "Capital humain et mobilité" and by NSERC Grant OGP0003017.

or, more generally, parametrices for these problems. Usual potential theory studies the case of an elliptic operator where there are n vector fields which are linearly independent. In this situation, the model case is the usual Laplace operator in $\mathbf{R}^n$

$$\frac{1}{2}\sum_{i=1}^{n}\frac{\partial^2}{\partial x_i^2}$$

and the Green function and heat kernel are explicitly given by

$$G(x,y) = \frac{2}{(n-2)\sigma_{n-1}} \cdot \frac{1}{|x-y|^{n-2}}, \qquad n > 2,$$

$$p_u(x,y) = \frac{1}{(2\pi u)^{n/2}} \exp\left(-\frac{|x-y|^2}{2u}\right),$$

(σ_{n-1} is the area of the unit sphere in $\mathbf{R}^n$).

More generally, if Δ is an elliptic operator of the form (1), the main singularity of the Green function is given by

$$(2) \qquad\qquad G(x,y) \sim \frac{1}{d(x,y)^{n-2}} \quad \text{for } x \sim y.$$

Here $d(x,y)$ is the riemannian distance associated to the riemannian metric $g_{ij}dx^i dx^j$, where $g^{ij}\xi_i\xi_j$ is the principal symbol of Δ. Moreover, it is well known that for u small,

$$(3) \qquad\qquad -\log p_u(x,y) \sim \frac{d(x,y)^2}{2u}.$$

For a general elliptic operator Δ one can also construct explicitly the Green function in terms of a convergent series (provided Δ has real analytic coefficients) of the form:

$$(4) \qquad G(x,y) = \sum_{k=0}^{\infty}\frac{V_k(x,y)}{d(x,y)^{n-2-k}} + \sum_{k\geq 0}\log d(x,y)\ d(x,y)^k W_k(x,y).$$

Here again $d(x,y)$ is the riemannian distance, $V_0(x,y)$ is esentially the square root of the Jacobian of the exponential map of center x and the $V_k(x,y)$ and $W_k(x,y)$ can be recursively constructed by integration, along the geodesics of the riemannian metric, of a first order transport equation. There exists also a similar formula for the heat kernel

$$(5) \qquad p_u(x,y) = \exp\left(-\frac{d(x,y)^2}{2u}\right)\frac{1}{u^{n/2}}\left[\sum_{k\geq 0}A_k(x,y)u^k\right],$$

where again the $A_k(x,y)$ are recursively defined by integration of the first order transport equations. From (5), one can recover (3). (See for example [1] among many references). As a consequence, we see that the theory of (second order) elliptic operators can be constructed using only riemannian geometric concepts. It is our purpose in this article to construct the Green function and the heat kernel for a class of subelliptic operators, the so called step 2 operators with one missing direction. Unfortunately, in this situation, the relation of these fundamental solutions to geometric concepts will be much less direct than in the elliptic situation. First, we state the hypotheses on the operator. We assume that we have $2n$ vector fields $X_1, \ldots, X_{2n}$ defined on a domain of $\mathbf{R}^{2n+1}$ and we consider the analogue of (1) in this situation

$$\Delta = \frac{1}{2} \sum_{j=1}^{2n} X_j^2.$$

We shall denote $(x_1, \ldots, x_{2n}, t)$ the $2n+1$ coordinates in $\mathbf{R}^{2n+1}$ and we shall assume that $X_1, \ldots, X_{2n}, \partial/\partial t$ is a basis of $\mathbf{R}^{2n+1}$ at each point. Moreover, we assume that the skew symmetric matrix of the commutators $[X_j, X_k]$ is non-degenerate along $\partial/\partial t$ everywhere and that the X_j are real analytic. In this situation, Δ is hypoelliptic in the sense of Hörmander but not elliptic and has a smooth fundamental solution outside its pole. Our purpose is to give a precise description of the singularity of the fundamental solution and of the heat kernel. In the next section we shall describe the model case and the answers. In Section 3 we shall interpret the answers and in Sections 4 and 5 we shall present a generalization to any operator Δ which has the structure described above.

2. The Model Operator

In this section and the next one, it is more convenient to change notation and use $(x_1, y_1, \ldots, x_n, y_n, t)$ to denote the coordinates in $\mathbf{R}^{2n+1}$ and X_i, Y_i to denote the vector fields. These vector fields are given explicitly by formulas of the type

$$X_i = \frac{\partial}{\partial x_i} + 2a_i y_i \frac{\partial}{\partial t}$$

$$Y_i = \frac{\partial}{\partial y_i} - 2a_i x_i \frac{\partial}{\partial t}$$

and we shall assume that the a_i are positive numbers and

$$\Delta = \frac{1}{2} \sum_{j=1}^{n} (X_j^2 + Y_j^2).$$

In the case where all a_j equal 1, the fundamental solution is explicitly given by a formula due to Folland [2], namely

$$(6) \qquad G(x, 0) = \frac{C_n}{\left(\sum_{j=1}^{n} (x_j^2 + y_j^2)^2 + t^2 \right)^{n/2}},$$

where C_n is an appropriate constant (in fact, it is well-known that the X_i, Y_i are left invariant vector fields for the Heisenberg group structure and it is sufficient to study the Green function of pole 0). The simplicity of formula (6) is unfortunately misleading. In fact, if the a_j are not equal, there is in general no such simple formula for the Green function and even if the a_j equal 1, there is no simple formula for the heat kernel. In fact, we have in general the following result.

The Green function of pole 0

$$\Delta G = \delta_0$$

is given by the integral formula

$$(7) \qquad G(x, y, t) = c_n \int_{-\infty}^{+\infty} \frac{V^{(0)}(\tau)}{f^n} d\tau$$

where

$$(8) \qquad f(x,y,t,\tau) = \sum_{j=1}^{n} a_j \tau r_j^2 \coth 2a_j\tau - it\tau$$

$$(9) \qquad V^{(0)}(\tau) = \prod_{j=1}^{n} \left(\frac{2a_j\tau}{\sinh 2a_j\tau} \right)$$

$$(10) \qquad c_n = -\frac{\Gamma(n)}{(2\pi)^{n+1}}$$

$$(11) \qquad r_j^2 = x_j^2 + y_j^2.$$

The heat kernel of pole 0

$$\frac{\partial}{\partial u} p_u(x,y,t) = \Delta p_u(x,y,t)$$

is given by the integral formula

$$(12) \qquad p_u(x,y,t) = \frac{c_n'}{u^{n+1}} \int_{+\infty}^{-\infty} e^{-(1/u)f} V^{(0)}(\tau)\, d\tau.$$

REMARK. It is easy to obtain formula (7) from (12) because

$$(13) \qquad G(x,y,t) = \int_0^\infty p_u(x,y,t)\, du$$

These results were obtained in [3] for the heat kernel (although with different normalizations) and [4] using different methods. In [3], the method was probabilistic and in [4], it used special function theory. Although both methods are quite respectable, they did not reveal the structure of these formulas, and in particular, it was not possible to generalize these formulas to more general situations, even in the step 2 case.

Moreover, probability theory was also misleading for another reason which we shall explain. Let us introduce the conjugate momenta $(\xi_1, \eta_1, \dots, \xi_n, \eta_n, \theta)$ of the coordinates $(x_1, y_1, \dots, x_n, y_n, t)$, respectively, and let us define the hamiltonian

$$(14) \qquad H(\xi, \eta, \theta, x, y, t) = \frac{1}{2} \sum_{j=1}^{n} [(\xi_j + 2a_j y_j \theta)^2 + (\eta_j - 2a_j x_j \theta)^2].$$

Thus H is exactly the symbol of the Heisenberg Laplace operator Δ. In a riemannian setting it would be the usual $\sum_{i,j} g^{ij} \xi_i \xi_j$ but here we are not in a riemannian setting because the hamiltonian is degenerate (the g^{ij} matrix is not invertible).

Nevertheless, we can construct a Hamilton-Jacobi equation

$$(15) \qquad \frac{\partial S_u}{\partial u} + H\left(\frac{\partial S}{\partial x}, \frac{\partial S}{\partial y}, \frac{\partial S}{\partial t}, x, y, t \right) = 0$$

and we can integrate this equation using the bicharacteristics of H

$$(16) \qquad \frac{dx_j}{ds} = H_{\xi_j}, \qquad \frac{dy_i}{ds} = H_{\eta_j}, \qquad \frac{dt}{ds} = H_\theta;$$

$$\frac{d\xi_j}{ds} = -H_{x_j}, \qquad \frac{d\eta_j}{ds} = -H_{y_j}, \qquad \frac{d\theta}{ds} = -H_t.$$

(The last term is identically 0 in this model). In riemannian geometry, if we take the bicharacteristics starting from z at time 0 and arriving at w at time u, we would obtain for S_u the function $d(z,w)^2/2u$.

Here we can do the same, namely start the bicharacteristic at time 0 from the origin and choose initial momenta so that the bicharacteristic arrive at point (x, y, t) at time u. We thus obtain a well defined function $S_u(x, y, t)$, which we shall call the classical action, by the usual formula

$$(17) \qquad S_u(x, y, t) = \int_0^u \left(\sum_j (\xi_j dx_j + \eta_j dy_j) + \theta dt - H du \right).$$

Let us assume that $n = 1$, so that there are only 3 coordinates (x, y, t). If you do these computations (they can be done explicitly), you discover that $S_u(x, y, t)$ is C^∞ everywhere for $u > 0$, except on the t axis where $S_u(x, y, t)$ is not even C^1 (it is only continuous). In fact, $\partial S_u / \partial x$ and $\partial S_u / \partial y$ on the t axis are not defined, because there is a whole manifold of bicharacteristics starting from 0 at time 0, arriving at a point $(x = y = 0, t)$ at a fixed time u, and having different tangents at their final point. In fact, the set of all possible tangents is the hyperplane $t = 0$. As a consequence S_u is not C^1 on the t axis and its level surfaces $S_u = \text{const}$ have a conical singularity on the t axis.

Thus we have a rather peculiar situation: all the points of the t axis are conjugate points (or focal points) of the origin. This situation never occurs in a riemannian context because in riemannian geometry, the exponential map is bijective in a small neighborhood of its origin (which is the same as to say that the geodesics flow has no conjugate point of a given point in a small ball).

This is only the first part of the story. The next part is the fact that nevertheless, one can define a metric on the so called *horizontal* space, which at a point (x, y, t) is the space $H_{(x,y,t)}$ generated by the X_j and the Y_j at that point. We define the metric by saying that the X_j and Y_j are an orthonormal basis of $H_{(x,y,t)}$. We can also say that the length of $\partial/\partial t$ is $+\infty$. Then we can ask the question: what are the curves which are always tangent to the horizontal space at any point, which go from 0 to (x, y, t) and minimize the length. In [3], [6] where all these concepts were introduced, it was proved that these curves are exactly the bicharacteristics, solutions of (16) starting from the origin and arriving at time $u = 1$ at (x, y, t) and that the squared length is $S_u(x, y, t)|_{u=1}$ (the proof is rather indirect and uses the end of our story (see [6])).

The third part of the story comes from probability theory and is also explained in [3], [6]. First of all, it was proved by a direct stationary phase estimation, that

$$(18) \qquad -\log p_u(x, y, t) \sim S_u(x, y, t)$$

for u small, where p_u is given by (12). The reason for this is the following: the probabilistic formula for the heat kernel is essentially

$$(19) \qquad p_u(x, y, t) = \int_{\Omega_h} \exp\left(-\frac{1}{2} \int_0^u \sum_{j=1}^n (\dot{x}_j^2 + \dot{y}_j^2) ds \right) \text{``}\mathcal{D} \text{ path''}$$

where Ω_h is the set of horizontal paths from 0 to (x, y, t) in time u and where $\dot{x}_j = dx_j/du$ as usual and "$\mathcal{D}$ path" is a short hand notation à la Feynman for the measure on this space of curves. This formula can be made precise (see [3]), but it immediately suggests that $\log p_u$ is for small u equivalent to the minimal length from 0 to (x, y, t) (renormalized by $1/u$) of horizontal curves, which in turn was proved to be equivalent to the classical action ([**3**]). Actually, it was proved in [**6**], in a very general setting that

$$- \log p_u \geq \text{Minimal length}$$

(this was done using probabilistic methods). Using the Malliavin calculus it is possible also to prove

$$- \log p_u \sim \text{Minimal length}$$

in the step 2 setting under some technical hypothesis and, using [**6**], to prove (18). Now, the final part of the story, and the reason why (18) was misleading, is the following. In a riemannian situation, we have

$$- \log p_u \sim \frac{d(x, y)^2}{2u}$$

and if we forget rigor, $p_u \sim 1/u^{n/2} e^{-d(x,y)^2/2u}$ (the power $n/2$ is there for reasons of homogeneity) so that from the identity

(20) $$G(x, y) = \int_0^\infty p_u(x, y) \, du$$

we deduce the correct Green function

$$G(x, y) \sim \frac{1}{d(x, y)^{n-2}}.$$

Now by the same reasoning we could say that in our subelliptic situation

$$G(x, y, t) \sim \frac{2}{S_u(x, y, t)^n}$$

(the power n is there for reasons of homogeneity), but this is completely wrong: S_u is not C^1 on the t axis, so the evaluation of Δ on the function $1/S_u(x, y, t)^n$ gives a Dirac mass all along the t axis.

In fact, it was proved in [**3**] for the case $n = 1$

$$p_u(x, y, t) \sim \frac{1}{u^{3/2}} e^{-S_u(x,y,t)},$$

$x_1^2 + y_1^2 \neq 0$ and also

$$p_u(x, y, t) \mid_{x=y=0} \sim \frac{1}{u^2} e^{-S_u(x,y,t)} \mid_{x=y=0} .$$

All these facts indicate that what is now known as "subriemannian geometry" is not well adapted to the study of subelliptic operators and their inverses.

The subriemannian geometric concepts are simply not the correct concepts contrary to the elliptic case and this is why we had to define new subelliptic geometric concepts. Moreover, the probabilistic arguments estimate the heat kernel as a function, but not as a kernel, which is of course not a surprise.

3. Interpretation of the formulas for the Green function and heat kernel

Let us now come back to the formulas for the Green function. Assume that we look for a Green function of Δ (always in the model case) under the form

$$(21) \qquad G = \int \frac{v^{(0)}(x, y, t, \tau)}{g(x, y, t, \tau)^n} d\tau,$$

where $v^{(0)}$ and g are unknown functions and let us compute Δ on G. It is easy to obtain

$$(22)$$

$$\Delta \left(\frac{v^{(0)}}{g^n} \right) = \frac{(\Delta v^{(0)})}{g^n} - \frac{n}{g^{n+1}} \left((\Delta g)v^{(0)} + \sum_{j=1}^{n} (X_j g \cdot X_j v^{(0)} + Y_j g \cdot Y_j v^{(0)}) \right)$$

$$+ \frac{n(n+1)}{g^{n+2}} H(\nabla g)v^{(0)},$$

where $H(\nabla g)$ is a short hand notation for

$$H(\nabla g) = H \left(\frac{\partial g}{\partial x}, \frac{\partial g}{\partial y}, \frac{\partial g}{\partial t}, x, y, t \right),$$

where H was defined as the symbol of Δ in (14).

Now let us assume that g satisfies a Hamilton Jacobi equation with a "time" τ, namely

$$(23) \qquad H(\nabla g) + \frac{\partial g}{\partial \tau} = 0$$

and rewrite the last term (the "most singular") term of (22) as

$$(24) \qquad \frac{n(n+1)}{g^{n+2}} \left(-\frac{\partial g}{\partial \tau} \right) v^{(0)} = n \frac{\partial}{\partial \tau} (g^{-(n+1)}) v^{(0)}$$

$$= n \frac{\partial}{\partial \tau} (g^{-(n+1)} v^{(0)}) - \frac{n}{g^{n+1}} \frac{\partial v^{(0)}}{\partial \tau}.$$

Then we obtain from (24) and (22)

$$\Delta \left(\frac{v^{(0)}}{g^n} \right) = \frac{\Delta v^{(0)}}{g^n} - \frac{n}{g^{n+1}} \left[\Delta(g)v^{(0)} + \left(\sum_{j=1}^{n} (X_j g)X_j + (Y_j g)Y_j \right) v^{(0)} + \frac{\partial v^{(0)}}{\partial \tau} \right]$$

$$+ n \frac{\partial}{\partial \tau} (g^{-(n+1)} v^{(0)}),$$

so if we assume that $v^{(0)}/g^{n+1}$ tends to 0 for large $|\tau|$, we obtain

$$(25) \qquad \Delta G = \int_{-\infty}^{+\infty} \frac{\Delta v^{(0)}}{g^n} d\tau - n \int_{-\infty}^{+\infty} \frac{T(v^{(0)})}{g^{n+1}} d\tau,$$

where we have defined $T(v)$ to be the first order operator:

$$(26) \qquad T(v) = (\Delta g)v + \left(\sum_{j=1}^{n} (X_j g)X_j + (Y_j g)Y_j \right) v + \frac{\partial v}{\partial \tau}.$$

The first problem is to integrate the Hamilton Jacobi equation (23). Again we use the bicharacteristics (16), but we shall assume the following boundary conditions: we consider a path $(x_j(s), y_j(s), t(s), \xi_j(s), \eta_j(s), \theta(s))$ for $0 \leq s \leq \tau$, solution of (16) such that:

$$(27) \qquad \begin{aligned} x_j(0) &= y_j(0) = 0, \quad \theta(0) = -i \equiv \theta_0; \\ x_j(\tau) &= x_j, \quad y_j(\tau) = y_j, \quad t_j(\tau) = t. \end{aligned}$$

Notice here that the initial conditions specify the x_j and y_j coordinates at their value 0 at the pole of the Green function, but they do not specify the value of t at $s = 0$. Instead, they fix the conjugate momentum $\theta(0)$ to be $-i$. On the other hand, the boundary conditions for the path at the final time τ are the usual ones. Because of these special conditions, the action satisfying the Hamilton Jacobi equation will be given by

$$(28) \qquad g(x,y,t,\tau) = -i\,t(0) + \int_0^\tau \left[\sum_{j=1}^n (\xi_j dx_j + \eta_j dy_j) + \theta dt - H ds \right],$$

where $t(0)$ is the initial value of the t-component of the path at time $s = 0$ and the integral is taken along the path satisfying (27).

It is easy to integrate explicitly the bicharacteristic system (16) and to prove that there is a unique path satisfying (27) and (16) for any x, y, t, τ. The action g can be also computed explicitly and is given by

$$(29) \qquad g(x,y,t,\tau) = \sum_{j=1}^n a_j(x_j^2 + y_j^2) \coth(2a_j\tau) - it,$$

so that f is simply given by $f = \tau g$ and is smooth everywhere even at $\tau = 0$. Moreover, g is a distance-like function in the sense that it vanishes for each fixed $0 \neq \tau \in (-\infty, \infty)$ if and only if $x_j = y_j = 0$ and $t = 0$.

The second problem is to cancel the singular term in $g^{-(n+1)}$ in (25). We do this by imposing

$$(30) \qquad T(v^{(0)}) = 0.$$

Now the definition of T in (26) tells that the first order part of T, namely

$$\frac{\partial}{\partial \tau} + \sum_{j=1}^n ((X_j g)X_j + (Y_j g)Y_j),$$

is exactly the differentiation d/ds along the path. This is the analogue of the transport equation along the geodesics. Moreover, because of the specific form of (29) and of the form of the vector fields X_j, Y_j, Δg is a function of τ only. In fact

$$(31) \qquad \Delta g = \sum_{j=1}^n a_j \coth(2a_j\tau)$$

and we can solve $T(v^{(0)}) = 0$ in a very simple way by a function $v^{(0)}(\tau)$ of τ only:

$$(32) \qquad v^{(0)}(\tau) = \prod_{j=1}^n \frac{1}{\sinh 2a_j t}.$$

But then, because $v^{(0)}$ does not depend on the spatial coordinate, $\Delta v^{(0)} = 0$ and we have proved that $\Delta G = 0$, at least outside 0.

Actually, there is one more small difficulty. Namely, for $t \neq 0$ and $x_j = y_j = 0$ for all j, the integral (7) is not convergent. The trick is to modify the contour of integration by adding to τ a small imaginary part

$$\tau \to \tau + i\varepsilon \operatorname{sgn} t$$

for a fixed positive ε (chosen small enough so that one does not reach the poles of $\coth 2a_j\tau$), and then the integral is absolutely convergent everywhere. Moreover, one can prove that $\Delta G = \delta_0$ using a regularization argument. This explains completely the intrinsic geometric meaning of formula (7). To summarize:

1°) $f = \tau g$ where g is an action function solution of the standard Hamilton Jacobi equation with time τ, but with non-standard initial conditions for the bicharacteristic paths (see (27));

2°) $V^{(0)}$ is essentially $\tau^n v^{(0)}$ where $v^{(0)}$ is obtained by solving a standard first order transport equation along the bicharacteristics. All details will appear in [**7**].

4. Generalization to step 2 subelliptic operators with one missing direction

The subelliptic geometric concepts introduced in the preceding section make sense in a general subelliptic situation. As in Section 1, we shall suppose that we have a real analytic operator

$$\Delta = \frac{1}{2} \sum_{j=1}^{2n} X_j^2$$

in a $(2n+1)$-dimensional space such that $X_1, \ldots, X_{2n}, \partial/\partial t$ is a basis of $\mathbf{R}^{2n+1}$ at any point and the matrix of commutator $[X_j, X_k]$ is non-degenerate $\partial/\partial t$. Let us now fix a point m_0. We can choose coordinates $(x_1, \ldots, x_{2n}, t)$ such that

1°) m_0 has coordinates 0.

2°) the X_j have the following form

$$(33) \qquad X_j = \frac{\partial}{\partial x_j} + (\Lambda x)_j \frac{\partial}{\partial t} + \sum_k h_{1jk} \frac{\partial}{\partial x_k} + h_{2j} \frac{\partial}{\partial t},$$

where Λ is a $2n \times 2n$ skew-symmetric matrix of the form

$$(34) \qquad \Lambda = \operatorname*{diag}_j \begin{pmatrix} 0 & 2a_j \\ -2a_j & 0 \end{pmatrix} \text{ (with } a_j > 0\text{)}$$

and h_{1jk} satisfy

$$(35) \qquad \begin{aligned} & h_{1jk}(0, t) = 0, \\ & h_{1jk}(x, t) = O(|x|) \end{aligned}$$

and h_{2j} satisfy

$$(36) \qquad \begin{aligned} & h_{2j}(0, t) = 0, \\ & h_{2j}(x, t) = O(|x|^2 + |x|\,|t|). \end{aligned}$$

We shall say that the X_j are in standard form and that the system of coordinates is a standard system. Although what we shall say is intrinsic, the proofs are done more easily in a standard system of coordinates.

We shall construct a Green function (or more precisely a parametrix) for Δ of pole 0 in the following form

(37) $$G(x,t) = \int_{-\infty}^{+\infty} d\tau \left(\sum_{k=0}^{\infty} \frac{V^{(k)}}{g^{n-k}} + \sum_{k=0}^{\infty} (\log g)\, g^k W^{(k)} \right),$$

where $g, V^{(k)}$ and $W^{(k)}$ are unknown functions of x, t, τ. This will give a formal series which will have the following property: if we stop the series at $k = N$ and call G_N the corresponding finite sum in (37) then

(38) $$\Delta G_N = \delta_0 + S_N,$$

where S_N is a smooth kernel of smoothing order $2N$.

We shall briefly sketch the argument. First we compute Δ on an integral like (37). We have to compute terms like

(39) $$\Delta \left(\frac{V^{(k)}}{g^{n-k}} \right) \quad \text{and} \quad \Delta((\log g)\, g^k W^{(k)}).$$

In particular, we first have the same formal computation as (22) for $k = 0$.

$$\Delta \left(\frac{V^{(0)}}{g^n} \right) = \frac{\Delta V^{(0)}}{g^n} - \frac{n}{g^{n+1}} \left((\Delta g) v^{(0)} + \sum_{j=1}^{2n} (X_j g)(X_j V^{(0)}) \right)$$

(40) $$+ \frac{n(n+1)}{g^{n+2}} H(\nabla g) V^{(0)},$$

where $H(\xi, \theta, x, t)$ is the full symbol of Δ

$$H(\xi, \theta, x, t) = \frac{1}{2} \sum_{j=1}^{2n} \zeta_j^2.$$

Here ζ_j is the full symbol of X_j given by (33):

$$\zeta_j = \xi_j + (\Lambda x)_j \theta + \sum_k h_{1jk} \xi_k + h_{2j} \theta.$$

The highest singularity is again the terms in $H(\nabla g)$ and we impose the Hamilton Jacobi equation

(41) $$H(\nabla g) + \frac{\partial g}{\partial \tau} = 0.$$

Using the same trick as in (24), we obtain

(42)

$$\Delta \left(\frac{V^{(0)}}{g^n} \right) = \frac{\Delta V^{(0)}}{g^n} - \frac{n}{g^{n+1}} \left((\Delta g) V^{(0)} + \sum_{j=1}^{2n} (X_j g)(X_j V^{(0)}) + \frac{\partial V^{(0)}}{\partial \tau} \right)$$

$$+ n \frac{\partial}{\partial \tau} \left(\frac{V^{(0)}}{g^{n+1}} \right).$$

The integral over τ of the total $d/d\tau$ derivation in (42) will give 0 provided that

$$\lim_{|\tau| \to \infty} \frac{V^{(0)}}{g^{n+1}} = 0.$$

We also get rid of the terms in $g^{-(n+1)}$ in (42) by imposing the transport equation

(43) $$T(V^{(0)}) = 0,$$

where

(44) $$T(v) \equiv (\Delta g)v + \left(\sum_{j=1}^{2n} (X_j g)X_j + \frac{\partial}{\partial \tau} \right) v.$$

At that point we have, at least formally:

$$\Delta \int \frac{V^{(0)}}{g^n} d\tau = \int \frac{\Delta V^{(0)}}{g^n} d\tau.$$

But in general, $\Delta V^{(0)}$ will not be 0. This is why we introduce a term $V^{(1)}/g^{n-1}$ (or $(\log g) W^{(0)}$ if $n = 1$) so that

(45) $$(n - 1)T(V^{(1)}) = \Delta V^{(0)}$$

and then

$$\Delta \int \left(\frac{V^{(0)}}{g^n} + \frac{V^{(1)}}{g^{n-1}} \right) d\tau = \int \frac{\Delta V^{(1)}}{g^{n-1}} d\tau, \ etc.$$

We see that the $V^{(k)}$ and $W^{(k)}$ can be recursively defined (using the previously forend $V^{(\ell)}$ or $W^{(\ell)}$ so that essentially, outside 0, we have

$$\Delta G_N = \int [(\log g)g^N \Delta W^{(N)} d\tau + g^N \Delta V^{(n+N)}] d\tau.$$

Until this point we have only done formal computations.

5. Integration of the Hamilton Jacobi and transport equations

To give a precise meaning to the previous computations, we must integrate for all τ the Hamilton Jacobi equation (41), the transport equations (43), (45) and the analogous equations, determining the $V^{(k)}$'s and $W^{(k)}$'s, prove that the integrals

$$\int_{-\infty}^{+\infty} \frac{V^{(k)}}{g^{n-k}} d\tau \quad \text{and} \quad \int_{-\infty}^{+\infty} (\log g)g^k W^{(k)} d\tau$$

are convergent and that the boundary terms at $|\tau| = +\infty$, $\lim_{|\tau|=\infty} V^{(k)}/g^{n-k}$ or $\lim_{|\tau|=\infty} (\log g) g^k W^{(k)}$ are all 0.

a) Trajectories.
We start again from the Hamiltonian system

(46) $$\dot{x} = H_\xi, \quad \dot{\xi} = -H_x \quad \dot{t} = H_\theta \quad \dot{\theta} = -H_t$$

for unknown trajectories $\{x(s), \xi(s), t(s), \theta(s)\}$ for $0 \leq s \leq \tau$ with the boundary conditons

(47)
$$x(0) = x_0, \quad x(\tau) = x_1;$$
$$\theta(0) = -i = \theta_0, \quad t(\tau) = t_1.$$

Let us denote by D a certain domain of the variables (x, t) in $\mathbf{C}^{2n+1}$ in which H is analytic as a function of (x, t). Then we can find a certain neighborhood N of the diagonal in $D \times D$, a neighborhood V of $D \times 0$ in $\mathbf{C}^{2n+1} \times \mathbf{C}^{2n+1}$ such that for $((x_0, t_0), (x_1, t_1)) \in N$, for any $\tau \in \mathbf{R} - \{0\}$, there exists a unique solution of the Hamilton equation (46) with boundary conditions (47) between 0 and τ so that $(x(s), t(s), (\tanh \alpha\tau)\zeta(s), (\tanh \alpha\tau)(\theta(s) - \theta_0))$, $s = \sigma\tau$ stay in V, are analytic functions of x_0, t_0, x_1, t_1, τ and extend analytically in a certain strip $|\mathrm{Im}\ \tau| < \delta_0$. Notice that the solution is globally defined for any τ.

b) Action.
Once this is done, we define the action

(48)
$$g(x_1, t_1, \tau \mid x_0, t_0) = \theta_0 t(0) + \int_0^\tau \left[\sum_j \xi_j \dot{x}_j + \theta t - H \right] ds,$$

where $t(0)$ is the initial value as $s = 0$ of the path defined above. This function g has the following properties.

(i) it vanishes if and only if $x_1 = x_0, t_1 = t_0$ and is analytic in the previously defined domains except at $\tau = 0$. Nevertheless, τg is analytic at $\tau = 0$.

(ii) $\partial g / \partial \tau + H(\nabla g, x, t) = 0$.

(iii) in the model case g is the action constructed in Section 2.

c) Transport equation.
First we consider the transport equation for $V^{(0)}$. We shall assume that $(x_0, t_0) = 0$ in $\mathbf{C}^{2n+1}$ and we write simply $V^{(0)}(x, t, \tau)$. The equation (43) is a transport equation along the trajectory $\{x(s), \xi(s), t(s), \theta(s)\}$ of the Hamilton system (46).

We can write
$$V^{(0)} = \tilde{v}^{(0)} v^{(0)},$$

where
$$\tilde{v}^{(0)} = \left(\frac{\det \Omega}{\det(\sin h\tau\Omega)} \right)^{1/2}$$

(Ω is the matrix $-i\Lambda$ of the standard form of the vector field).

Moreover, $v^{(0)}$ tends to 1 when $(x, t) \to 0$ uniformly for τ small and for any ε

(49)
$$|V^{(0)}| \leq C \exp\left(-\left(\frac{\mathrm{Tr}\,|\Omega|}{2} - \varepsilon \right) |\mathrm{Re}\ \tau| \right).$$

Then we can also write
$$V^{(k)} = V^{(0)} v^{(k)},$$
$$W^{(k)} = V^{(0)} w^{(k)},$$

where $v^{(k)}$ and $w^{(k)}$ are given by integrals along the trajectories of the Hamilton system (46).

d) Conclusion.

These results show that each term of the formal series for G make sense (the integrals over τ are convergent) and that the formal manipulation which were done in Section 4 are allowed. From this, we can construct by a standard method a parametrix, namely, we define

(50)

$$
\begin{aligned}
G_\infty = {}& \sum_{k=0}^{n} \int_{-\infty}^{+\infty} \frac{V^{(k)}}{g^{n-k}} d\tau + \int_{-\infty}^{+\infty} (\log g) W^{(0)} d\tau \\
& + \sum_{k \geq 1} \chi \left(\frac{|x|^4 + t^2}{\varepsilon_k^2} \right) \left[\int_{-\infty}^{+\infty} V^{(n+k)} g^k d\tau + \int_{-\infty}^{+\infty} (\log g) g^{(k)} W^{(k)} d\tau \right],
\end{aligned}
$$

where χ is a plateau function which is C^∞ with compact support and is 1 on a neighborhood of 0 and the ε_k are chosen decreasing sufficiently rapidly. Then

$$
\Delta G_\infty = \delta + S,
$$

where S is a C^∞ smoothing kernel.

6. The heat kernel

We consider now the heat equation

(51)
$$
\frac{\partial}{\partial u} P_u(x_1, t_1 \mid x_0, t_0) = \Delta P_u,
$$
$$
P_u(x_1, t_1 \mid x_0, t_0) \mid_{u=0} = \delta(x_1 - x_0) \delta(t_1 - t_0).
$$

Let us define

(52)
$$
f = \tau g,
$$

where g is the action introduced in the preceding section. Then we can define

$$
P_u(x_1, t_1 \mid x_0, t_0) = \sum_{k \geq 0} \frac{1}{u^{n+1-k}} \int_{-\infty}^{+\infty} e^{-(1/u)f} U^{(k)} d\tau
$$

as a formal series. The functions $U^{(k)}$ are recursively defined by integration over the trajectories $\{x(s), \xi(s), t(s), \theta(s)\}$ of the Hamiltonian system. Moreover, for $0 \leq k < n$

$$
U^{(k)} = (n-k)(n-k+1) \ldots (n-1) V^{(k)},
$$

where the $V^{(k)}$ have been defined in the previous sections.

As before the integrals can be made absolutely convergent for all x, t, $(u \neq 0)$ provided we displace the contour of integration in the complex τ space using a contour

$$
\operatorname{Im} \tau = \varepsilon \operatorname{sgn} t
$$

for $\varepsilon > 0$ small.

Using a regularization procedure, we define

(53)

$$
\begin{aligned}
P_\infty(x,t,u) = \sum_{k=0}^{n+1} \frac{1}{u^{n+1-k}} \int_{-\infty}^{+\infty} e^{-(1/u)f} U^{(k)} d\tau \\
+ \sum_{k \geq n+1} \chi\left(\frac{u + |x|^2 + |t|^2}{\varepsilon_k}\right) \int_{-\infty}^{+\infty} e^{-(1/u)f} \frac{U^{(k)}}{u^{n+1-k}} d\tau
\end{aligned}
$$

and we prove that

$$
\left(\Delta - \frac{\partial}{\partial u}\right) P_\infty = \delta(x)\delta(t)\delta(u) + S,
$$

where S is smoothing of any order.

7. Small time behavior of the heat kernel

Let us consider now any of the integrals appearing in the heat kernel as a function of the time variable u

(54)
$$
I_k(u) = \frac{1}{u^{n+1-k}} \int_{-\infty}^{+\infty} e^{-(1/u)f} U^{(k)} d\tau.
$$

The main singularity of the heat kernel is given by I_0. It is clear from the saddle point method applied to the integral (54) that the behavior of $I_k(u)$ is given by the critical points of f with respect to τ if such points exist.

By a rescaling argument, it is possible to prove that

$$
\frac{\partial f}{\partial \tau}(x,t,\tau_c,\theta_0) = 0
$$

is equivalent to

$$
\frac{\partial g}{\partial \theta_0}(x,t,1,\tau_c,\theta_0) = 0,
$$

where we have now indicated that g and $f \equiv \tau g$ are functions of the initial conjugate momentum θ_0 (which, until now was supposed to be $\theta_0 \equiv -i$). But standard Hamilton Jacobi theory tells also that

$$
\frac{\partial g}{\partial \theta_0} = t(0),
$$

so that the critical points τ_c of f as a function of τ satisfy $t(0) = 0$ which means that they correspond to trajectories of the hamiltonian system joining $(x_0, t = 0)$ to (x_1, t_1) in time τ. These are exactly the classical bicharacteristics of the subriemannian geometry associated to Δ. Moreover, one can prove that the classical action $S(x,t,u)$ is $1/u$ times the critical value of $f : (1/u)f(x,t,\tau_c)$.

We can in fact prove that if $x_0 = 0$ say, and $x \neq 0$, $f(x,t,\tau)$ has a unique critical point which is purely imaginary provided that $|x| + |t|$ is sufficiently small (the size of the neighborhood depends on the ratio $|t|/|x|^2$).

Then for generic point x, we have for $u \to 0^+$:

$$
P_u(x,t) \sim \frac{C_n}{u^{n+1/2}} e^{-S(x,t,u)} \theta(x,t),
$$

where $\theta(x,t)$ is a function which is smooth provided $|x|^2 \neq 0$.

The term "generic" means the following: let $0 < a_1 \leq a_2 \leq \ldots \leq a_p < a_{p+1} = \ldots = a_n$ the n positive eigenvalues of $i\Lambda$, the last ones being equal. A generic point is such that one of $x_{p+1}^2 + y_{p+1}^2, \ldots, x_n^2 + y_n^2$ is not 0.

References

[1] K. Kodaira, *Harmonic fields in a riemannian manifolds*, Annals of Math (1949), 587–665.
[2] G. Folland, *A fundamental solution for a subelliptic operator*, Bull. Am. Math. Soc. **79** (1973), 373-376.
[3] B. Gaveau, *Principe de moindre action, propagation de la chaleur et estimées sous elliptiques sur certains groupes nilpotents*, Acta Math vol 139 (1977), 95-153.
[4] R. Beals, P. Greiner, *Calculus on Heisenberg manifolds*, Annals of Math. Studies no. 119, Princeton University Press (1988).
[5] R. Beals, B. Gaveau, P. Greiner, J. Vauthier, *The Laguerre calculus on the Heisenberg group*, Bull. Sci. Math. (1986), 225–288.
[6] B. Gaveau, *Systèmes dynamiques associés à certains opérateurs hypoelliptiques*, Bull. Sci. Math. **102** (1978), 203-229.
[7] R. Beals, B. Gaveau, P. Greiner, *Complex hamiltonian mechanics and hypoelliptic operators*.

DEPARTMENT OF MATHEMATICS, YALE UNIVERSITY, P.O. BOX 208283, NEW HAVEN, CT 06520, U.S.A.
 E-mail address: `beals@math.yale.edu`

UNIVERSITÉ P. ET MARIE CURIE, MATHÉMATIQUES T 45–46, 5ÈME ÉTAGE, 4, PLACE JUSSIEU, 75252 PARIS CEDEX 05, FRANCE
 E-mail address: `gaveau@mathp6.jussieu.fr`

DEPARTMENT OF MATHEMATICS, UNIVERSITY OF TORONTO, TORONTO, ONTARIO M5S 1A1 CANADA
 E-mail address: `greiner@math.toronto.edu`

Contemporary Mathematics
Volume **212**, 1998

Higher Order Cauchy Pompeiu Operators

Heinrich Begehr and Gerald N. Hile

1. Introduction

The complex version of the Green formula leads to the Cauchy–Pompeiu representation formula, which for analytic functions coincides with Cauchy's formula. The additional term in the Cauchy-Pompeiu formula is an area integral with Cauchy kernel and the $\overline{z}$-derivative of the function under consideration as density. This area integral operator is denoted by T and is called the Pompeiu integral operator. If the density is chosen from $L_p(\overline{D})$, $1 < p$, D a domain in the complex plane $\mathbb{C}$, this area integral is a particular solution to the inhomogeneous Cauchy–Riemann system. The operator $\overline{T}$ is an integral operator analogous to T but with the Cauchy kernel replaced by its complex conjugate. Iterating the T-operator with itself and with $\overline{T}$ in a proper manner leads to a particular solution of an inhomogeneous equation of the form $\partial^{m+n}w/\partial\overline{z}^m\partial z^n = \rho$. The z- and $\overline{z}$-derivatives up to the order $m + n - 1$ of this weakly singular integral operator produce more weakly singular integral operators, while the derivatives of order $m + n$ yield strongly singular integrals.

In the particular case of the unit disc $\mathbb{D}$ the Pompeiu operator can be modified to handle the Dirichlet boundary condition. This $\widetilde{T}$-operator for $\mathbb{D}$ differs from the T-operator by an additional area integral defining an analytic function in $\mathbb{D}$ see [11]. The operator $\widetilde{T}\rho$ satisfies the homogeneous Dirichlet boundary condition $\operatorname{Re}\widetilde{T}\rho = 0$ on $\partial\mathbb{D}$. The $\widetilde{T}$-operator and its complex conjugate $\overline{\widetilde{T}}$ can also be iterated, to produce higher order integral operators which together with their derivatives satisfy certain Dirichlet boundary conditions.

The higher order integral operators are introduced and their main properties are explained. Details may be found in [7]. In order to show their usefullness two different kind of simple second order equations are treated. Representation formulas of Cauchy Pompeiu type are given for regular domains of the complex plane and especially for the unit disc expressing a function through a chain of its mixed derivatives with respect to z and $\overline{z}$ up to a certains order and kind.

1991 *Mathematics Subject Classification.* Primary 35J40, 30G20; Secondary 30E25, 35J25.

2. Pompeiu operator

Before development of the general operators, first the basic properties of the Pompeiu operator are listed and two fundamental examples of second order equations are explained.

DEFINITION 1. Let D be a domain in $\mathbb{C}$ and let $\rho \in L_1(\overline{D})$. Then the operator T given by

$$T\rho(z) := -\frac{1}{\pi} \int\limits_D \rho(\zeta) \frac{d\xi d\eta}{\zeta - z}, \quad z \in \mathbb{C},$$

is called the *Pompeiu operator*. By $\overline{T}$ we denote

$$\overline{T}\rho(z) := -\frac{1}{\pi} \int\limits_D \rho(\zeta) \frac{d\xi d\eta}{\overline{\zeta - z}}, \quad z \in \mathbb{C}.$$

The Pompeiu operator has the following properties; see [11].

LEMMA 1.
1. *For $\rho \in L_p(\overline{D})$, $2 < p$, we have $T\rho \in C^{\alpha_0}(\mathbb{C})$, $\alpha_0 := (p-2)/p$. T is a compact operator from $L_p(\overline{D})$ into $C^{\alpha_0}(\mathbb{C})$ satisfying*

$$\|T\rho\|_{C^{\alpha_0}(\mathbb{C})} \le M(p, D)\|\rho\|_{L_p(D)} \ .$$

2. *$T\rho$ has generalized first order derivatives, $(T\rho)_{\overline{z}} = \rho$, $(T\rho)_z = \Pi\rho$, where Π, given by*

$$\Pi\rho(z) := -\frac{1}{\pi} \int\limits_D \rho(\zeta) \frac{d\xi d\eta}{(\zeta - z)^2}, \quad z \in \mathbb{C},$$

is a singular integral operator of Calderon–Zygmund type. Π is an operator mapping $L_p(\overline{D})$ into $L_p(\overline{D})$, with norm $\Lambda_p := \|\Pi\|_p$ a continuous function of p for $1 < p$, and with $\Lambda_2 = 1$.

REMARK. $T\rho$ turns out to be a particular solution of the inhomogeneous Cauchy–Riemann equation $w_{\overline{z}} = \rho$. The operator T can be used also to solve more involved equations, such as the generalized Beltrami equation

$$w_{\overline{z}} + q_1 w_z + q_2 \overline{w_z} + aw + b\overline{w} + c == 0, \quad \|q_1\|_{L_\infty(\overline{D})} + \|q_2\|_{L_\infty(\overline{D})} \le q_0 < 1 \ ;$$

see e.g. [2, 4, 9, 12, 13]. We indicate the method by treating the inhomogeneous Beltrami equation

$$w_{\overline{z}} + q w_z = f, \quad \|q\|_{L_\infty(\overline{D})} \le q_0 < 1, \ f \in L_p(\overline{D}), \ 2 < p \ .$$

Looking for a particular solution in the form $w_0 = T\rho$ leads on the basis of Lemma 1 to the singular integral equation

$$\rho + q\,\Pi\rho = f$$

for the density ρ. This equation can be solved uniquely via a Neumann series provided that $\|q\Pi\|_{L_p} \le q_0\Lambda_p < 1$ when p is close enough to 2. The solution ρ satisfies the apriori estimate

$$\|\rho\|_{L_p(\overline{D})} \le \frac{\|f\|_{L_p(\overline{D})}}{1 - q_0\Lambda_p},$$

which leads to an estimate for w_0 in the form (see [2])

$$\|w_0\|_{C^{\alpha_0}(\overline{D})} + \|w_{0z}\|_{L_p(\overline{D})} + \|w_{0\overline{z}}\|_{L_p(\overline{D})} \leq M(p, q_0, D)\|f\|_{L_p(\overline{D})} \ .$$

If boundary conditions are introduced, an apriori estimate can be given for the solutions to the generalized Beltrami equation which then can be used to solve nonlinear problems; see [5, 6, 8, 13, 14].

3. Second order equations

There are two cases essentially different from one another:
(i) $w_{\overline{z}\,\overline{z}} = \rho$,
(ii) $w_{\overline{z}z} = \rho$, $\quad \rho \in L_p(\overline{D})$.
Obviously, particular solutions can be found in the form
(i) $w_0 = T^2\rho$,
(ii) $w_0 = T\overline{T}\rho$.
Observing

$$\overline{z} = \frac{1}{2\pi i} \int\limits_{\partial D} \frac{\overline{\zeta}d\zeta}{\zeta - z} - \frac{1}{\pi} \int\limits_{D} \frac{d\xi d\eta}{\zeta - z}$$

where

$$f(z) := \frac{1}{2\pi i} \int\limits_{\partial D} \frac{\overline{\zeta}d\zeta}{\zeta - z} , \quad z \notin \partial D$$

is an analytic function identically vanishing if D is a disc $|\zeta| < R$, we have

$$T^2\rho(z) = \frac{1}{\pi^2} \iint\limits_{DD} \frac{\rho(\widetilde{\zeta})}{(\widetilde{\zeta} - \zeta)(\zeta - z)} d\widetilde{\xi}d\widetilde{\eta}d\xi d\eta$$

$$= \frac{1}{\pi} \int\limits_{D} \frac{\overline{\zeta - z}}{\zeta - z}\rho(\zeta)d\xi d\eta + \frac{1}{\pi} \int\limits_{D} \frac{f(\zeta) - f(z)}{\zeta - z}\rho(\zeta)d\xi d\eta \ .$$

Here the last integral is an analytic function. Similarly, from

$$\log|z - \widetilde{\zeta}|^2 = \frac{1}{2\pi i} \int\limits_{\partial D} \log|\zeta - \widetilde{\zeta}|^2 \frac{d\zeta}{\zeta - z} - \frac{1}{\pi} \int\limits_{D} \frac{d\xi d\eta}{(\overline{\zeta - \widetilde{\zeta}})(\zeta - z)} , \quad z \neq \widetilde{\zeta} ,$$

where

$$g(z, \widetilde{\zeta}) := -\frac{1}{2\pi i} \int\limits_{\partial D} \log|\zeta - \widetilde{\zeta}|^2 \frac{d\zeta}{\zeta - z} , \quad z, \widetilde{\zeta} \in D ,$$

is analytic in z, we have

$$T\overline{T}\rho(z) = \frac{1}{\pi^2} \iint\limits_{DD} \frac{\rho(\widetilde{\zeta})}{(\overline{\widetilde{\zeta} - \zeta})(\zeta - z)} d\widetilde{\xi}d\widetilde{\eta}d\xi d\eta$$

$$= \frac{2}{\pi} \int\limits_{D} \log|\zeta - z|\rho(\zeta)d\xi d\eta + \frac{1}{\pi} \int\limits_{D} g(z, \zeta)\rho(\zeta)d\xi d\eta \ .$$

Another operator for (ii) is $\overline{T}T$, with the representation

$$\overline{T}T\rho(z) = \frac{2}{\pi} \int\limits_{D} \log|\zeta - z|\rho(\zeta)d\xi d\eta + \frac{1}{\pi} \int\limits_{D} \overline{g(z, \zeta)}\rho(\zeta)d\xi d\eta \ .$$

Hence, a particular solution in case (i) is

$$T_{0,2}\rho(z) := \frac{1}{\pi} \int_D \frac{\overline{\zeta - z}}{\zeta - z} \rho(\zeta) d\xi d\eta,$$

and in case (ii),

$$T_{1,1}\rho(z) := \frac{2}{\pi} \int_D \log|\zeta - z| \rho(\zeta) d\xi d\eta,$$

while the general solutions involve analytic functions ϕ and ψ,
   (i) $w = \psi + \overline{z}\phi + T_{0,2}\rho$,
   (ii) $w = \psi + \overline{\phi} + T_{1,1}\rho$.
   Obviously, instead of $T_{1,1}$ one can use

$$S_1\rho(z) := -\frac{2}{\pi} \int_D g(z,\zeta)\rho(\zeta) d\xi d\eta$$

where

$$g(z,\zeta) = -\log|\zeta - z| + h(z,\zeta)$$

is the Green function for the Laplace operator in the domain D. This operator additionally vanishes at the boundary ∂D. Without going into details we list some conclusions from the above representation formulas. We have in case (i)

$$w = \psi + \overline{z}\phi + T_{0,2}\rho, \quad T_{0,2}\rho(z) := \frac{1}{\pi} \int_D \frac{\overline{\zeta - z}}{\zeta - z} \rho(\zeta) d\xi d\eta \,,$$

$$w_{\overline{z}} = \phi + T_{0,1}\rho, \quad T_{0,1}\rho(z) := T\rho(z) = -\frac{1}{\pi} \int_D \frac{1}{\zeta - z} \rho(\zeta) d\xi d\eta \,,$$

$$w_z = \psi' + \overline{z}\phi' + T_{-1,2}\rho, \quad T_{-1,2}\rho(z) := \frac{1}{\pi} \int_D \frac{\overline{\zeta - z}}{(\zeta - z)^2} \rho(\zeta) d\xi d\eta \,,$$

$$w_{\overline{z}z} = \phi' + T_{-1,1}\rho, \quad T_{-1,1}\rho(z) := \Pi\rho(z) := -\frac{1}{\pi} \int_D \frac{1}{(\zeta - z)^2} \rho(\zeta) d\xi d\eta \,,$$

$$w_{zz} = \psi'' + \overline{z}\phi'' + T_{-2,2}\rho, \quad T_{-2,2}\rho(z) := \frac{2}{\pi} \int_D \frac{\overline{\zeta - z}}{(\zeta - z)^3} \rho(\zeta) d\xi d\eta \,,$$

$$w_{\overline{z}\,\overline{z}} = T_{0,0}\rho, \quad T_{0,0}\rho := \rho \,,$$

and in case (ii)

$$w = \psi + \overline{\phi} + T_{1,1}\rho, \quad T_{1,1}\rho(z) := \frac{2}{\pi} \int_D \log|\zeta - z|\rho(\zeta) d\xi d\eta \,,$$

$$w_{\overline{z}} = \overline{\phi'} + T_{1,0}\rho, \quad T_{1,0}\rho(z) := \overline{T}\rho(z) := -\frac{1}{\pi} \int_D \frac{1}{\overline{\zeta - z}} \rho(\zeta) d\xi d\eta \,,$$

$$w_z = \psi' + T_{0,1}\rho, \quad T_{0,1}\rho(z) := T\rho(z) = -\frac{1}{\pi} \int_D \frac{1}{\zeta - z} \rho(\zeta) d\xi d\eta \,,$$

$$w_{\overline{z}z} = T_{0,0}\rho, \quad T_{0,0}\rho = \rho \,,$$

$$w_{zz} = \psi'' + T_{-1,1}\rho, \quad T_{-1,1}\rho(z) := \Pi\rho(z) := -\frac{1}{\pi}\int\limits_D \frac{1}{(\zeta - z)^2}\rho(\zeta)d\xi d\eta\,,$$

$$w_{\bar{z}\bar{z}} = \overline{\phi''} + T_{1,-1}\rho, \quad T_{1,-1}\rho(z) := \overline{\Pi}\rho(z) := -\frac{1}{\pi}\int\limits_D \frac{1}{(\overline{\zeta - z})^2}\rho(\zeta)d\xi d\eta\,.$$

With these representations quite general elliptic second order equations can be treated; see [1, 3, 10]. The arbitrary analytic functions in the representation formulas can be determined by prescribing proper boundary values for w and $w_{\bar{z}}$.

4. Operators for higher order equations

By repeated iteration the following set of integral operators are produced; see [7].

DEFINITION 2. For integers m and $n, 0 \leq m+n, 0 < m^2 + n^2$, let

$$T_{m,n}\rho(z) := \int\limits_D K_{m,n}(z-\zeta)\rho(\zeta)d\xi d\eta$$

$$:= \begin{cases} \dfrac{(-m)!(-1)^{-m}}{(n-1)!\pi}\displaystyle\int\limits_D \dfrac{(\overline{z-\zeta})^{n-1}}{(z-\zeta)^{-m+1}}\rho(\zeta)d\xi d\eta\,, & \text{if } m \leq 0\,, \\[4mm] \dfrac{(-n)!(-1)^{-n}}{(m-1)!\pi}\displaystyle\int\limits_D \dfrac{(z-\zeta)^{m-1}}{(\overline{z-\zeta})^{-n+1}}\rho(\zeta)d\xi d\eta\,, & \text{if } n \leq 0\,, \\[4mm] \dfrac{1}{(m-1)!(n-1)!\pi}\displaystyle\int\limits_D (z-\zeta)^{m-1}(\overline{z-\zeta})^{n-1} \\[2mm] \qquad \times \left[\log|z-\zeta|^2 - \displaystyle\sum_{k=1}^{m-1}\dfrac{1}{k} - \sum_{l=1}^{n-1}\dfrac{1}{l}\right]\rho(\zeta)d\xi d\eta\,, & \text{if } 0 < m, 0 < n\,. \end{cases}$$

For convenience, set $T_{0,0}\rho := \rho$.

The essential properties of these singular integral operators are listed in the next lemma. For detailed proofs see [7]. Obviously, $T_{m,n}$ is a weakly singular operator for $0 < m + n$ and a singular operator of Calderon–Zygmund type for $m + n = 0 < m^2 + n^2$.

LEMMA 2. Let $\rho \in L_p(\overline{D})$, and let $m, n \in \mathbb{Z}$ with $0 \leq m+n$, $0 < m^2 + n^2$. Then

1. $\overline{T_{m,n}\rho} = T_{n,m}\overline{\rho}$.
2. If $2 \leq p$ then there exists a nonnegative constant M such that

$$|T_{m,n}\rho(z_1) - T_{m,n}\rho(z_2)| \leq M\|\rho\|_{L_p(\overline{D})}|z_1 - z_2|^{\alpha_0}$$

with

$$\alpha_0 := \begin{cases} 1, & \text{if } 2 \leq m+n, \\ (p-2)/p, & \text{if } m+n = 1\,, \end{cases}$$

for $|z_1|, |z_2| \leq R, 0 < R$. Here M depends on m, n and p, and also on D in the cases $2 \leq m+n \leq 3$ on D, and also on D and R for $4 \leq m+n$. For $3 \leq m+n$ the constant p is only restricted by $1 \leq p$.

3. *For $m + n = 0 < m^2 + n^2$ the operator $T_{m,n}$ maps $L_p(\mathbb{C})$, $1 < p$, into itself. $T_{m,n}\rho$ then has to be interpreted as a Cauchy principal value integral. Moreover,*

$$\|T_{m,n}\rho\|_{L_p(\mathbb{C})} \le M(p)\|\rho\|_{L_p(\mathbb{C})} \ .$$

$T_{m,n}\rho$ has generalized first order derivatives for $1 \le m + n$ given by

$$\frac{\partial}{\partial z}T_{m,n}\rho = T_{m-1,n}\rho, \quad \frac{\partial}{\partial \overline{z}}T_{m,n}\rho = T_{m,n-1}\rho \ ;$$

$$\frac{\partial^{k+l}}{\partial z^k \partial \overline{z}^l}T_{m,n}\rho = T_{m-k,n-l}\rho \quad \textit{if } k + l \le m + n \ .$$

4. *For $\rho \in L_2(\mathbb{C})$ we have*

$$T_{m,-m}T_{k,-k}\rho = T_{m+k,-m-k}\rho, \quad T_{m,-m}T_{-m,m}\rho = \rho \ .$$

$T_{m,-m}$ is a unitary operator from $L_2(\mathbb{C})$ into itself,

$$\|T_{m,-m}\rho\|_{L_2(\mathbb{C})} = \|\rho\|_{L_2(\mathbb{C})} \ ,$$

and $T_{-m,m}$ is both the inverse and adjoint operator to $T_{m,-m}$.

These operators are used to create higher order Cauchy–Pompeiu representations for $w \in C^{m+n}(\overline{D})$, $1 \le m+n$, which express w through $T_{m,n}(\partial^{m+n}w/\partial z^m \partial \overline{z}^n)$ and $m + n$ boundary integrals involving $\partial^{\mu+\nu}w/\partial z^\mu \partial \overline{z}^\nu$ for $0 \le \mu + \nu \le m + n - 1$.

5. Representation formulas of Cauchy–Pompeiu type

Integral representation formulas are usually obtained from the classical Gauß theorem. In complex form it reads as follows.

LEMMA 3. (Gauß Theorem). *Let $D \subset \mathbb{C}$ be a bounded domain with smooth boundary and let $u, v \in C^1(D; \mathbb{R} \cap C(\overline{D}; \mathbb{R})$, $w := u + iv$. Then*

$$\frac{1}{2\pi}\int_{\partial D} w(z)dz - \int_D w_{\overline{z}}(z)dxdy = 0 \ , \quad \frac{1}{2i}\int_{\partial D} w(z)d\overline{z} + \int_D w_z(z)dxdy = 0$$

where

$$2\frac{\partial}{\partial z} = \frac{\partial}{\partial x} - i\frac{\partial}{\partial y} \ , \quad 2\frac{\partial}{\partial \overline{z}} = \frac{\partial}{\partial x} + i\frac{\partial}{\partial y} \ .$$

In the same way that the Cauchy integral formula for analytic functions is attained from the Cauchy Theorem, we obtain the next result from the above complex forms of the Gauss Theorem.

LEMMA 4. (Pompeiu Representation). *Under the assumptions of Lemma 3 we have for $z \in D$ that*

$$w(z) = \frac{1}{2\pi i}\int_{\partial D} w(\zeta)\frac{d\zeta}{\zeta - z} - \frac{1}{\pi}\int_D w_{\overline{\zeta}}(\zeta)\frac{d\xi d\eta}{\zeta - z} \ ,$$

$$w(z) = -\frac{1}{2\pi i}\int_{\partial D} w(\zeta)\frac{d\overline{\zeta}}{\overline{\zeta} - \overline{z}} - \frac{1}{\pi}\int_D w_\zeta(\zeta)\frac{d\xi d\eta}{\overline{\zeta} - \overline{z}} \ .$$

By analogous arguments we get the following formulas.

LEMMA 5. *Under the assumptions of Lemma 3, for $0 < m + n$ and $z \in \mathbb{C} \backslash \partial D$ we have*

$$T_{m,n} w(z) = T_{m,n+1} w_{\bar{\zeta}}(z) - \frac{1}{2i} \int_{\partial D} K_{m,n+1}(z - \zeta) w(\zeta) d\zeta ,$$

$$T_{m,n} w(z) = T_{m+1,n} w_\zeta(z) + \frac{1}{2i} \int_{\partial D} K_{m+1,n}(z - \zeta) w(\zeta) d\bar{\zeta}.$$

Higher order Pompeiu formulas can more easily be described by using bi-indices.

DEFINITION 3. For nonnegative integers m and n let $\alpha := (m, n)$ denote the bi-index (m, n) and $z^\alpha = z^m \bar{z}^n$, $\partial_\alpha = \partial^{m+n}/\partial z^m \partial \bar{z}^n$, $T_\alpha := T_{m,n}$, $K_\alpha := K_{m,n}$, $\tilde{\alpha} = (n, m)$ and $|\alpha| := m + n$ for $\alpha = (m, n)$. Moreover, for $\alpha = (m, n)$, $\beta = (k, l)$ we write

$$\alpha \leq \beta \quad \text{if and only if } m \leq k \text{ and } n \leq l ,$$
$$\alpha < \beta \quad \text{if and only if } \alpha \leq \beta \text{ and } |\alpha| < |\beta| .$$

THEOREM 1. *Let $D \subset \mathbb{C}$ be a bounded smooth domain and let $w \in C^{m-1}(D; \mathbb{C})$ $\cap C^m(\overline{D}; \mathbb{C})$, where $1 \leq m$. Let*

$$(0, 0) =: \alpha_0 < \alpha_1 < \ldots < \alpha_m , \quad |\alpha_k| = k \text{ for } 0 \leq k \leq m$$

so that

$$\alpha_{k+1} = \alpha_k + \delta_k , \quad 0 \leq k \leq m - 1 ,$$

with

$$\delta_k \in \{(0, 1), (1, 0)\} , \quad 0 \leq k \leq m - 1 .$$

Then for $z \in D$,

$$w(z) = T_{\alpha_m} \partial_{\alpha_m} w(z) + \sum_{k=0}^{m-1} \frac{1}{2} \int_{\partial D} K_{\alpha_{k+1}}(z - \zeta) \partial_{\alpha_k} w(\zeta) d[(i\zeta)^{\tilde{\delta}_k}] .$$

PROOF. For $\alpha_m \in \{(0, 1), (1, 0)\}$ this equation is just the first or second formula in Lemma 4. By induction with respect to m, with use of Lemma 5, this generalized Pompeiu representation is shown. Assume the formula holds for some $m \in \mathbb{N}$ and that $w \in C^{m+1}(\overline{D}; \mathbb{C})$. Using the first or second formula in Lemma 5 applied to $\partial_{\alpha_m} w$ gives, for $\delta_m \in \{(1, 0), (0, 1)\}$,

$$T_{\alpha_m} \partial_{\alpha_m} w(z) = T_{\alpha_m + \delta_m} w(z) + \frac{1}{2} \int_{\partial D} K_{\alpha_m + \delta_m}(z - \zeta) \delta_{\alpha_m} w(\zeta) d[(i\zeta)^{\tilde{\delta}_m}] .$$

Inserting this equation into the representation formula for m immediately leads to the one for $m + 1$.

6. Operators for the unit disc

As the Cauchy formula for analytic functions in the unit disc $\mathbb{D}$ may easily be altered to obtain the Schwarz representation formula, also the Cauchy–Pompeiu formula may be modified in an analogous way; see e.g. [2].

LEMMA 6. (Schwarz–Poisson–Pompeiu formula). *Let* $w \in C^1(\mathbb{D}; \mathbb{C}) \cap C^0(\overline{\mathbb{D}}; \mathbb{C})$. *Then for* $z \in \mathbb{D}$,

$$w(z) = \frac{1}{2\pi i} \int_{\partial \mathbb{D}} \operatorname{Re} w(\zeta) \frac{\zeta + z}{\zeta - z} \frac{d\zeta}{\zeta} + \frac{1}{2\pi} \int_{\partial \mathbb{D}} \operatorname{Im} w(\zeta) \frac{d\zeta}{\zeta}$$

$$- \frac{1}{\pi} \int_{\mathbb{D}} \left[\frac{w_{\overline{\zeta}}(\zeta)}{\zeta - z} + \frac{z \overline{w_\zeta(\zeta)}}{1 - z\overline{\zeta}} \right] d\xi d\eta \,,$$

or

$$w(z) - i \operatorname{Im} w(0) = \frac{1}{2\pi i} \int_{\partial \mathbb{D}} \operatorname{Re} w(\zeta) \frac{\zeta + z}{\zeta - z} \frac{d\zeta}{\zeta} + S_1 w_{\overline{\zeta}}(z) \,,$$

where for $\rho \in L_1(\overline{\mathbb{D}})$,

$$S_1 \rho(z) := -\frac{1}{2\pi} \int_{\mathbb{D}} \left[\frac{\zeta + z}{\zeta - z} \frac{\rho(\zeta)}{\zeta} + \frac{1 + z\overline{\zeta}}{1 - z\overline{\zeta}} \frac{\overline{\rho(\zeta)}}{\overline{\zeta}} \right] d\xi d\eta \,.$$

LEMMA 7. S_1 *has the following properties.*

(1) *For* $\rho \in L_1(\overline{\mathbb{D}})$ *the function* $S_1\rho$ *has generalized first order derivatives* $(S_1\rho)_{\overline{z}} = \rho$,

$$(S_1\rho)_z(z) := \Pi_1 \rho(z) := -\frac{1}{\pi} \int_{\mathbb{D}} \left[\frac{\rho(\zeta)}{(\zeta - z)^2} + \frac{\overline{\rho(\zeta)}}{(1 - z\overline{\zeta})^2} \right] d\xi d\eta \,, \quad z \in \mathbb{D}.$$

The operator Π_1, *in* [11] *denoted by* $\widetilde{\Pi}$, *is an operator from* $L_p(\overline{\mathbb{D}})$ *into itself, satisfying*

$$\|\Pi_1\|_{L_2} = 1 \,.$$

(2) $S_1\rho$ *satisfies the Dirichlet condition*

$$\operatorname{Re} S_1\rho = 0 \quad \text{on } \partial\mathbb{D}$$

and

$$\operatorname{Im} S_1\rho(0) = 0 \,.$$

Iterating the S_1-operator is a little involved. The process leads to an integral operator which easily can be shown to satisfy a homogeneous Dirichlet condition as well as a side condition.

LEMMA 8. *For* $\rho \in L_1(\mathbb{D})$ *and for* $z \in \mathbb{D}$,

$$S_1^k \rho(z) = S_k \rho(z) := \frac{(-1)^k}{2\pi(k-1)!} \int_{\mathbb{D}} (2\operatorname{Re}(\zeta - z))^{k-1} \left[\frac{\zeta + z}{\zeta - z} \frac{\rho(\zeta)}{\zeta} + \frac{1 + z\overline{\zeta}}{1 - z\overline{\zeta}} \frac{\overline{\rho(\zeta)}}{\overline{\zeta}} \right] d\xi d\eta.$$

PROOF. The formula, obviously, holds for $k = 1$. For $k \geq 2$ we have $(\partial/\partial\overline{z}) S_k\rho$ $= S_{k-1}\rho$, $\operatorname{Re} S_k\rho = 0$ on $\partial\mathbb{D}$, $\operatorname{Im} S_k\rho(0) = 0$. Hence, by Lemma 6

$$S_k \rho(z) = -\frac{1}{2\pi} \int_{\mathbb{D}} \left[\frac{\zeta + z}{\zeta - z} \frac{S_{k-1}\rho(\zeta)}{\zeta} + \frac{1 + z\overline{\zeta}}{1 - z\overline{\zeta}} \frac{\overline{S_{k-1}\rho(\zeta)}}{\overline{\zeta}} \right] d\xi d\eta$$

$$= S_1 S_{k-1}\rho(z) \,.$$

COROLLARY 1. $S_k \rho$ is a particular solution to the Dirichlet problem

$$\frac{\partial^k}{\partial \bar{z}^k} S_k \rho = \rho, \quad \text{Re}\, \frac{\partial^\kappa}{\partial \bar{z}^\kappa} S_k \rho = 0 \quad \text{on } \partial \mathbb{D}, \ \text{Im}\, \frac{\partial^\kappa}{\partial \bar{z}^\kappa} S_k \rho(0) = 0, \ 0 \le \kappa < k - 1 .$$

Moreover, $\dfrac{\partial^\kappa}{\partial \bar{z}^\kappa} S_k \rho$ is a weakly singular integral if $0 \le \kappa \le k - 1$. For $\kappa = k$ we have

$$\frac{\partial^k}{\partial \bar{z}^k} S_k \rho$$

$$= \frac{(-1)^k k}{\pi} \int\limits_{\mathbb{D}} \left[\left(\frac{\overline{\zeta - z}}{\zeta - z} \right)^{k-1} \frac{\rho(\zeta)}{(\rho - z)^2} + \left(\frac{|\zeta|^2 + \zeta(\zeta - z)}{1 - z\bar{\zeta}} \right)^{k-1} \frac{\overline{\rho(\zeta)}}{(1 - z\bar{\zeta})^2} \right] d\xi d\eta .$$

This last integral operator is singular. It is not known whether $\dfrac{\partial^k}{\partial \bar{z}^k} S_k$ has L^2-norm equal to 1 or not. This information would be necessary to treat more general k-th order partial differential equations with principal part of the form

$$\frac{\partial^k w}{\partial \bar{z}^k} + \sum_{l=1}^{k-1} q_l \frac{\partial^k w}{\partial \bar{z}^{k-l} \partial z^l},$$

as was done in the case $k = 2$ in [1, 10].

References

[1]   H. Begehr., *Elliptic second order equations*, Second workshop on functional-analytic methods in complex analysis and applications to partial differential equations, ICTP, Trieste, 1993, ed. A. S. Mshimba, W. Tutschke, World Sci., Singapore, 1995, pp. 115–152.

[2]   H. Begehr, *Complex analytic methods for partial differential equations. An introductory text*, World Sci., Singapore, 1994.

[3]   H. Begehr and A. Dzhuraev, *On boundary value problems for overdetermined elliptic systems*, Complex Analysis in Several Complex Variables (Th. M. Rassias, ed.) (to appear).

[4]   H. Begehr and R. P. Gilbert, *Transformations, transmutations and kernel functions, I; II*, Longman, Harlow, 1992; 1993.

[5]   H. Begehr and G. N. Hile, *Nonlinear Riemann boundary value problems for a nonlinear elliptic system in the plane*, Math. Z. **179** (1982), 241–261.

[6]   H. Begehr and G. N. Hile, *Riemann boundary value problems for nonlinear elliptic systems*, Complex Variables, Theory Appl. **1** (1983), 239–261.

[7]   H. Begehr and G. N. Hile, *A hierarchy of integral operators*, Rocky Mountain J. Math., to appear.

[8]   H. Begehr and G. C. Hsiao, *The Hilbert boundary value problem for nonlinear elliptic systems*, Proc. Roy. Soc. Edinburgh **94A** (1983), 97–112.

[9]   B. Bojarski, *Generalized solutions of a system of first order elliptic equations with discontinuous coefficients*, Mat. Sbornik **43(85)** (1957), 451–503 (Russian).

[10] A. Dzhureav, *Methods of singular integral equations*, Longman, Harlow, 1992.

[11] I. N. Vekua, *Generalized analytic functions*, Pergamon Press, Oxford, 1962.

[12] V. S. Vinogradov, *On a boundary value problem for linear elliptic systems of first order in the plane*, Dokl. Akad. Nauk SSSR **118** (1958), 1059–1062 (Russian).

[13] G. C. Wen and H. Begehr, *Boundary value problems for elliptic equations and systems*, Longman, Harlow, 1990.

[14] W. Wendland, *Elliptic systems in the plane*, Pitman, London, 1979.

FREIE UNIVERSITÄT BERLIN, I. MATHEMATISCHES INSTITUT, ARNIMALLEE 3, 14195 BERLIN, GERMANY
E-mail address: `begehr@math.fu-berlin.de`

DEPARTMENT OF MATHEMATICS, UNIVERSITY OF HAWAII AT MANOA, 2565 THE MALL, HONOLULU, HAWAII 96822, U.S.A.
E-mail address: `hile@uhunix.uhcc.hawaii.edu`

Contemporary Mathematics
Volume **212**, 1998

A Polydisk Version of Beurling's Characterization for Invariant Subspaces of Finite Multi-Codimension

Mischa Cotlar and Cora Sadosky

ABSTRACT. The characterization of invariant subspaces of finite codimension in $H^2(\mathbf{T})$ as those of the form $\mathbf{b}H^2(\mathbf{T})$, for $\mathbf{b}$ a finite Blaschke product, fails in product spaces. A multidimensional analogue is given, but this becomes possible only by replacing finite-codimensional subspaces by a new class of appropriately defined "subspaces of finite multi-codimension".

1. Introduction

The invariant subspaces $\mathbf{I} \subset H^2(\mathbf{T})$ of the Hardy space on the circle, were characterized by Beurling in [3], and those of finite codimension N are of the form $\mathbf{I} = \mathbf{b}H^2(\mathbf{T})$, where $\mathbf{b}$ is a Blaschke product of degree N (i.e., of N factors). Their orthocomplements, $H^2(\mathbf{T}) \ominus \mathbf{b}H^2(\mathbf{T})$, are the finite-dimensional model subspaces which play an important role in a number of questions in harmonic analysis (cf. [7]).

In the product space $\mathbf{T}^d$, the subspaces of the form $\mathbf{b}H^2(\mathbf{T}^d)$, where $\mathbf{b}(x) = b_1(x_1)\cdots b_d(x_d)$, for b_j, $j = 1,\ldots d$, one-dimensional finite Blaschke products, are not, in general, of finite codimension, as follows from a theorem of Ahern and Clark [1]. Still, their orthocomplements are of interest in problems of interpolation in product spaces, as in the following context.

The classical Pick-Nevanlinna theorem gives a criterion for bounded analytic interpolation to be possible to a finite number of values assigned at a finite number of points in the unit disk; the condition is the positivity of a finite matrix constructed from the data (Pick condition) [9]. The corresponding finite matrix condition for the analoguous problem in several variables (either in the polydisk or in the unit ball of $\mathbf{C}^d$) fails to be sufficient [2].

It is well-known that, because of the Nehari theorem [8], the Pick condition is equivalent to the boundedness of the norm of a Hankel operator constructed from the data. And there is a version of the Nehari theorem valid in $H^2(\mathbf{T}^d)$, $d > 1$ [5]. But the derivation of the (finite) Pick condition from the boundedness of the norm of an (infinite) Hankel operator in one dimension, involves projecting on a model subspace of $H^2(\mathbf{T})$ that turns out to be of finite dimension by the

---

1991 *Mathematics Subject Classification*. Primary 30D50, 47A15; Secondary 42B30, 47B35.

Beurling characterization (cf. [5], Remark 3). If the model subspaces involved in several dimensions are not finite-dimensional, in what terms can a finite criterion be obtained?

In this paper, we introduce a notion of *subspace of multi-finite type* in a product space. These will be seen to be the subspaces capable of playing the role of those of finite dimension in the Beurling characterization for functions of one variable.

In fact, a Beurling type characterization of the invariant subspaces of $H^2(\mathbf{T}^d)$, $d > 1$, of the form $\mathbf{I} = \mathbf{b}H^2(\mathbf{T}^d)$, $\mathbf{b} = b_1 \otimes \cdots \otimes b_d$, $b_1, \ldots, b_d$, one-variable Blaschke products, is given in Theorem 1 below, as those *of finite multi-codimension*, i.e., those for which $H^2(\mathbf{T}^d) \ominus \mathbf{I}$ are of multi-finite type. A related result, necessary in applications to interpolation theory, is given as Theorem 2. We begin with a Proposition on subspaces of multi-finite type, on which the other results are based.

Applications of the Beurling type results obtained here to Hankel operators and to Pick-Nevalinna interpolation problems in several variables are included in a forthcoming paper [6]. In a related matter, in [4] it was shown that for $d > 1$ all Hankel operators of finite rank in $\mathbf{T}^d$ are zero, so there cannot be a meaningful theory of singular numbers for those operators based on the notion of finite-dimensional subspaces. Again, the notion of a subspace of multi-finite type provides a substitute for the singular numbers of operators acting in product spaces, to be developed elsewhere.

2. Subspaces of bi-finite type in the torus

Fixing $d \geq 1$, let $H^2(\mathbf{T}^d) = \{f \in L^2(\mathbf{T}^d) : \hat{f}(n) = \hat{f}(n_1, \ldots, n_d) = 0$ if $n_k < 0$ for some $k = 1, \ldots, d\}$, where $\hat{f}$ is the Fourier transform of f. The d shifts in $L^2(\mathbf{T}^d)$, $S_k = S_{x_k}$, $k = 1, \ldots, d$, are defined by

$$(1) \qquad\qquad S_k f(x) = \exp(ix_k)\, f(x).$$

For simplicity, in what follows the definitions and results are written for the case of $\mathbf{T}^2$, but are valid for $\mathbf{T}^d$, $d > 1$, with obvious modifications.

Given a subspace $L \subset L^2(\mathbf{T})$ and two integers $N \geq 0$, $M \geq 0$, a subspace $E \subset L^2(\mathbf{T}^2)$ is called of *finite type* $(L; N, M)$ if there exist two sets of linearly independent functions $\{\Phi_1, \ldots, \Phi_N\} \subset H^2(\mathbf{T})$ and $\{\Psi_1, \ldots, \Psi_M\} \subset H^2(\mathbf{T})$, such that $f \in E$ iff f can be written as

$$(2) \qquad f(x,y) = \sum_1^N A_n(x)\Phi_n(y) + \sum_1^M B_n(y)\Psi_n(x)$$

for some arbitrary functions $A_n, B_n \in L$.

Observe that when $L = \mathbf{C}$ and $M = 0$, E reduces to an N-dimensional subspace of $H^2(\mathbf{T})$.

Calling W_N the linear span of $\{\Phi_1, \ldots, \Phi_N\}$, and V_M that of $\{\Psi_1, \ldots, \Psi_M\}$, if $\{\Phi'_1, \ldots, \Phi'_N\}$ and $\{\Psi'_1, \ldots, \Psi'_M\}$ are other bases for W_M, V_N, respectively, then $(L, \{\Phi'_1, \ldots, \Phi'_N\}, \{\Psi'_1, \ldots, \Psi'_M\})$ gives rise to the same subspace E as $(L, \{\Phi_1, \ldots, \Phi_N\}, \{\Psi_1, \ldots, \Psi_M\})$. Thus, E is said to be given by L and W_N, V_M, and we write

$$(3) \qquad E = L \otimes W_N + V_M \otimes L, \quad \text{where } \dim W_N = N \text{ and } \dim V_M = M.$$

Here we are primarily interested in the cases $L = H^2(\mathbf{T})$ and $L = L^2(\mathbf{T})$.

3. Beurling type characterizations of invariant subspaces of finite bi-codimension

In this section we write $S^\perp = \mathrm{H}^2(\mathbf{T}^d) \ominus S$, for every subspace $S \subset \mathrm{H}^2(\mathbf{T}^d)$, $d = 1, 2$.

PROPOSITION. Let $E = \mathrm{H}^2(\mathbf{T}) \otimes W_N + V_M \otimes \mathrm{H}^2(\mathbf{T}) \subset \mathrm{H}^2(\mathbf{T}^2)$ be a subspace of bi-finite type. The following conditions are equivalent.

(a) $E^\perp$ is invariant under the two shifts of $\mathrm{H}^2(\mathbf{T}^2)$.

(b) $W_N^\perp$ and $V_M^\perp$ are both invariant under the shift of $\mathrm{H}^2(\mathbf{T})$.

(c) $E^\perp = \mathbf{b}\mathrm{H}^2(\mathbf{T}^2)$, where $\mathbf{b} = b_1 \otimes b_2$, and b_1 and b_2 are one-dimensional Blaschke products of degrees N and M, respectively, and

(4) $$V_M = (b_1 \mathrm{H}^2(\mathbf{T}))^\perp, \quad W_N = (b_2 \mathrm{H}^2(\mathbf{T}))^\perp.$$

PROOF. (a) $\Rightarrow$ (b). Let $f(x) \in V_M^\perp$, then $e^{ix} f(x) \in V_M^\perp$. In fact, take $g \in W_N^\perp$, $g \neq 0$, so that, for some k, $\int g(y)e^{iky}dy \neq 0$. Then $f(x)g(y) \in E^\perp$ and, by assumption (a), $e^{ix}e^{iky}f(x)g(y) \perp E$, which implies $0 = \int(\int e^{iky} g(y)dy)e^{ix} f(x)\overline{h(x)}dx$, $\forall h \in V_M$. Hence, for all such h, $\int e^{ix} f(x)\overline{h(x)}dx = 0$ and $e^{ix} f(x) \perp V_M$. Similarly for W_N.

(b) $\Rightarrow$ (c). By the Beurling characterization of finite codimensional invariant subspaces of $\mathrm{H}^2(\mathbf{T})$, there exist two Blaschke products, $b_1(x)$ and $b_2(y)$, such that (4) holds. Let $h \in E^\perp$, so that $\langle h, A(x)\Phi_n(y)\rangle = 0$ for $A \in \mathrm{H}^2(\mathbf{T})$, arbitrary, $\{\Phi_1, \ldots, \Phi_N\}$ basis for W_N, and $\langle h, B(y)\Psi_n(x)\rangle = 0$, for $B \in \mathrm{H}^2(\mathbf{T})$, arbitrary, $\{\Psi_1, \ldots, \Psi_M\}$ basis for V_M. Hence, for any $A, B \in \mathrm{H}^2(\mathbf{T})$,

$$\int \left(\int h(x,y)\overline{A(x)}dx \right) \overline{\Phi_n(y)}dy = \int \left(\int h(x,y)\overline{B(y)}dy \right) \overline{\Psi_n(x)}dx = 0,$$

that is,

$$\int h(x,y)\overline{A(x)}dx \perp \Phi_n(y) \quad \text{for } n = 1, \ldots, N$$

and

$$\int h(x,y)\overline{B(y)}dy \perp \Psi_n(x) \quad \text{for } n = 1, \ldots, M$$

which, by (4), implies

$$\int h(x,y) \overline{A(x)}dx = b_2(y)\eta_a(y), \quad \int h(x,y) \overline{B(y)}dy = b_1(x)\eta_b(x)$$

where η_a and η_b are in $\mathrm{H}^2(\mathbf{T})$. For a zero of b_2 at y_0, we have $\int h(x,y_0) \overline{A(x)}dx = 0$ and, taking $A(x) = h(x,y_0)$, we get $\int |h(x,y_0)|^2 dx = 0$. Thus, $h(x,y_0) = 0$ wherever $b_2(y_0) = 0$, so that $h(x,y) = b_2(y)h'(x,y)$, for $h' \in \mathrm{H}^2(\mathbf{T}^2)$. Similarly, $h(x,y) = b_1(x)h''(x,y)$, for $h'' \in \mathrm{H}^2(\mathbf{T}^2)$. Thus, $h(x,y) = b_1(x)b_2(y)h_0(x,y)$ for $h_0 \in \mathrm{H}^2(\mathbf{T}^2)$, which is (c).

(c) $\Rightarrow$ (a) is immediate. $\square$

THEOREM 1. (Beurling type result for invariant subspaces of finite bi-codimension in $\mathrm{H}^2(\mathbf{T}^2)$). Given a subspace $\mathbf{I} \subset \mathrm{H}^2(\mathbf{T}^2)$, invariant under both shifts of $\mathrm{H}^2(\mathbf{T}^2)$, the following are equivalent.

(a) $\mathbf{I} = \mathbf{b}\mathrm{H}^2(\mathbf{T}^2)$, where $\mathbf{b} = b_1 \otimes b_2$, b_1 and b_2 one-dimensional Blaschke products, of degree N and M, respectively.

(b) $\mathbf{I}$ *is of finite bi-codimension, i.e.,* $E = \mathrm{H}^2(\mathbf{T}^2) \ominus \mathbf{I}$ *is of bi-finite type* $(\mathrm{H}^2(\mathbf{T}); N, M)$.

PROOF. (a) $\Rightarrow$ (b). Setting $W_M = \mathrm{H}^2(\mathbf{T}) \ominus b_2 \mathrm{H}^2(\mathbf{T})$ and $V_M = \mathrm{H}^2(\mathbf{T}) \ominus b_1 \mathrm{H}^2(\mathbf{T})$, the subspace $E = \mathrm{H}^2(\mathbf{T}) \otimes W_M + V_N \otimes \mathrm{H}^2(\mathbf{T}) \subset \mathrm{H}^2(\mathbf{T}^2)$ is of bi-finite type. Furthermore, by Beurling theorem, V_M and W_N satisfy condition (b) of the Proposition, so that E has to satisfy its condition (c). Thus, $\mathbf{I} = \mathrm{H}^2(\mathbf{T}^2) \ominus E$ and (b) follows.

(b) $\Rightarrow$ (a). Since $\mathbf{I}$ is an invariant subspace of $\mathrm{H}^2(\mathbf{T}^2)$, this is immediate from (a) implies (c) in the Proposition. $\qquad\square$

A related result is needed for applications to interpolation problems and other questions.

In the one-dimensional case, $\overline{\mathrm{H}^2(\mathbf{T})^\perp} = H_0^2(\mathbf{T}) = e^{it}H^2(\mathbf{T})$, so that the Beurling theorem applies also to $\overline{H^{2\perp}} \ominus b\overline{H^{2\perp}}$, b a Blaschke product. The situation is different in $\mathbf{T}^2$, where

$$(5) \qquad \overline{\mathrm{H}^2(\mathbf{T}^2)^\perp} = H_x^2 + H_y^2 = \mathrm{H}^2 \oplus H_{x,-y}^2 \oplus H_{-x,y}^2 \underset{\ne}{\supsetneq} \mathrm{H}^2(\mathbf{T}^2),$$

for $H_x^2 = \{f \in L^2(\mathbf{T}^2) : \hat{f}(m,n) = 0 \text{ if } m < 0\}$, $H_{-x,y}^2 = \{f \in L^2(\mathbf{T}^2) : \hat{f}(m,n) = 0 \text{ if } m \ge 0 \text{ or } n < 0\}$ and H_y^2, $H_{x,-y}^2$ similarly defined. Thus, the Beurling type results of Theorems 1 and 1a do not apply to $\overline{\mathrm{H}^2(\mathbf{T}^2)^\perp} \ominus b\overline{\mathrm{H}^2(\mathbf{T}^2)^\perp}$. However, $\overline{\mathrm{H}^2(\mathbf{T}^2)^\perp}$ is still invariant under both shifts in $\mathbf{T}^2$, and the following result, necessary for the multidimensional version of Pick's interpolation, holds.

THEOREM 2. *For a subspace* $\mathbf{E} \subset \overline{\mathrm{H}^2(\mathbf{T}^2)^\perp}$, *invariant under both shifts in* $\mathbf{T}^2$, *the following is equivalent.*

(a)

$$(6) \qquad \mathbf{E} = b_1 \otimes b_2 \, \mathrm{H}^2(\mathbf{T}^2) + b_1 H_{x,-y}^2 + b_2 H_{-x,y}^2$$

where b_1 *and* b_2 *are one-dimensional Blaschke products of degrees* M *and* N, *respectively.*

(b) $E = \overline{\mathrm{H}^2(\mathbf{T}^2)^\perp} \ominus \mathbf{E}$ *is a subspace of* $\overline{\mathrm{H}^2(\mathbf{T}^2)^\perp}$ *of bi-finite type* $(L^2(\mathbf{T}); N, M)$.

PROOF. Let us first show that if

$$(7) \qquad W_N = \mathrm{H}^2(\mathbf{T}) \ominus b_2 \mathrm{H}^2(\mathbf{T}), \quad V_M = \mathrm{H}^2(\mathbf{T}) \ominus b_1 \mathrm{H}^2(\mathbf{T})$$

for b_1, b_2, two Blaschke products of degrees M and N, respectively, and if

$$(7a) \qquad E' = L^2(\mathbf{T}) \otimes W_N + V_N \otimes L^2(\mathbf{T}),$$

then $\overline{\mathrm{H}^2(\mathbf{T}^2)^\perp} \ominus E'$ is of the form (6).

Let $g = h_{x,y} + h_{x,-y} + h_{-x,y} \in \overline{\mathrm{H}^2(\mathbf{T}^2)^\perp}$ be such that $g \perp E'$, i.e., $g(x,y) \perp A(x)\Phi_n(y)$, $n = 1, \ldots, N$, and $g(x,y) \perp B(x)\Psi_m(y)$, $m = 1, \ldots, M$, where $A, B \in L^2(\mathbf{T})$ are arbitary, and $\{\Phi_n\}$, $\{\Psi_m\}$ are bases of W_N and V_M, respectively. We claim that g can be decomposed as in (6). In fact, since $h_{x,-y}$ is antianalytic in y, it is orthogonal to all $A(x)\Phi_n(y)$, which implies $h_{x,y} + h_{-x,y} = g - h_{x,-y} \perp A(x)\Phi_n(y)$, for $n = 1, \ldots, N$, i.e., $\int(h_{x,y} + h_{-x,y}) \, A(x)dx \perp W_N$, for any arbitrary $A \in L^2(\mathbf{T})$. By (7), then $\int(h_{x,y} + h_{-x,y})A(x)dx = b_2(y)\eta(y)$, $\eta \in \mathrm{H}^2(\mathbf{T})$. With an argument as in the proof of the Proposition, this implies $h_{x,y} + h_{-x,y} = b_2(y)h_{x,y}' + b_2(y)h_{-x,y}'$, and, both sides being orthogonal sums, $h_{x,y} = b_2(y) \, h_{x,y}'$ and $h_{-x,y} = b_2(y) \, h_{-x,y}'$.

Similarly, from $g(x,y) \perp B(y)\Psi_m(x)$, $m = 1, \ldots, M$, we obtain $h_{x,y} = b_1(x)h''_{x,y}$ and $h_{x,-y} = b_1(x)h''_{x,-y}$. Therefore, we can write

$$g = b_1(x)b_2(y)h^{\circ}_{x,y} + b_2(y)h'_{-x,y} + b_1(x)h''_{x,-y}$$

as asserted. Now we can proceed with proving the thesis' equivalence.

(a) $\Rightarrow$ (b). By the previous remark, it is enough to prove that $\mathbf{E} \perp E'$, for $\mathbf{E}$ as in (a). By the definition (7a) of E', we have, then, to prove that every $g \in \mathbf{E}$, $g = b_1(x)b_2(y)h_{x,y} + b_1(x) \, h_{x,-y} + b_2(y)h_{-x,y}$ is orthogonal to every $A(x)\Phi_n(y)$, $B(y)\Psi_m(x)$, for $A, B \in L^2(\mathbf{T})$, arbitrary, and $\{\Phi_n : n = 1, \ldots, ; N\}$, $\{\Phi_m : m = 1, \ldots, M\}$ bases of W_M and V_N, as given in (7).

Observe that, since $\mathrm{h}_{x,-y}$ is antianalytic in y, by (7), $b_1(x)b_2(y)h_{x,y}$ and $b_2(y)h_{-x,y}$ are orthogonal to $A(x)\Phi_n(y)$, $n = 1, \ldots, N$, since $h_{x,y}$ and $h_{-x,y}$ are analytic in y. In turn, $b_1(x)h_{x,-y}$ is orthogonal to $A(x)\Phi_n(y)$ since $h_{x,-y}$ is antianalytic and $\Phi_n(y)$ is analytic, both in y. Thus $g \in \mathbf{E}$ is orthogonal to all $A(x)\Phi_n(y)$ and $B(y)\Psi_m(x)$ as wanted.

(b) $\Rightarrow$ (a). By the first remark, it is enough to show that the subspace E of finite type $(L^2(\mathbf{T}); N, M)$ in (b) can be written $E = L^2(\mathbf{T}) \otimes W'_N + V'_M \otimes L^2(\mathbf{T})$, for $W'_N = \mathrm{H}^2(\mathbf{T}) \ominus b_1\mathrm{H}^2(\mathbf{T})$, and $V'_M = \mathrm{H}^2(\mathbf{T}) \ominus b_2\mathrm{H}^2(\mathbf{T})$, for some Blaschke products b_1 and b_2 of degree N and M, respectively. In fact, if f and g are two functions in $\mathrm{H}^2(\mathbf{T})$ satisfying $f \perp W'_N$ and $g \perp V'_M$, then $f(x)g(y) \in \overline{\mathrm{H}^2(\mathbf{T}^2)^{\perp}} \ominus E = \mathbf{E}$, and, by the invariance of $\mathbf{E}$ with respect to both shifts, $f(x)g(y)\mathrm{e}^{ix}$ and $f(x)g(y)\mathrm{e}^{iy}$ are both orthogonal to E. As in the proof of the Proposition, this implies that W'_N and V'_M are equal to $b_1\mathrm{H}^2$ and $b_2\mathrm{H}^2$, for some pair of Blaschke products b_1 and b_2, and the rest of the thesis follows. $\qquad\square$

REMARK. After the above was written, Charles Neville pointed out to us that the proof of the Proposition leads to a characterization of the invariant subspaces in $\mathrm{H}^2(\mathbf{T}^2)$ containing Theorem 1 as a particular case. The precise statement will appear elsewhere.

References

[1] P. Ahern and D. N. Clark, *Invariant subspaces and analytic continuation in several variables*, J. Math. Mech. **19** (1969/70), 963–969.

[2] E. Amar, *Ensembles d'interpolation dans le spectre d'une algèbre d'opérateurs*, Thèse, Université de Paris-Sud, 1977.

[3] A. Beurling, *On two problems concerning linear transformations in Hilbert space*, Acta Math. **83** (1949), 239–255.

[4] M. Cotlar and C. Sadosky, *Abstract, weighted, and multidimensional AAK theorems, and the singular numbers of Sarason commutants*, Integr. Equat. and Oper. Th. **17** (1993), 169–201.

[5] M. Cotlar and C. Sadosky, *Nehari and Nevanlinna-Pick problems and holomorphic extensions in the polydisk in terms of restricted BMO*, J. Func. Anal. **124** (1994), 205–210.

[6] M. Cotlar and C. Sadosky, *Two distinguished subspaces of product BMO and Nehari-AAK theory for Hankel operators on the torus*, Integr. Equat. and Oper. Th. **26** (1996), 273–304.

[7] H. Helson, *Lectures on Invariant Subspaces*, Academic Press, New York, 1964.

[8] Z. Nehari, *On bilinear forms*, Annals of Math. **68** (1957), 153–162.

[9] G. Pick, *Über die Beschränkungen analytischer Funktionen, welche durch vorgegebene Funktionswerte bewirkt werden*, Math. Ann. **77** (1916), 7–23.

[10] N. Salinas, *Product of kernel functions and module tensor products*, Operator Theory: Advances and Applications, vol. 32, 1988, pp. 219–241.

[11] S. Treil, *The theorem of Adamyan-Arov-Krein: Vector variant*, Publ. Seminar LOMI Leningrad **141** (1985), 56–72 (In Russian).

FACULTAD DE CIENCIAS, UNIVERSIDAD CENTRAL DE VENEZUELA, CARACAS 1040, VENEZUELA
E-mail address: mcotlar@euler.ciens.ucv.edu

DEPARTMENT OF MATHEMATICS, HOWARD UNIVERSITY, WASHINGTON, DC 20059, U.S.A.
E-mail address: cs@scs.howard.edu

Contemporary Mathematics
Volume **212**, 1998

A Representation of Solutions with Singularities

Bert Fischer and Nikolai Tarkhanov

ABSTRACT. Let P be a homogeneous elliptic differential operator of order $p < n$ with constant coefficients in $\mathbb{R}^n$. Denote by Φ the fundamental solution of convolution type to P which vanishes at infinity. Let f be a solution to $Pf = 0$ away from a d-plane S in $\mathbb{R}^n$, and let $q \in [1, q_0)$, where $q_0 = \frac{d}{d+p-n}$ if $d > n - p$ and $q_0 = \infty$ if $d \le n - p$. We prove that there exists a solution f_e to this equation on the whole space $\mathbb{R}^n$ and a sequence $\{f_\alpha\}$ in $L^q(S)$ satisfying $\|\alpha! \, f_\alpha\|_{L^q(S)}^{1/|\alpha|} \to 0$ such that $f(x) = f_e(x) + \sum_{\alpha \in \mathbb{Z}_+^n} \int_S D^\alpha \Phi(x - y) \, f_\alpha(y) \, ds(y)$, where ds is the induced Lebesgue measure on S. This complements an earlier result of the second author on representation of solutions to $Pf = 0$ outside a compact subset of $\mathbb{R}^n$. As an immediate consequence we get a representation theorem for hyperfunctions of $n - 1$ variables.

1. Introduction

In 1972 Baernstein [1] proved that every function holomorphic off the real axis can be represented in the form

$$f(z) = f_e(z) + \sum_{j=0}^{\infty} \int_{-\infty}^{\infty} (z - t)^{-1-j} f_j(t) \, dt \quad (z \notin \mathbb{R}),$$

where f_e is an entire function and $\{f_j\}$ a sequence in $L^q(\mathbb{R})$ ($q \in [1, \infty)$) satisfying the condition $\lim_{j \to \infty} \|f_j\|_{L^q(\mathbb{R})}^{1/|\alpha|} = 0$.

In this article we prove a similar result for solutions of a homogeneous elliptic differential equation with constant coefficients on $\mathbb{R}^n$, with $(z - t)^{-1-j}$ replaced by the derivatives of the standard fundamental solution of the equation.

In Section 2 we set up notation and terminology and state our main result. The converse statement is quite easy to prove; we give the proof in Section 3. Section 4 contains a brief exposition of abstract framework for proving the main theorem. In Section 5 we treat the duality between the spaces of solutions to the equation on a subset of $\mathbb{R}^n$ and solutions to the transposed equation off the subset. This duality is a principal tool in the proof. It is also of independent interest and extends the

---

1991 *Mathematics Subject Classification.* Primary 35C10; Secondary 35A20.

The first author was supported by the Max-Planck-Gesellschaft.

The second author was supported by the Alexander-von-Humboldt Foundation.

results of Grothendieck [2] to the case of noncompact sets. In Section 6 we recall a standard fact on the dual to the space of sequences which decrease like Taylor coefficients of an analytic function. Using both the dualities we describe in Section 7 the transpose to a mapping whose surjectivity is equivalent to the statement of the theorem. Section 8 establishes then the invariance of equicontinuous sets under the inverse to the transposed mapping; this suffices for surjectivity. In the final two sections some applications are indicated. Namely, in Section 9 we derive an interesting formula for functions harmonic outside a plane in $\mathbb{R}^n$. Section 10 indicates how these results may be used to represent hyperfunctions on $\mathbb{R}^d$.

2. Notation and the main theorem

Assume that P is a homogeneous elliptic operator of order $p < n$ with constant coefficients on $\mathbb{R}^n$.

Let us fix a fundamental solution $\Phi(z)$ of convolution type for P which vanishes at infinity. The existence and uniqueness of such a fundamental solution on $\mathbb{R}^n$ is well-known (see, for instance, Hörmander [3]). Moreover, Φ is a homogeneous distribution of degree $p - n$ on $\mathbb{R}^n$.

If U is an open subset of $\mathbb{R}^n$, then denote by $sol(U, P)$ the vector space of all C^∞ solutions to the equation $Pf = 0$ on U. We will write it simply $sol(U)$ when no confusion can arise.

Note that all the solutions in $sol(U)$ are in fact analytic functions on U, by the *Petrovskii Theorem*.

We endow the space $sol(U)$ with the topology of uniform convergence on compact subsets of U. This topology is generated by the family of seminorms

$$\|f\|_{C(K)} = \sup_{x \in K} |f(x)|,$$

where K runs over compact subsets of U.

LEMMA 2.1. *If $U \subset \mathbb{R}^n$ is open, then $sol(U)$ is a Fréchet-Schwartz space.*

PROOF. By a priori estimates for solutions of elliptic equations, if K' and K'' are compact subsets of U and K' lies in the interior of K'', then

$$(2.1) \qquad \sup_{|\alpha| \le j} \|D^\alpha f\|_{C(K')} \le c \, \|f\|_{C(K'')} \quad \text{for all } f \in sol(U),$$

with c a constant depending only on K', K'' and j. Hence it follows that the original topology on $sol(U)$ coincides with that induced by $C^\infty_{loc}(U)$. To finish the proof we use the fact that $C^\infty_{loc}(U)$ is a Fréchet-Schwartz space. □

We shall use the so-called one-point, or Aleksandrov, compactification of $\mathbb{R}^n$, denoted by $\widehat{\mathbb{R}^n}$. This means that $\widehat{\mathbb{R}^n}$ is the union of $\mathbb{R}^n$ and the symbolic point ∞, and the topology of $\widehat{\mathbb{R}^n}$ is given by the following neighborhoods bases:

1. If $x \in \mathbb{R}^n$, then we take the usual basis of neighborhoods of x (for example, the family of all balls centered at x).
2. If $x = \infty$, then the basis of neighborhoods of x is defined to be the family $\{U \cup \infty\}$, where U is an open subset of $\mathbb{R}^n$ with compact complement.

We shall also need the concept of a solution to $Pf = 0$ in a neighborhood $\{U \cup \infty\}$ of ∞. By this, is meant a solution to $Pf = 0$ on the "finite part" U of this neighborhood such that $f(\infty) = 0$.

Given any open set U with compact complement in $\mathbb{R}^n$, we denote by $sol(U \cup \infty)$ the subspace of $sol(U)$ consisting of those solutions which vanish at infinity. This subspace is closed (see [6, 5.4.4]), and so, when equipped with the induced topology, it is a Fréchet-Schwartz space.

If σ is a closed subset of $\widehat{\mathbb{R}^n}$, then $sol(\sigma)$ stands for the space of (equivalence classes of) solutions to $Pf = 0$ in some neighborhood of σ. Two such solutions are called *equivalent* if there is a neighborhood of σ where they are equal. In $sol(\sigma)$, a sequence $\{f_\nu\}$ is said to converge if there exists a neighborhood $\mathcal{N}$ of σ such that all the solutions are defined at least in $\mathcal{N}$ and converge uniformly on compact subsets of $\mathcal{N}$.

Alternatively, $sol(\sigma)$ can be described as the inductive limit of the spaces $sol(U_\nu)$, where $\{U_\nu\}$ is any decreasing sequence of open sets containing σ such that each neighborhood of σ contains some U_ν and such that each component of each U_ν meets $U_{\nu+1}$.

LEMMA 2.2. *The space $sol(\sigma)$ is separated, a subset is bounded if and only if it is contained and bounded in some $sol(U_\nu)$, and each closed bounded set is compact.*

PROOF. This follows by the same method as in Köthe [4, p. 379]. $\qquad\square$

Given an integer d with $0 \le d \le n$, we consider $\mathbb{R}^d$ as the subspace of $\mathbb{R}^n$ defined by

$$\mathbb{R}^d = \{x \in \mathbb{R}^n : x_{d+1} = \cdots = x_n = 0\}.$$

The induced Lebesgue measure on $\mathbb{R}^d$ is $ds = dx_1 \cdot \cdots \cdot dx_d$.

In this paper we consider solutions to $Pf = 0$ in $\mathbb{R}^n \setminus \mathbb{R}^d$. The singular set is now $\mathbb{R}^d \cup \{\infty\}$. There are very few such f having a representation $f = D^\alpha \Phi * m$, where m is a complex Borel measure of finite total variation on $\mathbb{R}^d$ and $\alpha \in \mathbb{Z}^n_+$, since those that do satisfy

$$|f(x)| \le const(\alpha) \, dist(x, \mathbb{R}^d)^{p-n-|\alpha|} \, \|m\|.$$

Thus, they are bounded in every region $dist(x, \mathbb{R}^d) \ge \varepsilon > 0$ and tend to zero as $dist(x, \mathbb{R}^d) \to \infty$.

On the other hand, we are going to prove the following.

THEOREM 2.3. *Let q be any real number with $1 \le q < \frac{d}{d+p-n}$ if $d > n - p$, and with $1 \le q < \infty$ if $d \le n - p$. Then for each solution $f \in sol(\mathbb{R}^n \setminus \mathbb{R}^d)$ there exist a solution $f_e \in sol(\mathbb{R}^n)$ and a sequence $\{f_\alpha\}_{\alpha \in \mathbb{Z}^n_+} \subset L^q(\mathbb{R}^d)^l$, satisfying $\|\alpha! \, f_\alpha\|_{L^q(\mathbb{R}^d)}^{1/|\alpha|} \to 0$ when $|\alpha| \to 0$, such that*

$$(2.2) \quad f(x) = f_e(x) + \sum_{\alpha \in \mathbb{Z}^n_+} \int_{\mathbb{R}^d} D^\alpha \Phi \, (x - y) \, f_\alpha(y) \, ds(y) \quad \text{for all } x \in \mathbb{R}^n \setminus \mathbb{R}^d.$$

The theorem asserts that any solution to $Pf = 0$ in $\mathbb{R}^n \setminus \mathbb{R}^d$ can be split into two parts, one of which has the only singularity at infinity (of $\widehat{\mathbb{R}^n}$), and the other part (the sum of potentials $D^\alpha \Phi * (f_\alpha \, ds)$) carries all the finite singularities and has a "weak singularity" at infinity (it is bounded outside of the tubular neighborhoods

$$\{x \in \mathbb{R}^n : dist(x, \mathbb{R}^d) < \varepsilon\}$$

of $\mathbb{R}^d$).

S. Simonova (1991, *unpublished*) obtained an analogous representation theorem for functions harmonic off a hyperplane.

We now turn to the proof of Theorem 2.3. Throughout the proof, we let $q_0 = \frac{d}{d+p-n}$ if $d > n - p$, and $q_0 = \infty$ if $d \le n - p$. To shorten notation, we write $\| \cdot \|_q$ for the norm in $L^q(\mathbb{R}^d)$.

3. The converse theorem

The converse statement to Theorem 2.3 is quite easy to prove.

LEMMA 3.1. *Let $q \in [1, q_0)$. For every sequence $\{f_\alpha\}_{\alpha \in \mathbb{Z}_+^n} \subset L^q(\mathbb{R}^d)$ satisfying $\|\alpha! f_\alpha\|_q^{1/|\alpha|} \to 0$ when $|\alpha| \to \infty$, the series $\sum_\alpha \int_{\mathbb{R}^d} D^\alpha \Phi(x-y) f_\alpha(y) \, ds(y)$ converges for $x \in \mathbb{R}^n \setminus \mathbb{R}^d$ and defines an element in $sol(\mathbb{R}^n \setminus \mathbb{R}^d)$.*

PROOF. First note that for $x \in \mathbb{R}^n \setminus \mathbb{R}^d$ we have

$$P \int_{\mathbb{R}^d} D^\alpha \Phi(x - y) \, f_\alpha(y) \, ds(y) = \int_{\mathbb{R}^d} D^\alpha (P\Phi)(x - y) \, f_\alpha(y) \, ds(y) = 0.$$

Thus the proof will be complete if we show that the series we look at converges uniformly on compact subsets of $\mathbb{R}^n \setminus \mathbb{R}^d$.

It is well-known that a C^∞ function g on an open set $U \subset \mathbb{R}^n$ is real analytic if and only if for every compact set $K \subset U$ there are constants $b = b(g, K)$ and $c = c(g, K)$ such that

$$\sup_{z \in K} |D^\alpha g(z)| \le c \, b^{|\alpha|} |\alpha|! \quad \text{for all} \ \alpha \in \mathbb{Z}_+^n.$$

We apply this estimate to the fundamental solution Φ; recall that $\Phi(z)$ is a real analytic function away from the origin in $\mathbb{R}^n$. It follows that there exist constants b and c depending only on P such that

(3.1) $$\sup_{|z|=1} |D^\alpha \Phi(z)| \le c \, b^{|\alpha|} |\alpha|! \quad \text{for all} \ \alpha \in \mathbb{Z}_+^n.$$

So, by Hölder's inequality, we have

$$\left| \int_{\mathbb{R}^d} D^\alpha \Phi(x - y) \, f_\alpha(y) \, ds(y) \right|$$

$$\le \left| \int_{\mathbb{R}^d} D^\alpha \Phi \left(\frac{x - y}{|x - y|} \right) |x - y|^{p-n-|\alpha|} f_\alpha(y) \, ds(y) \right|$$

$$\le c \, b^{|\alpha|} |\alpha|! \left(\int_{\mathbb{R}^d} |x - y|^{(p-n-|\alpha|)q'} \, ds(y) \right)^{\frac{1}{q'}} \|f_\alpha\|_q,$$

with $\frac{1}{q} + \frac{1}{q'} = 1$.

It is not hard to see that

$$\int_{\mathbb{R}^d} |x - y|^{(p-n-|\alpha|)q'} \, ds(y)$$

$$= dist(x, \mathbb{R}^d)^{d+(p-n-|\alpha|)q'} \int_{\mathbb{R}^d} (|z|^2 + 1)^{\frac{(p-n-|\alpha|)q'}{2}} \, dz$$

$$\le dist(x, \mathbb{R}^d)^{d+(p-n-|\alpha|)q'} \int_{\mathbb{R}^d} (|z|^2 + 1)^{\frac{(p-n)q'}{2}} \, dz,$$

provided that $(p-n)q' < -d$. As the latter condition is equivalent to $q \in [1, q_0)$ (and hence it is fulfilled by assumption), we get

(3.2)
$$\left| \int_{\mathbb{R}^d} D^\alpha \Phi(x-y) \, f_\alpha(y) \, ds(y) \right|$$
$$\leq C \, b^{|\alpha|} \, |\alpha|! \, dist(x, \mathbb{R}^d)^{\frac{d}{q'} + (p-n-|\alpha|)} \, \|f_\alpha\|_q$$

with C a constant independent of α and x.

Let $dist(x, \mathbb{R}^d) \geq \varepsilon > 0$. Then (3.2) implies

(3.3)
$$\sum_{\alpha \in \mathbb{Z}_+^n} \left| \int_{\mathbb{R}^d} D^\alpha \Phi(x-y) \, f_\alpha(y) \, ds(y) \right|$$
$$\leq C \, \varepsilon^{\frac{d}{q'} + (p-n)} \sum_{\alpha \in \mathbb{Z}_+^n} \left(b \, \varepsilon^{-1} \right)^{|\alpha|} |\alpha|! \, \|f_\alpha\|_q$$
$$\leq C \, \varepsilon^{\frac{d}{q'} + (p-n)} \sum_{j=0}^{\infty} \left(b \, \varepsilon^{-1} \, n \sup_{|\alpha|=j} \|\alpha! \, f_\alpha\|_q^{1/|\alpha|} \right)^j,$$

where we used the equality $\sum_{|\alpha|=j} \frac{|\alpha|!}{\alpha!} = n^j$.

Now, since $\sup_{|\alpha|=j} \|\alpha! \, f_\alpha\|_q^{1/|\alpha|} \to 0$ when $j \to \infty$, the last sum can be majorized by a geometric sum. Hence

$$\sum_{\alpha \in \mathbb{Z}_+^n} \left| \int_{\mathbb{R}^d} D^\alpha \Phi(x-y) \, f_\alpha(y) \, ds(y) \right| \leq const(\varepsilon) < \infty,$$

as desired.   □

4. An abstract framework

The main idea in the proof of Theorem 2.3 is the duality between the space of solutions to $Pf = 0$ on an open set $U \subset \mathbb{R}^n$ and the space of solutions to the transposed equation $P'g = 0$ on the complement of U in $\widehat{\mathbb{R}^n}$, which will be developed in the next section. We will also use some facts from the theory of locally convex spaces. In particular, we will use the following criterion for a linear map between Fréchet spaces to be onto.

PROPOSITION 4.1. *Let K and L be Fréchet spaces and let M be a continuous linear mapping from L to K. If $(M')^{-1}(B)$ is equicontinuous in K' whenever B is an equicontinuous set in L', then $M(L) = K$.*

Here K' and L' denote the topological dual spaces and M' denotes the transposed mapping. Note that "equicontinuous" can be interchanges with "weakly bounded," since Fréchet spaces are barreled.

PROOF. See Baernstein [1].   □

Now we begin with the proof of Theorem 2.3. To this end, we introduce three Fréchet spaces, K, L_1, and L_2.

K is the space $sol(\mathbb{R}^n \setminus \mathbb{R}^d)$, and L_1 is the space $sol(\mathbb{R}^n)$. These two spaces are given the usual topology of uniform convergence on compact subsets (of $\mathbb{R}^n \setminus \mathbb{R}^d$ or $\mathbb{R}^n$, respectively), as above.

L_2 is the space of all sequences $\{f_\alpha\}_{\alpha \in \mathbb{Z}_+^n}$ in $L^q(\mathbb{R}^d)$ such that $\|\alpha! \, f_\alpha\|_q^{1/|\alpha|} \to 0$ as $|\alpha| \to \infty$. This space is topologized by the sequence of norms $\| \cdot \|_R$ defined by

$$\|\{f_\alpha\}\|_R = \sup_\alpha R^{|\alpha|} \, \|\alpha! \, f_\alpha\|_q \quad (R = 1, 2, \dots).$$

Then L_2 is a locally convex metrizable space, and, as the reader may easily check, it is complete (i.e., a Fréchet space).

We define the linear mapping $M : \; L_1 \oplus L_2 \to K$ by
(4.1)
$$M(f_e \oplus \{f_\alpha\})(x) = f_e(x) + \sum_{\alpha \in \mathbb{Z}_+^n} \int_{\mathbb{R}^d} D^\alpha \Phi(x - y) \, f_\alpha(y) \, ds(y) \quad \text{for } x \in \mathbb{R}^n \setminus \mathbb{R}^d.$$

It follows from Lemma 3.1 that M is well-defined.

LEMMA 4.2. *The mapping M defined in (4.1) is continuous.*

PROOF. Given a compact set K in $\mathbb{R}^n \setminus \mathbb{R}^d$, we let $\varepsilon > 0$ be the distance from K to $\mathbb{R}^d$. If R is a natural number with

$$R \geq 2 \, b \, \varepsilon^{-1} \, n,$$

then we deduce from (3.3) that

$$\|M(f_e \oplus \{f_\alpha\})\|_{C(K)}$$

$$\leq \|f_e\|_{C(K)} + C \, \varepsilon^{\frac{d}{q'} + (p-n)} \left(\sum_{j=0}^\infty 2^{-j} \right) \|\{f_\alpha\}\|_R$$

$$= \|f_e\|_{C(K)} + C' \, \|\{f_\alpha\}\|_R;$$

the constant C' depends only on K and q. This is the desired conclusion. $\qquad\square$

The conclusion of Theorem 2.3 is that M is onto. We want to deduce this from the proposition, so we look for concrete representations of the dual spaces.

5. Grothendieck duality

Fix a Green operator G_P for the differential operator P. By definition, G_P is a bidifferential operator of order $p-1$ with values in the space of differential forms of degree $n-1$ on $\mathbb{R}^n$ such that $d \, G_P(g, f) = (g \, Pf - P'g \, f) \, dx$ for all smooth functions g and f.

Let U be an open set in $\mathbb{R}^n$. Given any solution $g \in sol(\widehat{\mathbb{R}^n} \setminus U, P')$, we define a linear functional $\mathcal{F}_g$ on $sol(U, P)$ as follows.

There is an open set $\mathcal{O}_g \subset\subset U$ with piecewise smooth boundary such that g still satisfies $P'g = 0$ in a neighborhood of $\mathbb{R}^n \setminus \mathcal{O}_g$. Put

(5.1)
$$\langle \mathcal{F}_g, f \rangle = \int_{\partial \mathcal{O}_g} G_P(g, f) \quad (f \in sol(U, P)).$$

It follows from Stokes' formula that the value $\langle \mathcal{F}_g, f \rangle$ is independent of the particular choice of $\mathcal{O}_g$ with the properties previously mentioned.

LEMMA 5.1. *As defined by (5.1), $\mathcal{F}_g$ is a continuous linear functional on the space $sol(U, P)$.*

PROOF. Use the estimate (2.1) with $K' = \partial \mathcal{O}_g$ and $j = p - 1$. □

The following result goes back at least as far as the work of Grothendieck [2] who proved it in the case where U is relatively compact.

LEMMA 5.2. [Grothendieck Duality]. *For each open set U in $\mathbb{R}^n$, the correspondence $g \mapsto \mathcal{F}_g$ induces the topological isomorphism*

$$sol(U, P)' \overset{\text{top}}{\cong} sol(\widehat{\mathbb{R}^n} \setminus U, P').$$

PROOF. Pick a continuous linear functional $\mathcal{F}$ on $sol(U, P)$. Since $sol(U, P)$ is a subspace of $C_{loc}(U)$, the space of continuous functions on U, this functional can be extended, by the Hahn-Banach Theorem, to a measure m with compact support in U. Set $K = supp\, m$.

Let $\mathcal{O} \subset\subset U$ be any open set with piecewise smooth boundary such that $K \subset \mathcal{O}$. For each solution $f \in sol(U, P)$, we have, by Green's formula,

$$f(x) = - \int_{\partial \mathcal{O}} G_P(\Phi(x - y), f(y)) \quad (x \in \mathcal{O}).$$

Therefore

$$\langle \mathcal{F}, f \rangle = \int_K f \, dm$$
$$= \int_{\partial \mathcal{O}} G_P(g, f),$$

where $g(y) = - \int_K \Phi(x - y) \, dm(x)$.

Now we look more closely at the properties of this function g called the "Fantappiè indicatrix" of $\mathcal{F}$. Since $P\Phi = \delta$, where δ is the Dirac functional, we deduce that $P'g = 0$ away from K. Moreover, $g(\infty) = 0$, because the fundamental solution $\Phi(x)$ behaves like $|x|^{p-n}$ at infinity.

From what has already been proved, it follows that $g \in sol(\widehat{\mathbb{R}^n} \setminus U, P')$ and $\mathcal{F} = \mathcal{F}_g$. Our next claim is that such a solution g is unique.

To this end, we let $g \in sol(\widehat{\mathbb{R}^n} \setminus U, P')$ satisfy

(5.2)
$$\int_{\partial \mathcal{O}_g} G_P(g, f) \quad \text{for all } f \in sol(U, P),$$

where $\mathcal{O}_g \subset\subset U$ is an open set with piecewise smooth boundary such that g still satisfies $P'g = 0$ in a neighborhood of $\mathbb{R}^n \setminus \mathcal{O}_g$.

We represent g in the complement of $\mathcal{O}_g$ by Green's formula. This is possible because of $g(\infty) = 0$. We get

$$g(y) = - \int_{\partial \mathcal{O}_g} G_P(g(x), \Phi(x - y)) \quad \text{for } y \in \mathbb{R}^n \setminus \overline{\mathcal{O}_g}.$$

For any fixed $y \in \mathbb{R}^n \setminus U$, we have $\Phi(\cdot - y) \in sol(U, P)$, and so $g(y) = 0$ by condition (5.2). But this conclusion is vacuous if $U = \mathbb{R}^n$, say. We would like to have more, namely that $g = 0$ in some neighborhood of $\widehat{\mathbb{R}^n} \setminus U$. For this purpose, we use the *Runge Theorem* for solutions of the equation $Pf = 0$.

There exists an open set $\mathcal{O} \subset\subset U$ with the following properties:
1. $\mathcal{O}_g \subset\subset \mathcal{O}$, and

2. the complement of $\mathcal{O}$ has no compact connected components in U.
(The second property can always be achieved by adding all compact connected components of $U \setminus \mathcal{O}$ to $\mathcal{O}$.)

Fix $y \in \mathbb{R}^n \setminus \mathcal{O}$. Then $\Phi(\cdot - y)$ is in $sol(\mathcal{O}, P)$. By the *Runge Theorem*, there is a sequence $\{f_\nu\}$ in $sol(U, P)$ such that $f_\nu \to \Phi(\cdot - y)$ uniformly on compact subsets of $\mathcal{O}$. Applying (2.1) we can assert that the derivatives up to order $p - 1$ of f_ν also converge to the corresponding derivatives of $\Phi(\cdot - y)$ uniformly on compact subsets of $\mathcal{O}$. Therefore,

$$
\begin{aligned}
g(y) &= -\lim_{\nu \to \infty} \int_{\partial \mathcal{O}_g} G_P(g, f_\nu) \\
&= -\lim_{\nu \to \infty} 0 \\
&= 0.
\end{aligned}
$$

Thus, $g = 0$ in $\mathcal{O}$, i.e., g is the zero element of $sol(\widehat{\mathbb{R}^n \setminus U}, P')$.

We have proved that the correspondence $g \mapsto \mathcal{F}_g$ induces the isomorphism of vector spaces

$$
sol(\widehat{\mathbb{R}^n \setminus U}, P') \overset{\cong}{\to} sol(U, P)'.
$$

That this algebraic isomorphism is in fact a topological one follows from a version of the *Open Mapping Theorem*. This completes the proof. $\qquad\square$

6. Duality in the space of sequences

In this section we describe the dual space to L_2.

Let $\{g_\alpha\}_{\alpha \in \mathbb{Z}_+^n}$ be a sequence in $L^{q'}(\mathbb{R}^d)$ satisfying $\left\| \frac{g_\alpha}{\alpha!} \right\|_{q'} = O(R^{|\alpha|})$ when $|\alpha| \to \infty$, for some positive integer R. We define a linear functional $\mathcal{F}_{\{g_\alpha\}}$ on L_2 by

$$
(6.1) \qquad \langle \mathcal{F}_{\{g_\alpha\}}, \{f_\alpha\} \rangle = \sum_{\alpha \in \mathbb{Z}_+^n} \int_{\mathbb{R}^d} g_\alpha(y)\, f_\alpha(y)\, ds(y) \quad (\{f_\alpha\} \in L_2).
$$

LEMMA 6.1. *As defined by (6.1), $\mathcal{F}_{\{g_\alpha\}}$ is a continuous linear functional on the space L_2.*

PROOF. If $\{f_\alpha\} \in L_2$, then an easy computation shows that

$$
\begin{aligned}
\langle \mathcal{F}_{\{g_\alpha\}}, \{f_\alpha\} \rangle &\leq \sup_\alpha \left(R^{-|\alpha|} \left\| \frac{g_\alpha}{\alpha!} \right\|_{q'} \right) \left(\sum_{j=0}^{\infty} 2^{-j} \right) \|\{f_\alpha\}\|_{2nR} \\
&= C\, \|\{f_\alpha\}\|_{2nR};
\end{aligned}
$$

the constant C depends only on $\{g_\alpha\}$. This is precisely the assertion of the lemma. $\qquad\square$

The following result is certainly not new. It has been frequently used, beginning with the original paper of Baernstein [1].

LEMMA 6.2. *The dual space to L_2 can be identified with the set of all sequences $\{g_\alpha\}_{\alpha \in \mathbb{Z}_+^n}$ in $L^{q'}(\mathbb{R}^d)$ satisfying $\left\| \frac{g_\alpha}{\alpha!} \right\|_{q'} = O(R^{|\alpha|})$ when $|\alpha| \to \infty$, for some positive integer R. Moreover, the duality is given by (6.1).*

PROOF. Fix a continuous linear functional $\mathcal{F}$ on L_2. By a property of countable-normed spaces, there is a number R such that $\mathcal{F}$ is continuous on L_2 equipped merely with the norm $\|\cdot\|_R$.

For a multi-index $\alpha \in \mathbb{Z}_+^n$ and a function $f \in L^q(\mathbb{R}^d)$, denote by $i_\alpha(f)$ the element in L_2 which is f in the ith entry and 0 in all other entries.

Let $\{f_\alpha\}_{\alpha \in \mathbb{Z}_+^n} \in L_2$. Then $\{f_\alpha\} = \sum_{\alpha \in \mathbb{Z}_+^n} i_\alpha(f_\alpha)$ and the series converges with respect to the norm $\|\cdot\|_R$. Since $\mathcal{F}$ is continuous, we have $\langle \mathcal{F}, \{f_\alpha\}\rangle = \sum_\alpha \langle F, i_\alpha(f_\alpha)\rangle$.

Given any fixed multi-index α, we consider the linear functional on $L^q(\mathbb{R}^d)$ defined by $f \mapsto \langle \mathcal{F}, i_\alpha(f)\rangle$, for $f \in L^q(\mathbb{R}^d)$. This functional is obviously continuous, so by the *Riesz Theorem* there is a function $g_\alpha \in L^{q'}(\mathbb{R}^d)$ such that $\langle \mathcal{F}, i_\alpha(g)\rangle = \int_{\mathbb{R}^d} g_\alpha(y) f(y)\, ds(y)$ for all $f \in L^q(\mathbb{R}^d)$.

Hence

$$\langle \mathcal{F}, \{f_\alpha\}\rangle = \sum_\alpha \int_{\mathbb{R}^d} g_\alpha(y) f_\alpha(y)\, ds(y) \quad \text{for all } \{f_\alpha\} \in L_2.$$

The only point remaining concerns the behavior of $\|g_\alpha\|_{q'}$ in dependence on α. To evaluate these norms we again use the *Riesz Theorem* which states that $\|g_\alpha\|_{q'}$ is equal to the norm of the functional $f \mapsto \langle \mathcal{F}, i_\alpha(f)\rangle$ on $L^q(\mathbb{R}^d)$. So, letting $\|\mathcal{F}\|$ be the norm of $\mathcal{F}$ on the space $(L_2, \|\cdot\|_R)$, we get

$$\|g_\alpha\|_{q'} = \sup_{f \in L^q(\mathbb{R}^d)} \frac{\langle \mathcal{F}, i_\alpha(f)\rangle}{\|f\|_q}$$
$$\leq \|\mathcal{F}\| \, R^{|\alpha|} \, \alpha!,$$

as desired.

The rest of the proof is obvious. $\qquad\square$

7. Transpose

First of all, for any pair of locally convex spaces L_1 and L_2, we can identify $(L_1 \oplus L_2)'$ and $L_1' \oplus L_2'$ by the rule

$$\langle \mathcal{F}, f_1 \oplus f_2 \rangle = \langle \mathcal{F}_1, f_1\rangle + \langle \mathcal{F}_2, f_2\rangle,$$

where $\mathcal{F} \in (L_1 \oplus L_2)'$, $f_i \in L_i$, and $\mathcal{F}_i \in L_i'$ is the restriction of $\mathcal{F}$ to $L_i \oplus \{0\}$.

Our next goal is to describe the operator transposed to M under the foregoing identifications. To this end, define the linear mapping $T: K' \to L_1' \oplus L_2'$ by

(7.1) $$Tg = g \oplus \{-(-1)^{|\alpha|} D^\alpha g\},$$

where $D^\alpha g$ denotes the αth derivative of g restricted to $\mathbb{R}^d$.

LEMMA 7.1. *For each $g \in sol(\mathbb{R}^d \cup \{\infty\}, P')$ there is a positive integer R such that*

$$\left\| \frac{D^\alpha g}{\alpha!} \right\|_{q'} = O(R^{|\alpha|}) \quad \text{when } |\alpha| \to \infty.$$

PROOF. We take a function $f \in L^q(\mathbb{R}^d)$ with $\|f\|_q \leq 1$ and evaluate the integral $\int_{\mathbb{R}^d} D^\alpha g(y) f(y) ds(y)$.

Select an open set $\mathcal{O} \subset\subset \mathbb{R}^n \setminus \mathbb{R}^d$ with piecewise smooth boundary such that g still satisfies $P'g = 0$ in a neighborhood of $\mathbb{R}^n \setminus \mathcal{O}$. As g vanishes at ∞, we have, by Green's formula,

$$(7.2) \qquad (-1)^{|\alpha|} D^\alpha g(y) = -\int_{\partial\mathcal{O}} G_P(g(x), D^\alpha \Phi(x-y)) \quad (y \in \mathbb{R}^d).$$

Substituting (7.2) into the above integral and applying Fubini's theorem yield

$$\left| \int_{\mathbb{R}^d} D^\alpha g(y) f(y) ds(y) \right|$$

$$= \left| \int_{\partial\mathcal{O}} G_P \left(g(x), \int_{\mathbb{R}^d} D^\alpha \Phi(x-y) f(y) ds(y) \right) \right|$$

$$\leq const(P) \left(\sum_{|\beta| \leq p-1} \|D^\beta g\|_{L^1(\partial\mathcal{O})} \right)$$

$$\cdot \sup_{|\gamma| \leq p-1} \left\| \int_{\mathbb{R}^d} D^{\alpha+\gamma} \Phi(\cdot - y) f(y) ds(y) \right\|_{C(\partial\mathcal{O})}.$$

So, letting $\varepsilon > 0$ be the distance from $\partial\mathcal{O}$ to $\mathbb{R}^d$ and using estimate (3.2), we find

$$\left| \int_{\mathbb{R}^d} D^\alpha g(y) f(y) ds(y) \right|$$

$$\leq C \, \varepsilon^{d/q'+(p-n)} \left(\sum_{|\beta| \leq p-1} \|D^\beta g\|_{L^1(\partial\mathcal{O})} \right) R^{|\alpha|} \alpha!,$$

where R is any integer with $R > b\varepsilon^{-1} n$ and the constant C is independent of α and f.

Hence it follows, by the *Riesz Theorem*, that $\|\frac{D^\alpha g}{\alpha!}\|_{q'} = O(R^{|\alpha|})$, which completes the proof. $\qquad \square$

It follows from this lemma that, given any $g \in K'$, the sequence $\{-(-1)^{|\alpha|} D^\alpha g\}$ satisfies the growth condition for L_2'. Thus, the mapping T is well-defined.

LEMMA 7.2. *The mapping T defined by (7.1) can be identified with the mapping transposed to M.*

PROOF. Take $f_e \in L_1$, $\{f_\alpha\} \in L_2$, and $g \in K'$. Then, writing $\langle \cdot, \cdot \rangle_L$ to denote $\langle L', L \rangle$ duality, we have

$$(7.3) \quad \begin{aligned} &\langle Tg, f_e \oplus \{f_\alpha\} \rangle_{L_1 \oplus L_2} \\ &= \langle g, f_e \rangle_{L_1} + \langle \{-(-1)^{|\alpha|} D^\alpha g\}, \{f_\alpha\} \rangle_{L_2} \\ &= \int_{\partial\mathcal{O}} G_P(g, f_e) - \sum_\alpha (-1)^{|\alpha|} \int_{\partial\mathcal{O}} D^\alpha g(y) f_\alpha(y) ds(y), \end{aligned}$$

where $\mathcal{O}$ depends on g and is the same as in the proof of Lemma 7.1. (This open set is a suitable one for effecting the Grothendieck duality in both $\langle L_1', L_1 \rangle$ and $\langle L_2', L_2 \rangle$.)

On the other hand, setting

$$F_\alpha(x) = \int_{\mathbb{R}^d} D^\alpha \Phi(x-y) \, f_\alpha(y) \, ds(y),$$

we get

(7.4)
$$\begin{aligned}
\langle g, M(f_e \oplus \{f_\alpha\}) \rangle_K &= \langle g, f_e + \sum_\alpha F_\alpha \rangle_K \\
&= \langle g, f_e \rangle_K + \sum_\alpha \langle g, F_\alpha \rangle_K \\
&= \int_{\partial \mathcal{O}} G_P(g, f_e) + \sum_\alpha \int_{\partial \mathcal{O}} G_P(g, F_\alpha),
\end{aligned}$$

the second equality being a consequence of Lemma 3.1.

From Fubini's theorem and (7.2) it follows that

$$\begin{aligned}
\int_{\partial \mathcal{O}} G_P(g, F_\alpha) &= \int_{\mathbb{R}^d} \left(\int_{\partial \mathcal{O}} G_P(g(x), D^\alpha \Phi(x-y)) \right) f_\alpha(y) \, ds(y) \\
&= -(-1)^{|\alpha|} \int_{\mathbb{R}^d} D^\alpha g(y) \, f_\alpha(y) \, ds(y),
\end{aligned}$$

hence the right-hand sides of (7.3) and (7.4) are the same.

Thus T and M' can be identified, as asserted. $\square$

8. Conclusion of proof

We are now in a position to finish the proof. It suffices, by Proposition 4.1, to show that $T^{-1}(B)$ is equicontinuous in K' whenever B is equicontinuous in $L_1' \oplus L_2'$. We can certainly assume that $B = B_1 \oplus B_2$, where B_i is an equicontinuous subset of L_i'.

The task is now to find a proper description of equicontinuous sets in L_1' and K'.

LEMMA 8.1. *Under the isomorphism of Lemma 5.2, a set $b \subset sol(\widehat{\mathbb{R}^n} \setminus U, P')$ is equicontinuous if and only if it is contained in one of the form*

$$\{g \in sol(\widehat{\mathbb{R}^n} \setminus K, P') : \ |g(x)| \le C \quad for \ all \ \ x \in \mathbb{R}^n \setminus K\},$$

with K a compact subset of U and C a positive constant.

PROOF. Since $sol(U, P)$ is a Fréchet space, "equicontinuous" can be interchanged with "weakly bounded." Combining Lemmas 2.1 and Lemma 2.2 we can assert that a set $b \subset sol(\widehat{\mathbb{R}^n} \setminus U, P')$ is weakly bounded iff it is contained and bounded in one of the spaces $sol(\mathcal{N}, P')$, where $\mathcal{N}$ is a neighborhood of $\widehat{\mathbb{R}^n} \setminus U$ in the compactified space. From this the desired conclusion follows immediately. $\square$

Thus, there is no loss of generality in assuming that

$$B_1 = \{g \in sol(\widehat{\mathbb{R}^n} \setminus B(0, r), P') : \ |g(x)| \le 1 \quad for \ all \ \ x \in \mathbb{R}^n \setminus B(0, r)\},$$

where $B(0, r)$ is the ball of center 0 and radius $r > 0$ in $\mathbb{R}^n$.

Moreover, a set b in L_2' is equicontinuous if and only if it is contained in the polar of a neighborhood of zero in L_2. So, without restriction of generality we can assume that B_2 is the polar of $\{\{f_\alpha\} \in B_2 : \|\{f_\alpha\}\|_R < 1\}$, for some fixed $R = 1, 2, \ldots$. We choose

$$B_2 = \left\{ \{g_\alpha\} : \left\| \frac{g_\alpha}{\alpha!} \right\| \le R^{|\alpha|} \quad \text{for all} \quad \alpha \in \mathbb{Z}_+^n \right\}.$$

If $g \in K'$, then Tg is in $B_1 \oplus B_2$ if and only if g lies in B_1 and

(8.1)
$$\left\| \frac{D^\alpha g}{\alpha!} \right\| \le R^{|\alpha|} \quad \text{for all} \quad \alpha \in \mathbb{Z}_+^n.$$

Let N be a positive integer satisfying $N q' > d$ (to have satisfied this condition it suffices to take $N = n - p$). By the *Sobolev Embedding Theorem*,

$$\sup_{y \in B(0,r) \cap \mathbb{R}^d} |g(y)| \le c \sup_{|\beta| \le N} \|D^\beta g\|_{q'},$$

the constant c being independent of g. Moreover, we get

$$\sup_{y \in B(0,r) \cap \mathbb{R}^d} |D^\alpha g(y)| \le c \sup_{|\beta| \le N} \|D^{\beta+\alpha} g\|_{q'}$$

$$\le c \sup_{|\gamma| \le N+|\alpha|} \|D^\gamma g\|_{q'}$$

for all $\alpha \in \mathbb{Z}_+^n$. Combining these with (8.1) yields

(8.2)
$$\sup_{y \in B(0,r) \cap \mathbb{R}^d} |D^\alpha g(y)| \le c \sup_{|\gamma| \le N+|\alpha|} R^{|\gamma|} \gamma!$$

$$\le c N! (2R)^{N+|\alpha|} |\alpha|!$$

for all $\alpha \in \mathbb{Z}_+^n$. (The last estimate is a consequence of the obvious inequality $(N + |\alpha|)! \le 2^{N+|\alpha|} N! |\alpha|!$.)

Given any point $y \in B(0,r) \cap \mathbb{R}^d$, we now consider the Taylor series of g at y. It follows from (8.2) that

$$\left| \sum_\alpha \frac{D^\alpha g(y)}{\alpha!} (x-y)^\alpha \right| \le c (2R)^N N! \sum_{j=0}^\infty (2R)^j \sum_{|\alpha|=j} \frac{j!}{\alpha!} |(x-y)^\alpha|$$

$$= c (2R)^N N! \sum_{j=0}^\infty (2R)^j (|x_1 - y_1| + \cdots + |x_n - y_n|)^j,$$

hence the series converges whenever $|x_1 - y_1| + \cdots + |x_n - y_n| < (2R)^{-1}$. Since its sum coincides with g near $B(0,r) \cap \mathbb{R}^d$, it is independent of y and provides a single valued analytic continuation of g to the polyhedron

$$\Pi = \left\{ x \in B(0,r) : |x_{d+1}| + \cdots + |x_n| < \frac{1}{2} (2R)^{-1} \right\}.$$

It is a simple matter to see that this continuation also satisfies $P'g = 0$ in Π. (Indeed, $P'g$ is analytic in Π and vanishes near $B(0,r) \cap \mathbb{R}^d$.)

Moreover, g satisfies $P'g = 0$ in a neighborhood of the closure of Π, and $|g| \le C'$ in Π, where $C' = 2c (2R)^N N!$ depends only on R, q, and r. Thus we have proved that $T^{-1}(B_1 \oplus B_2)$ is contained in the set

$$\{g \in sol(\widehat{\mathbb{R}^n} \setminus K, P') : |g(x)| \le C \quad \text{for all} \quad x \in \mathbb{R}^n \setminus K\},$$

with $K = \overline{B(0,r)} \setminus \Pi$ and $C = \max(1, C')$. By Lemma 8.1, $T^{-1}(B_1 \oplus B_2)$ is equicontinuous in K', as desired.

This proves Theorem 2.3.   □

REMARK 8.2. The proof above gives more, namely Theorem 2.3 remains valid when we replace the "singular set" $\mathbb{R}^d$ by any smooth surface S of dimension d in $\mathbb{R}^n$ which is flat at infinity (an asymptotic flatness is also sufficient).

9. An example for harmonic functions

For the Laplace operator in $\mathbb{R}^n$ $(n > 2)$ we obtain the following result.

COROLLARY 9.1. *Let q be any real number with $1 \leq q < \frac{d}{d+2-n}$ if $d > n-2$, and with $1 \leq q < \infty$ if $d \leq n - 2$. Then every harmonic function f in $\mathbb{R}^n \setminus \mathbb{R}^d$ has a representation*

$$f(x) = f_e(x) + \sum_{j=0}^{\infty} \int_{\mathbb{R}^d} \frac{h_j(y, x-y)}{|x-y|^{n+2(j-1)}} \, dm(y) \quad (x \notin \mathbb{R}^d),$$

where f_e is a harmonic function on $\mathbb{R}^n$, and $h_j(y, z)$ are homogeneous harmonic polynomials of degree j in z with coefficients in $L^q(\mathbb{R}^d)$ with respect to y, such that $\lim_{j \to \infty} \left(\frac{1}{j!} \int_{\mathbb{R}^d} |h_j(y, D_z) h_j(y, z)|^{q/2} \, ds(y) \right)^{\frac{1}{qj}} = 0$.

PROOF. It suffices to transform formula (2.2) by means of an identity of Hecke (cf. Stein [5]).   □

10. Hyperfunctions

Sato hyperfunctions generalize Schwartz distributions. Loosely speaking, a hyperfunction on $\mathbb{R}^{n-1}$ is supposed to be like a "boundary jump"

$$f(x_1, \ldots, x_{n-1}, 0+) - f(x_1, \ldots, x_{n-1}, 0-)$$

of some solution $f \in sol(\mathbb{R}^n \setminus \mathbb{R}^{n-1})$.

Precisely, the vector space of hyperfunctions on $\mathbb{R}^{n-1}$ is defined to be the quotient of the space of all solutions in $\mathbb{R}^n \setminus \mathbb{R}^{n-1}$ modulo the space of all solutions in $\mathbb{R}^n$. Usually, the differential operator P is chosen to be the Laplace operator in $\mathbb{R}^n$.

From Theorem 2.3 we obtain the following result concerning the representation of hyperfunctions by integrals over $\mathbb{R}^{n-1}$.

COROLLARY 10.1. *Every hyperfunction on $\mathbb{R}^{n-1}$ has a representing solution of the form*

$$\sum_{\alpha \in \mathbb{Z}_+^n} \int_{\mathbb{R}^{n-1}} D^\alpha \Phi(x-y) \, f_\alpha(y) \, ds(y) \quad (x \in \mathbb{R}^n \setminus \mathbb{R}^{n-1}),$$

with $\{f_\alpha\}_{\alpha \in \mathbb{Z}_+^n} \subset L^q(\mathbb{R}^{n-1})$ $(q < \frac{n-1}{p-1})$ satisfying $\|\alpha! \, f_\alpha\|_{L^q(\mathbb{R}^{n-1})}^{1/|\alpha|} \to 0$ when $|\alpha| \to 0$.

PROOF. This is obvious because of (2.2).   □

References

[1] A. Baernstein, *A representation theorem for functions holomorphic off the real axis*, Trans. Amer. Math. Soc. **165** (1972), 159–165.

[2] A. Grothendieck, *Sur les espaces de solutions d'une classe generale d'equations aux derivees partielles*, J. Anal. Math. **2** (1952–1953), 243–280.

[3] L. Hörmander, *The Analysis of Linear Partial Differential Operators. Vol. 1: Distribution Theory and Fourier Analysis*, Springer-Verlag, Berlin et al., 1983.

[4] G. Köthe, *Topologische lineare Räume. I*, Springer-Verlag, Berlin et al., 1960.

[5] E. M. Stein, *Singular Integrals and Differentiability Properties of Functions*, Princeton Univ. Press, Princeton, 1970.

[6] N. N. Tarkhanov, *Complexes of Differential Operators*, Kluwer Academic Publishers, Dordrecht, NL, 1995.

MAX-PLANCK-ARBEITSGRUPPE "PARTIELLE DIFFERENTIALGLEICHUNGEN UND KOMPLEXE ANALYSIS," FB MATHEMATIK, UNIVERSITÄT POTSDAM, POSTFACH 60 15 53, 14415 POTSDAM, GERMANY

E-mail address: `tarkhan@mpg-ana.uni-potsdam.de`

Contemporary Mathematics
Volume **212**, 1998

Bounded Monogenic Functions on Unbounded Domains

Edwin Franks and John Ryan

ABSTRACT. A minor modification of the Clifford Cauchy kernel is introduced. This kernel works extremely well at transposing results from Clifford analysis over bounded domains to unbounded domains lying in a half space. The functions considered here are either bounded, or have slow decay at infinity. Kelvin inversion is used to obtain another reproducing kernel of Cauchy type, which describes the behaviour of monogenic functions defined over other domains lying in a half space. These domains have the origin on their boundary, and the functions blow up at a specific rate near the origin.

1. Introduction

The idea of adding on extra terms to the Cauchy kernel in the complex plane, or to the Newtonian potential in R^n, with $n \geq 3$, in order to develop analysis over unbounded domains is not new. See for instance [9], and references there in. This idea is also used in the framework of Clifford analysis in [22] and [21] to study boundary value problems over half space.

We show here that the idea can be effectively used in the context of Clifford analysis to study bounded monogenic functions on an unbounded domain U satisfying $U \cap -U = \phi$, the empty set. Such a domain U lies in a half space of R^n. First we set up the Cauchy integral formula for bounded monogenic functions over such domains. In fact this formula works for monogenic functions defined on these domains, and satisfying a slow divergence at infinity. We illustrate this point here. We also use a Cauchy transform type argument to establish the existence of such functions.

We investigate the transform of this new Cauchy kernel over unbounded domains lying in a half space. We illustrate that this transform is well defined for Lipschitz continuous functions with slow decay at infinity.

1991 *Mathematics Subject Classification*. Primary 30G35; Secondary 35C15, 53A50.

The research covered in this paper was initiated while the second author was the recipient of a research award from the Arkansas Science and Technology Authority. It was completed while he was the recipient of a von Humboldt Research Fellowship, visiting the Bergakademie in Freiberg, Sachsen, Germany. The main ideas were obtained while the second author was a visitor at Macquarie University, Sydney, Australia in August 1994. He is grateful to Alan McIntosh for the support that he and his analysis research group has given during this visit.

We conclude by applying Kelvin inversion to our analysis. This transforms the modified Cauchy kernel to a kernel acting over domains defined in a half space, and containing the origin on the boundary. This kernel acts on monogenic functions defined on the domain and satisfying the inequality $\|f(x)\| \leq C\|x\|^{-n+1}$. Though the types of domains considered here do not include the sector domains used in [10], [11] and elsewhere, it is worth noting that these functions do satisfy the basic inequality employed for monogenic functions on sector domains in those papers. This link will be investigated elsewhere. Here we simply show that results obtained for the first modification of the Cauchy kernel that we introduce here also carry over to the new class of functions introduced here.

Clifford analysis over unbounded domains has been developed and applied by a number of authors using different techniques, see for instance [2, 16, 11, 13, 21, 22]. The methods employed here differ from those used in [2, 16, 11, 13].

2. Preliminaries

We shall use the real Clifford algebra, A_n, generated from R^n, such that for each $x \in R^n$ we have that $x^2 = -\|x\|^2$. Consequently, we are assuming that R^n is embedded in the algebra A_n. Though in the previous formula we have used the Clifford algebra generated from a negative definite inner product over R^n we could have just as easily taken the Clifford algebra generated from the positive definite inner product on R^n.

For the algebra A_n we have the anticommutation relationship

$$e_i e_j + e_j e_i = -2\delta_{i,j},$$

where $\delta_{i,j}$ is the Kroneker delta function, and $e_1 \ldots, e_n$ is an orthonormal basis for R^n. As no other relations are assumed then it may be deduced that the algebra A_n has dimension 2^n, and the algebra has as a basis $1, e_1 \ldots, e_n, \ldots, e_{j_1} \ldots e_{j_r}, \ldots e_1 \ldots e_n$, where $1 \leq r \leq n$ and $j_1 < \ldots j_r$. For a general element $A = a_0 + \ldots + a_{1\ldots n} e_1 \ldots e_n$, with $a_0, \ldots, a_{1\ldots n} \in R$ we denote the norm, $(a_0^2 + \ldots a_{1\ldots n}^2)^{1/2}$ by $\|A\|$. For $A, B \in A_n$ it follows that as A_n is a finite dimensional algebra that there exists dimensional constants $C(n)$ such that $\|AB\| \leq \|A\|\|B\|$. Sharp estimates for these dimensional constants are worked out in [6] using matrix representations for the Clifford algebras.

We shall also need the antiautomorphism

$$\sim: A_n \to A_n : e_{j_1} \ldots e_{j_r} \to e_{j_r} \ldots e_{j_1}.$$

For a general element $A \in A_n$ we shall write $\overset{A}{\sim}$ for $\sim (A)$.

When $n \geq 3$ the algebra is no longer a division algebra. However, from an analysis viewpoint one extremely crucial property of the algebra A_n is that each non-zero vector $x \in R^n$ has a multiplicative inverse given by $-x/\|x\|^2$. Up to a sign this inverse corresponds to the Kelvin inverse of a vector in euclidean space.

We now introduce the Dirac operator $D = \sum_{j=1}^{n} e_j (\partial/\partial x_j)$. This operator is a direct analogue over euclidean space of the Dirac operator used over Minkowski space, see for instance [8]. In particular we have that $D^2 = -\triangle_n$, where $\triangle_n$ is the Laplacian over R^n.

DEFINITION 1. Suppose that U is a domain in R^n, then a differentiable function $f : U \to A_n$ is said to be left monogenic if it satisfies the equation $Df = 0$.

A similar definition can be given for right monogenic functions. The equation $Df = 0$ is a generalization of the Cauchy-Riemann equations. Basic properties of monogenic functions are described in the book of Brackx, Delanghe and Sommen, [4]. In particular, there is the following Cauchy integral formula:

THEOREM 1. *Suppose that $f : U \to A_n$ is a left monogenic function, and V is a closed bounded region in U with a sufficently smooth boundary, ∂V. Then for each y lying in the interior of V we have that $f(y) = \int_{\partial V} G(x-y)n(x)f(x)d\sigma(x)$, where $G(x) = (1/\omega_n)(x/\|x\|^n)$ and ω_n is the surface area of the unit sphere in R^n, $n(x)$ is the outward pointing normal vector to ∂V at x, wherever this vector is defined, and σ is the Lebesgue measure on ∂V.*

In the previous theorem it is sufficent to assume that ∂V is piecewise C^1, though one can easily get the same result if one assumes that ∂V is Lipschitz continuous.This generalized Cauchy integral formula was apparantly first introduced for the special case $n = 3$ by Dixon, [5] in 1904. The function $G(x)$ appearing in Theorem 1 is the gradient of the Newtonian potential.It is both left and right monogenic, and it is the generalization to euclidean space of the Cauchy kernel from one variable complex analysis.

3. Cauchy integral formula on unbounded domains

In this section we shall assume that all domains lie in a half space of R^n, so that for each such domain we have $U \cap -U = \phi$, the empty set. Furthermore, we will assume that each domain is unbounded and has a Lipschitz continuous boundary.

THEOREM 2. *Suppose that U is a domain satisfying the conditions outlined in the previous paragraph, and f is a left monogenic function defined in a neighbourhood of U. Moreover, f is bounded on the closure of U. Then*

$$f(y) = \int_{\partial V} K(x,y)n(x)f(x)d\sigma(x),$$

for each $y \in U$, where $K(x,y) = G(x-y) - G(x+y)$.

PROOF. Consider a point $y \in U$, and the sphere, $S(0,r)$, centred at 0 and of radius r. For r sufficently large y will lie in the domain $U \cap D(0,r)$, where $D(0,r)$ is the open disc centred at 0 and of radius r. We denote this domain by $U(r)$. As $G(x+y)$ has no singularity in U the Cauchy integral formula now gives

$$f(y) = \int_{\partial U(r)} K(x,y)n(x)f(x)d\sigma(x).$$

This expression can be rewriten as

$$\int_{\partial U \cap D(0,r)} K(x,y)n(x)f(x)d\sigma(x) = f(x) - \int_{S(0,r) \cap U} K(x,y)n(x)f(x)d\sigma(x).$$

Consequently,

$$\lim_{r \to \infty} \int_{\partial U \cap D(0,r)} K(x,y)n(x)f(x)d\sigma(x)$$

$$= f(x) - \lim_{r \to \infty} \int_{S(0,r) \cap U} K(x,y)n(x)f(x)d\sigma(x).$$

Now

$$\lim_{r \to \infty} \int_{\partial U \cap D(0,r)} K(x,y)n(x)f(x)d\sigma(x) = \int_{\partial U} K(x,y)n(x)f(x)d\sigma(x).$$

Also, following arguments given in [7] it may be observed that $\|K(x,y)\| \leq C(n)(\|y\|/\|x\|^n)$, for $\|x\| \geq 2\|y\|$, where $C(n)$ is a dimensional constant. Moreover, f is bounded on the closure of U. So there is a constant C such that $\|f(x)\| \leq C$ for each x in the closure of U.

Consequently,

$$\left\| \int_{U \cap S(0,r)} K(x,y)n(x)f(x)d\sigma(x) \right\| \leq D(n)r^{-1},$$

where $D(n)$ is a dimensional constant. The result now follows on letting r tend to infinity.

For the type of domain, U, considered in Theorem 2 we can consider a homotopy $H : \partial U \times [0,1] \to \partial U \cup U$ such that (i) $H(x,0) = x$ for each $x \in \partial U$, (ii) for each $t \in [0,1]$ the set $H(\partial U, t)$ is a Lipschitz continuous surface bounding an unbounded subdomain of U, (iii) for $t > 0$ we have that $H(\partial U, t) \cap U = \phi$. We shall call such a homotopy a homotopy of Hardy type. By applying Theorem 2 to each surface $H(\partial U, t)$, with $t > 0$ we obtain:

PROPOSITION 1. *Suppose that the domain U is as in Theorem 2, and we have a homotopy of Hardy type acting over the boundary of U. Suppose also that $f : \partial U \cup U \to A_n$ is such that (i) f is monogenic on U, (ii) $f \mid \partial U$ is an essentially bounded function, and (iii) the function*

$$F : [0,1] \to L^\infty(\partial U, A_n) : F(t) = f \mid H(\partial U, t),$$

where $L^\infty(\partial U, A_n)$ is the A_n module of essentially bounded functions on ∂U, is continuous. Then for each $y \in U$ we have

$$f(y) = \int_{\partial U} K(x,y)n(x)f(x)d\sigma(x).$$

By minor adaptations of continuity arguments given in [15] it may be observed that the integral formula given in Proposition 1 is independant of the choice of homotopy of Hardy type acting over $\partial U \cap U$. Also, the set of functions satisfying the conditions layed out in Proposition 1 is a right module over A_n. This module is the Hardy space of essentially bounded functions on ∂U which have extensions to left monogenic functions on U. We can denote this space by $H^\infty(U)$. Proposition 1 gives a Cauchy integral formula for each member of $H^\infty(U)$. Examples of domains to consider here are the cases where U is a half space, or $U = \{x_1 e_1 + \ldots + x_n e_n : 0 < x_1 < 1, \text{and } x_2, \ldots x_n \in R\}$, the infinite strip, or U is a Lipschitz domain of the type described in [10], and elsewhere, and lies in a half space.

The proof of Theorem 2 easily modifies to yield the following extension of Proposition 1.

THEOREM 3. *Suppose that the domain U is as in Theorem 2, that H is a homotopy of Hardy type acting over the boundary of U. Suppose also that $f : \partial U \cup U \to A_n$ satisfies (i) f is left monogenic on U, (ii) there is a real, positive constant C such that $\|f(x)\| \leq C\|x\|^s$ for some $s \in [0,1)$, and $x \in \partial U \cup U$, (iii) $f \mid \partial U$ is* *measurable, and* (iv) *the function* $F : [0,1] \to L^\infty(U, \|x\|^{-s}) : F(t) = f \mid H(\partial U, t)$ *is continuous, where* $L^\infty(U, \|x\|^{-s})$ *is the space of* A_n *valued functions on* ∂U *which are essentially bounded on* ∂U *with respect to the weight function* $\|x\|^{-s}$. *Then,*

$$f(y) = \int_{\partial U} K(x,y)n(x)f(x)d\sigma(x)$$

for each $y \in U$.

Theorem 3 gives a Cauchy integral formula for the weighted Hardy space, $H^\infty(U, \|x\|^{-s})$, of monogenic functions satisfying conditions (i)-(iv) from Theorem 3, via an arbitrary homotopy of Hardy type over ∂U. Following arguments given in [15], and elsewhere, it is a simple matter to see that the spaces $H^\infty(U, \|x\|^{-s})$ is a complete Banach space for each choice of $s \in [0,1)$.

The existence of left mongenic functions defined on such domains and satisfying the inequality given in Theorem 3 is partially provided by the following proposition.

PROPOSITION 2. *Suppose that* U *is an unbounded domain with Lipschitz continuous boundary, and whose closure lies in a cone,* T, *in open half space,* Q. *Moreover, the tip of this cone lies at the origin. Suppose also* $g : \partial Q \to A_n$ *is a measurable function which satisfies* $\|g(x)\| \leq C\|x\|^s$ *for some* $s \in [0,1)$ *and all* $x \in \partial Q$. *Then* $\int_{\partial Q} K(x,y)n(x)g(x)dx^{n-1}$ *defines a left monogenic function* f *on* U, *which satisfies* $\|f(x)\| \leq C_1\|x\|^s$ *for all* $x \in U$, *and some positive number* C_1.

PROOF. Consider the integral $\int_{\partial Q \setminus D(0,2\|y\|)} K(x,y)n(x)g(x)dx^{n-1}$ for some $y \in U$. From arguments layed out in the proof of Theorem 2 this integral is bounded by $\|y\|^{1+s}C\int_{2\|y\|}^\infty r^{-2}dr$, for some constant C. This integral is bounded by $C\|y\|^s$ for some $C \in R^+$.

Now consider the integral

$$\int_{\partial Q \cap D(0,2\|y\|)} K(x,y)n(x)g(x)dx^{n-1} = \int_{\partial Q \cap D(0,2\|y\|)} G(x-y)n(x)g(x)dx^{n-1}$$
$$+ \int_{\partial Q \cap D(0,2\|y\|)} G(x+y)n(x)g(x)dx^{n-1}.$$

Let us first consider the term

$$\int_{\partial Q} G(x-y)n(x)g(x)dx^{n-1}.$$

This integral is bounded by

$$C\|y\|^s \int_{\partial Q \cap D(0,2\|y\|)} \|G(x-y)\|dx^{n-1}.$$

By translating the origin to the point $u \in \partial Q \cap D(0,2\|y\|)$, which is the projection of y to ∂Q, it may be observed that this term is dominated by

$$C\|y\|^s \int_0^{8\|y\|} r^{n-2}((\|y\|\tan\mu)^2 + r^2)^{-n/2}dr,$$

where μ is the angle the cone T subtends with the space ∂Q. On changing variable to $R = r/(\|y\|\tan\mu)$ we obtain the desired estimate for this term. A similar estimate may be obtained by the same techniques for the remaining term.

In [17] we introduce special types of manifolds in C^n. These manifolds are natural generalizations of domains in R^n. Each such manifold is a real n-dimensional manifold. If M is such a manifold and is unbounded, with $M \cap -M = \phi$, then one can easily adapt the arguments given so far to produce a Cauchy integral formula for solutions f of the Dirac operator in C^n which satisfy the inequality $\|f(z)\| \leq C\|z\|^s$ for some $s \in [0, 1)$ and each $z \in M$.

4. A modified Cauchy transform

In this section we look at some basic properties of the transform of the modified Cauchy kernel $K(x, y)$ over the types of domains described in the previous section. We begin with:

THEOREM 4. *Suppose that U is an unbounded domain in a half space, and that U does not contain 0 on its boundary. Suppose also that $h : U \to A_n$ is a Lipschitz continuous function and for some $s \in (0, \infty)$ we have $\|h(x)\| \leq C\|x\|^{-s}$ for each $x \in U$. Then the integral*

$$\int_U K(x, y) h(x) dx^n$$

defines a differentiable function on U.

PROOF. Let us consider a point $y \in U$. For r large enough we have that $y \in D(0, r)$. So

$$\int_U K(x, y) h(x) dx^n = \int_{U \cap D(0,r)} K(x, y) h(x) dx^n + \int_{U \backslash D(0,r)} K(x, y) h(x) dx^n.$$

By a minor adaptation of arguments given in [7] it may be observed that

$$\int_{U \cap D(0,r)} K(x, y) h(x) dx^n$$

defines a differentiable function on $U \cap D(0, r)$.

Moreover, if r is chosen to be greater than $2\|y\|$ then

$$\left\| \int_{U \backslash D(0,r)} K(x, y) h(x) dx^n \right\| \leq C \int_r^\infty R^{-1-s} dR,$$

for some positive constant C. It follows by elementary uniform convergence arguments that that the integral $\int_{U \backslash D(0,r)} K(x, u) h(x) dx^n$ defines a left monogenic function in a neighbourhood of y.

COROLLARY 1. *The function $K(h)(y) = \int_U K(x, y) h(x) dx^n$ is a solution on U of the equation $Df = h$ for each h defined on U and satisfying $\|h(x)\| \leq C\|x\|^{-s}$ for some $s \in (0, \infty)$ and all $x \in U$.*

For the type of domain considered here Theorem 4 and its corollary gives an improvement on similar results obtained in [18] using Moebius transformations.

5. Kelvin inversion

From the identities $u^{-1} - v^{-1} = u^{-1}(v-u)v^{-1}$ and $u^{-1} + v^{-1} = u^{-1}(v+u)v^{-1}$, where $u, v \in R^n \backslash \{0\}$, it may be deduced that

$$K(u^{-1}, v^{-1}) = -G(u)^{-1}(G(u-v) + G(u+v))G(v)^{-1}.$$

A similar identity is deuced by the same technique in [14]. We shall denote the kernel $-(G(u-v) + G(u+v))$ by $L(u,v)$. In [19, 20] it is shown that if the function f is left monogenic in the variable u^{-1} then the function $G(u)f(u^{-1})$ is left monogenic in the variable u. Also, the Kelvin inverse of a half space is also a half space. Moreover, [15],

$$\int_{\partial U} f(u^{-1}) n(u^{-1}) g(u^{-1}) d\sigma(u^{-1}) = \int_{\partial U^{-1}} f(u^{-1}) G(u) n(u) G(u) g(u^{-1}) d\sigma(u)$$

for any pair of measurable functions f, g defined on ∂U.

Combining these facts with Theorem 2 we obtain:

THEOREM 5. *Suppose that U is a domain in a half space and has 0 sitting on its boundary. Suppose that the boundary of U is locally Lipschitz, and f is a left monogenic function defined on a neighbourhood of U in $R^n \backslash \{0\}$. Moreover, $\|f(u)\| \leq C \|u\|^{-n+1}$. Then for each $v \in U$ we have*

$$f(v) = \int_{\partial U} L(u,v) n(u) f(u) d\sigma(u).$$

It is now also possible to use Theorem 3 to deduce:

THEOREM 6. *Suppose that U is as in Theorem 5 and $f : U \cup \partial U \to A_n$ is such that f is measurable on ∂U, f is monogenic on U, $\|f(u)\| \leq C\|u\|^{-n+1+s}$ for some $C \in R^+$ and $s \in [0,1)$, and there is a homotopy of Hardy type $H : \partial U \times [0,1] \to U$ such that $F : [0,1] \to L^\infty(\partial U, \|u\|^{-n+1+s}) : F(t) = f \mid H(\partial U, t)$ is continuous. Then*

$$f(v) = \int_{\partial U} L(u,v) n(u) f(u) d\sigma(u)$$

for each $u \in U$.

As observed in the introduction Theorems 5 and 6 give Cauchy integral formulae for monogenic functions of the type considered in [10, 11, 12] and elsewhere, though the types of domains considered here do not include the sector domains used in those papers.

Also,

$$\int_U K(x,y) h(x) dx^n = G(v)^{-1} \int_{U^{-1}} L(u,v) G(u)^{-1} h(u^{-1}) \|u\|^{-2n} du^n,$$

where $x = u^{-1}$ and $y = v^{-1}$. It follows that

$$\int_U K(x,y) h(x) dx^n = G(v)^{-1} \int_{U^{-1}} L(u,v) G_{-1} h(u^{-1}) du^n,$$

where $G_{-1}(u) = u\|u\|^{-n-1}$.

Applying these calculations to Theorem 4 we obtain:

THEOREM 7. *Suppose that U is a bounded domain satisfying the conditions layed out in Theorem 5 and $h : U \to A_n$ is Lipschitz continuous and satisfies $\|h(u)\| \leq C\|u\|^{-n+s}$ for some constant C and $s \in [0, \infty)$. Then the integral*

$$\int_U L(u, v)h(u)du^n$$

gives a well defined, differentiable function.

In [3, 14] it is shown that the functions G and G_{-1} act as intertwining operators for the Dirac operator D under Kelvin inversion. It follows that under Kelvin inversion the corollary to Theorem 4 transforms to give:

LEMMA 1. *Suppose that U and h are as in Theorem 7, then the expression $\int_U L(u, v)h(u)du^n$ gives a solution to the equation $Dg = h$.*

The results obtained so far in this section are special cases of results that one can obtain under general conformal transformations. In [1] it is shown that any Moebius transformation over the one point compactification of R^n can be expressed as $y = \phi(x) = (ax + b)(cx + d)^{-1}$, where the coefficents a, b, c and d are elements of A_n satisfying certain constraints which are spelt out in [1] and elsewhere. Also,in [16, 3, 14] and elsewhere it is shown that if $f(\phi(x))$ is left monogenic with respect to the variable $y = \phi(x)$ then the function $J(\phi, x)f(\phi(x))$ is left monogenic with respect to the variable x, where $J(\phi, x) = \overset{(cx+d)}{\sim} \|cx+d\|^{-n}$. It follows that the kernel $K(\phi(u), \phi(v))$ conformally transforms itself to $M(u, v) = J(cv + d)K(\phi(u), \phi(u)) \overset{J}{\sim}$ $(cu + d)$. This kernel is left monogenic in the variable v and right monogenic with respect to the variable u. Consequently, the results developed in this paper can also be obtained for suitable function spaces under inverse Moebius transformations of domains lying in a half space.

References

[1] L. V. Ahlfors, *Moebius transformations in R^n expressed through 2×2 matrices of Clifford numbers*, Complex Variables **5** (1986), 215–224.

[2] S. Bernstein, *Elliptic boundary value problems in unbounded domains*, Clifford Algebras and their Applications in Mathematical Physics edited by F. Brackx, H. Serras and R. Delanghe, Kluwer, Dordrecht, 1993, pp. 45–53.

[3] B. Bojarski, *Conformally covariant differential operators*, Proc. of the XXth Iranian Math. Congress, Tehran, 1989.

[4] F. Brackx, R.Delanghe and F.Sommen, *Clifford Analysis,* Research Notes in Mathematics, No 76, Pitman, London, 1982.

[5] A. C. Dixon, *On the Newtonian potential*, Quaterly J. of Math. **35** (1904), 283–296.

[6] G. N. Hile and P. Lounesto, *Inequalities for spinor norms in Clifford algebras*, Spinors in Physics and Geometry, World Scientific Publishing, Singapore, 1988, pp. 285–297.

[7] V. Iftimie, *Fonctions hypercomplexes*, Bull. Math. de la Soc. Sci. Math. de la R. S. Roumanie **4** (1965), 279–332.

[8] H. Jakobsen and M. Vergne, *Wave and Dirac operators and representations of the conformal group*, J. Funct. Anal. **24** (1977), 52–106.

[9] L. Karp, *Generalized Newton potential and its Applications*, Journal of Mathematical Analysis and its Applications **174** (1993), 480–497.

[10] C. Li, A. McIntosh and S. Semmes, *Convolution singular integrals on Lipschitz surfaces*, Journal of the American Mathematical Society **5** (1992), 455–481.

[11] C. Li, A. McIntosh, and T. Qian, *Clifford algebras, Fourier transforms, and singular convolution operators on Lipschitz surfaces*, Revista Mathematica Iberoamericana, to appear.

[12] A. McIntosh, *Clifford algebras, Fourier theory, singular integral opertors, and partial differential equations on Lipschitz domains*, Clifford Algebras in Analysis and Related Topics, edited by J. Ryan, CRC Press, Boca Raton, 1995, pp. 33–88.

[13] M. Mitrea, *Singular Integrals, Hardy Spaces, and Clifford Wavelets*, Lecture Notes in Mathematics, No 1575, Springer Verlag, 1994.

[14] J. Peetre and T. Qian, *Moebius invariance of iterated Dirac operators*, Journal of the Australian Mathematical Society, Series A, to appear.

[15] T. Qian and J. Ryan, *Conformal transformations and Hardy spaces arising in Clifford analysis,* to appear.

[16] J. Ryan, *Iterated Dirac operators and conformal transformations in R^n*, Proc of the XV International Conference on Differential Geometric Methods in Theoretical Physics, World Scientific, Singapore, 1987, pp. 390–399.

[17] J. Ryan, *Cells of harmonicity and generalized Cauchy integral formulae*, Proc. of the London Math. Soc. **60** (1990), 295–318.

[18] J. Ryan, *Conformally covariant operators in Clifford analysis, to appear.*

[19] F. Sommen, *Spherical monogenic functions and analytic functionals on the unit sphere,* Tokyo J. Math. **4** (1981), 427–456.

[20] A. Sudbery, *Quaternionic analysis*, Mathematical Proceedings of the Cambridge Philosophical Society **86** (1979), 199–225.

[21] Z. Xu and C. Zhou, *On boundary value problems of Riemann-Hilbert type for monogenic functions in a half space of R^n (m $\geq$ 2)*, Complex Variables.

[22] C. Zhou, *On boundary value problems of Neumann type for the Dirac operator in a half space of R^n (m $\geq$ 2)*, Complex Variables.

SCHOOL OF MATHEMATICS, PHYSICS, COMPUTING AND ENGINEERING, MACQUARIE UNIVERSITY, NORTH RYDE, NSW 2109, AUSTRALIA
E-mail address: `edwin@macadam.mpce.mq.edu.au`

DEPARTMENT OF MATHEMATICS, UNIVERSITY OF ARKANSAS, FAYETTEVILLE, AR 72701, USA
E-mail address: `jryan@comp.uark.edu`

Contemporary Mathematics
Volume **212**, 1998

L^2 Holomorphic Functions on Pseudo-Convex Coverings

M. Gromov, G. Henkin, and M. Shubin

0. Preliminaries and main results

1. Let M be a complex manifold with a non-empty smooth boundary which will be denoted by bM, $\dim_{\mathbb{C}} M = n$. Let us assume that a real-valued C^∞-function $\rho = \rho(z)$ is given in a complex neighbourhood $\tilde{M}$ of $\overline{M} = M \cup bM$, $\dim_{\mathbb{C}} \tilde{M} = n$, so that

$$(0.1) \quad M = \{z \mid \rho(z) < 0\}, \quad bM = \{z \mid \rho(z) = 0\}; \qquad \nabla \rho(z) \neq 0 \text{ for all } z \in bM .$$

For any $z \in bM$ denote by $T_z^c(bM)$ the complex tangent space to bM: the maximal complex subspace in the real tangent space $T_z(bM)$, $\dim_{\mathbb{C}} T_z^c(bM) = n - 1$. If $z_1, \ldots, z_n$ are complex local coordinates in $\tilde{M}$ near $z \in bM$, then $T_z \tilde{M}$ is identified with $\mathbb{C}^n$ and

$$(0.2) \qquad T_z^c(bM) = \left\{ w = (w_1, \ldots, w_n) \Big| \sum_{j=1}^n \frac{\partial \rho}{\partial z_j}(z) w_j = 0 \right\}.$$

The *Levi form* is an hermitian form on $T_z^c(bM)$ defined in the local coordinates as follows:

$$(0.3) \qquad L_z(w, \bar{w}) = \sum_{j,k=1}^n \frac{\partial^2 \rho}{\partial z_j \partial \bar{z}_k}(z) w_j \bar{w}_k.$$

The manifold M is called *pseudoconvex* if $L_z(w, \bar{w}) \geq 0$ for all $z \in bM$ and $w \in T_z^c(bM)$. It is called *strongly pseudoconvex* if $L_z(w, \bar{w}) > 0$ for all $z \in bM$ and all $w \neq 0$, $w \in T_z^c(bM)$. In this case replacing ρ by $e^{\lambda \rho} - 1$ with sufficiently large $\lambda > 0$ we can assume that $L_z(w, \bar{w}) > 0$ for all $w \neq 0$ (not only for w satisfying the condition in (0.2)).

Denote by $\mathcal{O}(M)$ the set of all holomorphic functions on M.

A point $z \in bM$ is called a *peak point* for $\mathcal{O}(M)$ if there exists a function $f \in \mathcal{O}(M)$ such that f is unbounded on M but bounded outside $U \cap M$ for any neighbourhood U of z in $\tilde{M}$. A point $z \in bM$ is called a *local peak point* for $\mathcal{O}(M)$

1991 *Mathematics Subject Classification.* Primary 32F30, 32F20; Secondary 46L99.

The third author was partially supported by NSF grant DMS-9222491.

if there exists a function $f \in \mathcal{O}(M)$ such that f is unbounded in $U \cap M$ for any neighbourhood U of z in $\tilde{M}$ and there exists a neighbourhood U of z in $\tilde{M}$ such that for any neighbourhood V of z in $\tilde{M}$ the function f is bounded in $U - V$.

The Oka-Grauert theorem ([9], see also [7], [11]) states that if M is strongly pseudoconvex and $\overline{M}$ is compact, then every point $z \in bM$ is a peak point for $\mathcal{O}(M)$. (Moreover for every $z \in bM$ there exist functions $f_1, \ldots, f_n \in \mathcal{O}(M)$ which are local complex coordinates in $U \cap M$ for a neighbourhood U of z in $\tilde{M}$.) It follows in particular that the space $\mathcal{O}(M)$ is infinite-dimensional.

One of the goals of this paper is to extend this result to the case when $\overline{M}$ is not necessarily compact but admits a free holomorphic action of a discrete group Γ such that the orbit space $\overline{M}/\Gamma$ is compact (or in other words $\overline{M}$ is a regular covering of a compact complex manifold with a strongly pseudoconvex boundary). In this case we shall use the von Neumann Γ-dimension $\dim_\Gamma$ to measure Hilbert spaces of holomorphic functions (or some exterior forms) which are in L^2 with respect to a Γ-invariant smooth measure on $\overline{M}$. In case when the group Γ is trivial (i.e. has only one element) the Γ-dimension is just the usual dimension $\dim_{\mathbb{C}}$. We shall prove that in general case the space of L^2-holomorphic functions on a strongly pseudoconvex regular covering of a compact manifold has an infinite Γ-dimension and every point $z \in bM$ is a local peak point.

Let us choose a boundary point x for a strongly pseudoconvex manifold M and describe the classical E. Levi construction of a locally defined holomorphic function on $U \cap M$ (here U is a neighbourhood of x in $\tilde{M}$) with the peak point x. Let us consider the Taylor expansion of ρ at x:

$$(0.4) \qquad \rho(z) = \rho(x) + 2\operatorname{Re} f(x, z) + L_x(z - x, \bar{z} - \bar{x}) + O(|z - x|^3),$$

where L_x is the Levi form at x and $f(x, z)$ is a complex quadratic polynomial with respect to z:

$$f(x, z) = \sum_{1 \leq \nu \leq n} \frac{\partial \rho}{\partial z_\nu}(x)(z_\nu - x_\nu) + \frac{1}{2} \sum_{1 \leq \mu, \nu \leq n} \frac{\partial^2 \rho}{\partial z_\mu \partial z_\nu}(x)(z_\mu - x_\mu)(z_\nu - x_\nu).$$

The complex quadric hypersurface $S_x = \{z \mid f(x, z) = 0\}$ has $T_x^c(bM)$ as its tangent plane at x. Therefore the strict pseudoconvexity implies that $\rho(z) > 0$ if $f(x, z) = 0$ and $z \neq x$ is close to x. This means that near x the intersection of the hypersurface S_x with $\overline{M}$ consists of one point x. Hence the function $1/f(x, \cdot)$ is holomorphic in $U \cap M$ (where U is a neighbourhood of x in $\tilde{M}$) and x is its peak point.

The technique which allows to pass from locally defined holomorphic functions to global ones is $\bar{\partial}$-cohomologies on complex manifolds. For any integers p, q with $1 \leq p, q \leq n$ denote by $\Lambda^{p,q}(M)$ the space of all C^∞ forms on M which can be written in local complex coordinates as

$$\omega = \sum_{|I|=p, |J|=q} \omega_{I,J} dz^I \wedge d\bar{z}^J,$$

where $dz^I = dz^{i_1} \wedge \cdots \wedge dz^{i_p}$, $d\bar{z}^J = d\bar{z}^{j_1} \wedge \cdots \wedge d\bar{z}^{j_q}$, $I = (i_1, \ldots, i_p)$, $J = (j_1, \ldots, j_q)$, $i_1 < \cdots < i_p$, $j_1 < \cdots < j_q$, and $\omega_{I,J}$ are C^∞ functions in local coordinates. For such a form ω its $\bar{\partial}$ differential is written as

$$\bar{\partial}\omega = \sum_{|I|=p, |J|=q} \sum_{k=1}^n \frac{\partial \omega_{I,J}}{\partial \bar{z}^k} d\bar{z}^k \wedge dz^I \wedge d\bar{z}^J,$$

so $\bar{\partial}$ defines a linear map $\bar{\partial} : \Lambda^{p,q}(M) \longrightarrow \Lambda^{p,q+1}(M)$. All these maps constitute a complex of vector spaces

$$\Lambda^{p,\bullet} : \quad 0 \longrightarrow \Lambda^{p,0} \longrightarrow \Lambda^{p,1} \longrightarrow \cdots \longrightarrow \Lambda^{p,n} \longrightarrow 0.$$

Its cohomologies are denoted $H^{p,q}(M)$.

An important part of the Grauert theorem is the fact that $\dim_{\mathbb{C}} H^{p,q}(M) < \infty$ for all p, q with $q > 0$ provided M is strictly pseudoconvex and $\overline{M}$ is compact. This fact is used in constructing global holomorphic functions on M with a peak point $x \in bM$ as follows. We start with a locally defined function $g \in \mathcal{O}(U \cap M)$ (here U is a neighbourhood of x in $\tilde{M}$), multiply it by a cut-off function $\chi \in C_0^\infty(U)$ which equals 1 in a neighbourhood of x, then solve the equation $\bar{\partial}f = \bar{\partial}(\chi g)$ on M in appropriate function spaces consisting of bounded functions on M. If we can do this then the function $\chi g - f$ is holomorphic on M and x is its peak point. The solvability of the equation $\bar{\partial}f = \alpha \in \Lambda^{0,1}(M)$ for all forms α with $\bar{\partial}\alpha = 0$ is equivalent to the vanishing of $H^{0,1}(M)$. If we only know that the latter space has a finite dimension then we still can solve the equation $\bar{\partial}f = \alpha$ for all $\bar{\partial}$-closed forms α in the space of finite codimension in the space of all $\bar{\partial}$-closed forms. This is sufficient to construct holomorhpic functions on M with the peak point x because it is easy to provide an infinite-dimensional space of holomorphic functions in a neighbourhood of x having x as its peak point (e.g. we can take a linear space spanned by all powers of one function with the peak point x).

We should be also able to provide bounded solutions of the equation $\bar{\partial}f = \alpha$ provided α is $\bar{\partial}$-closed and sufficiently regular up to the boundary. Therefore we should consider cohomologies $H^{p,q}(M)$ with estimates.

2. Now we shall give a very brief description of the Γ-dimension. It will be used to measure Γ-invariant spaces (of functions and forms) which are infinite-dimensional in the usual sense. It is also convenient to use the Γ-trace. For more details we refer the reader to [1], [4] and textbooks on von Neumann algebras (e.g. [6], [15], [21]).

We shall denote the Γ-dimension by $\dim_\Gamma$. It is defined on the set of all (projective) Hilbert Γ-modules and takes values in $[0, \infty]$. The simplest Hilbert Γ-module is given by a left regular representation of Γ: it is the Hilbert space $L^2\Gamma$ consisting of all complex-valued L^2-functions on Γ. The group Γ acts unitarily on $L^2\Gamma$ by $\gamma \mapsto L_\gamma$ where L_γ is defined as follows:

$$L_\gamma f(x) = f(\gamma^{-1}x), \ x \in \Gamma; \qquad f \in L^2\Gamma.$$

By definition $\dim_\Gamma L^2\Gamma = 1$.

For any (complex) Hilbert space $\mathcal{H}$ define a free Hilbert Γ-module $L^2\Gamma \otimes \mathcal{H}$. Its Γ-dimension equals $\dim_{\mathbb{C}} \mathcal{H}$. The action of Γ in $L^2\Gamma \otimes \mathcal{H}$ is defined by $\gamma \mapsto L_\gamma \otimes I$.

A general Hilbert Γ-module is a closed Γ-invariant subspace in a free Hilbert Γ-module. It would be natural to call such subspaces *projective* Hilbert modules, but the word "projective" is usually omitted, so only projective Hilbert modules are considered.

For any Hilbert space $\mathcal{H}$ denote by $\mathcal{A}_\Gamma$ a von Neumann algebra which consists of all bounded linear operators in $L^2\Gamma \otimes \mathcal{H}$ which commute with the action of Γ there. This algebra is in fact generated by the operators of the form $R_\gamma \otimes B$, $B \in \mathcal{B}(H)$, $\gamma \in \Gamma$, where $\mathcal{B}(H)$ is the algebra of all bounded linear operators in $\mathcal{H}$, R_γ

is the operator of the right translation in $L^2\Gamma$ i.e.

$$R_\gamma f(x) = f(x\gamma), \ x \in \Gamma; \qquad f \in L^2\Gamma.$$

This means that the algebra $\mathcal{A}_\Gamma$ is the weak closure of all finite linear combinations of the operators of the form $R_\gamma \otimes B$. So in fact $\mathcal{A}_\Gamma$ is a tensor product (in the sense of von Neumann algebras) of $\mathcal{R}_\Gamma$ and $\mathcal{B}(H)$ where $\mathcal{R}_\Gamma$ is the von Neumann algebra generated by the operators R_γ in $L^2\Gamma$ (it consists of all operators in $L^2\Gamma$ which commute with all operators L_γ, $\gamma \in \Gamma$).

There is a natural trace on $\mathcal{R}_\Gamma$. It is denoted by tr_Γ and defined as the diagonal matrix element (all of them are equal) in the δ-functions basis. For example we can define it by

$$\mathrm{tr}_\Gamma S = (S\delta_e, \delta_e), \qquad S \in \mathcal{R}_\Gamma,$$

where e is the neutral element of Γ, $\delta_e \in L^2\Gamma$ is the "Dirac delta-function" at e, i.e. $\delta_e(x) = 1$ if $x = e$ and 0 otherwise. There is also a natural trace on $\mathcal{A}_\Gamma$ too: $\mathrm{Tr}_\Gamma = \mathrm{tr}_\Gamma \otimes \mathrm{Tr}$ where Tr is the usual trace on $\mathcal{B}(H)$.

Now for any Hilbert Γ-module which is a closed Γ-invariant subspace L in $L^2\Gamma \otimes \mathcal{H}$, its Γ-dimension is defined by the natural formula

$$\dim_\Gamma L = \mathrm{Tr}_\Gamma P_L,$$

where P_L is the orthogonal projection on L in $L^2\Gamma \otimes \mathcal{H}$.

3. Let us describe the spaces of reduced Dolbeault cohomologies on a complex (generally non-compact) manifold M with a given hermitian metric. Denote the Hilbert space of all (measurable) square-integrable (p,q)-forms on M by $L^2\Lambda^{p,q} = L^2\Lambda^{p,q}(M)$. The operator

$$\bar\partial : L^2\Lambda^{p,q}(M) \longrightarrow L^2\Lambda^{p,q+1}(M)$$

is defined as the maximal operator i.e. its domain $D^{p,q} = D^{p,q}(\bar\partial; M)$ is the set of all $\omega \in L^2\Lambda^{p,q}$ such that $\bar\partial\omega \in L^2\Lambda^{p,q+1}$ where $\bar\partial\omega$ is applied in the sense of distributions. Obviously $\bar\partial^2 = 0$ on $D^{p,q}$ and we can form a complex

$$L^2\Lambda^{p,\bullet}: \quad 0 \longrightarrow D^{p,0} \longrightarrow D^{p,1} \longrightarrow \cdots \longrightarrow D^{p,n} \longrightarrow 0.$$

Its cohomologies are denoted $L^2 H^{p,q}(M)$ and called L^2 Dolbeault cohomologies of M:

$$L^2 H^{p,q}(M) = \mathrm{Ker}\,(\bar\partial : D^{p,q} \to D^{p,q+1})/\mathrm{Im}\,(\bar\partial : D^{p,q-1} \to D^{p,q}).$$

We actually need reduced cohomologies

$$L^2 \bar{H}^{p,q}(M) = \mathrm{Ker}\,(\bar\partial : D^{p,q} \to D^{p,q+1})/\overline{\mathrm{Im}\,(\bar\partial : D^{p,q-1} \to D^{p,q})},$$

where the line over $\mathrm{Im}\,\bar\partial$ means its closure in the corresponding L^2 space. Since $\mathrm{Ker}\,\bar\partial$ is a closed subspace in L^2, the reduced cohomology space $L^2\bar{H}^{p,q}(M)$ is a Hilbert space.

Note that the space $L^2\bar{H}^{0,0}(M)$ coincides with the space $L^2\mathcal{O}(M)$ of all square-integrable holomorphic functions on M.

4. Let us assume now that M is a complex manifold (with boundary) with a free action of a discrete group Γ on $\overline{M}$ such that $\overline{M}/\Gamma$ is compact and the action is holomorphic on M. (Here $\overline{M} = M \cup bM$.) Let us assume that an hermitian Γ-invariant metric is given on $\overline{M}$. Then the reduced L^2 Dolbeault cohomologies become Hilbert Γ-modules. Hence they have a well defined Γ-dimension (possibly infinity).

Now we will formulate our main results. We will always assume that we are in the situation described above, M is strongly pseudoconvex and bM is nonempty.

THEOREM 0.1. $\dim_\Gamma L^2 \bar{H}^{p,q}(M) < \infty$ for all p, q provided $q > 0$.

THEOREM 0.2. $\dim_\Gamma L^2 \mathcal{O}(M) = \infty$ and each point in bM is a local peak point for $L^2 \mathcal{O}(M)$.

Under the same conditions it is also possible to construct holomorphic functions which have stronger local singularities (not L^2) but are in L^2 in a generalized sense. For any $s \in \mathbb{R}$ denote by $W^s = W^s(M)$ the uniform (Γ-invariant) Sobolev space of distributions on M, based on the space $W^0 = L^2(M)$ constructed with the use of a smooth Γ-invariant measure on $\overline{M}$ (see e.g. [18] for the details on the Sobolev spaces). The space W^{-s} for large $s > 0$ contains in particular holomorphic functions on M with power singularities at the boundary. For any $s \in \mathbb{R}$ the space W^s is a Hilbert Γ-module with respect to the natural action of Γ. Denote by $W^s\mathcal{O}(M)$ the space of all elements in W^s which are actually holomorphic functions on M. Now we can formulate another version of Theorem 0.2.

THEOREM 0.3. For any $x \in bM$ and any integer $N > 0$ there exists $s > 0$ and a closed Γ-invariant subspace $L \subset W^{-s}\mathcal{O}(M)$ such that
   (i) $\dim_\Gamma L = N$;
   (ii) $L \cap L^2(M) = \{0\}$ but for any $f \in L$ and any Γ-invariant neighbourhood U of x in $\overline{M}$ we have $f \in L^2(M - U)$.

It is also possible to construct L^2-holomorphic functions on M which are in $C^\infty(\overline{M})$:

THEOREM 0.4. For any integer $N > 0$ there exists a Γ-invariant subspace $L \subset L^2\mathcal{O}(M) \cap C^\infty(\overline{M})$ such that $\dim_\Gamma \bar{L} = N$ where $\bar{L}$ is the closure of L in $L^2(M)$.

EXAMPLES. 1) Let X be a compact real-analytic manifold with an infinite fundamental group $\Gamma = \pi_1(X)$. Assume that X is imbedded into its complexification Y and a Riemannian metric is chosen on Y. Let X_ε be a ε-neighbourhood of X in Y where ε is sufficiently small. It is known ([13], [9]) that then X_ε is strongly pseudoconvex. Let M be the universal covering of X_ε. Theorems 0.1–0.4 can be applied to M and we conclude in particular that there are sufficiently many L^2 holomorphic functions on M.

A particular case: strip $\{z \mid |\operatorname{Im} z| < 1\}$ in $\mathbb{C}$ with the action of $\Gamma = \mathbb{Z}$ by translations along $\mathbb{R}$. Of course in this case L^2 holomorphic functions can be obtained by the Fourier transform or explicitly (e.g. take $1/(a^2 + z^2)$ where $a > 1$).

2) Let X be a compact complex manifold with a holomorphic positive vector bundle E on X. The positivity means that E is supplied with an hermitian metric and ε-neighbourhood X_ε of X in the total space of E is strongly pseudoconvex (for some $\varepsilon > 0$ or, equivalently, for any $\varepsilon > 0$).

Note that X_ε is not a Stein manifold because it has a non-trivial compact complex submanifold X (the zero section of E). But we are again in the situation of the Theorems 0.1–0.4 and these theorems give extensions of some results of Gromov [10] and Napier [16]. Namely let M be the universal covering of X_ε. Theorems 0.2–0.4 garantee that there are many L^2 holomorphic functions on M. In particular,

$\dim_\Gamma L^2 \mathcal{O}(M) = \infty$. Spaces of L^2 holomorphic functions with a finite positive Γ-dimension were constructed in [10], [16] from functions which are polynomial along the fibers.

REMARKS. 1) If in the assumptions of the theorems above M/Γ is a Stein manifold then Stein [20] proved that M is also a Stein manifold. It follows from this result that there are sufficiently many holomorphic functions on M then, but it does not follow that there exist non-trivial L^2 holomorphic functions. On the other hand it can happen that M/Γ is not Stein (see Example 2 above). Then even the existence of any holomorphic function on M which is not constant along orbits of Γ is not obvious.

2) If $bM = \emptyset$ then it follows from the arguments of Atiyah [1] that $\dim_\Gamma L^2 \bar{H}^{p,q}(M) < \infty$ for all p, q (including $q = 0$). In this case in fact $L^2 \mathcal{O}(M) = \{0\}$. If E is a positive Γ-invariant holomorphic line bundle, then the Atiyah index theorem [1] implies that $\dim_\Gamma L^2 \mathcal{O}(M, E^k) > 0$ for large k; in particular, the space of all holomorphic sections of E^k is infinite-dimensional in the usual sense. (See [10] and [16] for further results.)

3) Theorems 0.2–0.4 remain valid if we replace holomorphic functions by holomorphic $(p, 0)$-forms. More generally all Theorems 0.1–0.4 are true for sections of arbitrary holomorphic vector Γ-bundles over M.

4) Theorems 0.1, 0.3, 0.4 can be extended to the case when M is strongly pseudoconvex but with possibly non-smooth boundary i.e. we can drop the requirement $\nabla\rho \neq 0$ on bM in (0.1) but require instead that the Levi form (0.3) is positive for all $z \in bM$ and all $w \neq 0$, $w \in \mathbb{C}^n$.

5) Let us assume that the Levi form (0.3) is non-degenerate on $T_z^c(bM)$ for all $z \in bM$ and the boundary bM is connected (this is authomatically true if bM is strongly pseudoconvex). Let r be the number of negative eigenvalues of the Levi form in $T_z^c(bM)$. Then

$$\dim_\Gamma L^2 \bar{H}^{0,r} = \infty$$

and

$$\dim_\Gamma L^2 \bar{H}^{p,q} < \infty, \qquad q \neq r.$$

This is a generalization to the covering case of the classical theorems by Andreotti-Grauert and Andreotti-Vesentini (see [2], [7], [12]).

6) First applications of von Neumann algebras to constructions of non-trivial spaces of L^2-holomorphic functions or sections of holomorphic vector bundles are due to M. Atiyah [1] and A. Connes [5]. J.Roe proved existence of an infinite-dimensional space of L^2 holomorphic sections of a power E^k for a uniformly positive holomorphic line bundle E over a complete Kähler simply connected manifold of non-positive curvature without any action of a discrete group (see [17] for further results and references).

1. $\bar{\partial}$-cohomologies of pseudoconvex coverings

1. In this section we will prove Theorem 0.1. We will start by extending the Kohn-Morrey estimates ([7],[14]) to our case. We will always assume that M is strictly pseudoconvex.

First we will consider a general Γ-invariant analytic situation. Namely let M be a C^∞-manifold (possibly with boundary) with a free action of a discrete group Γ such that $\overline{M}/\Gamma$ is compact. Let E be a (complex) vector Γ-bundle on $\overline{M}$ with a Γ-invariant hermitian metric in the fibers of E. We shall use Γ-invariant Sobolev spaces W^s of sections of E over M. The scale of the Hilbert spaces $W^s = W^s(M, E)$ is based on the Hilbert space $L^2(M, E)$ which is taken with respect to a smooth Γ-invariant measure on $\overline{M}$ and the given Γ-invariant hermitian metric on E over $\overline{M}$. Let $\tilde{M}$ be a Γ-invariant complex neighbourhood of $\overline{M}$. Assume that E and the measure on M are extended to $\tilde{M}$ in a smooth Γ-invariant way. For any $s \in \mathbb{R}$ the space $W^s = W^s(M, E)$ is a Hilbert space which consists of all restrictions to M of finite linear combinations of all sections Au where $u \in L^2(\tilde{M}, E)$ and A is a properly supported Γ-invariant pseudodifferential operator of order $-s$ on $\tilde{M}$ (see e.g. [1] or [18]). The norm in W^s is denoted $\| \cdot \|_s$.

In particular Γ-invariant Sobolev spaces $W^s \Lambda^{p,q}$ of (p, q)-differential forms on M are well defined.

Let us consider $\bar{\partial}$ as the maximal operator in L^2 and let $\bar{\partial}^*$ be the Hilbert space adjoint operator. We shall also use the corresponding Laplacian

$$\Box_{p,q} = \bar{\partial}\bar{\partial}^* + \bar{\partial}^*\bar{\partial} \quad \text{on} \quad L^2\Lambda^{p,q}(M).$$

We shall denote the domain of any operator A by $D(A)$. Let $\Lambda_c^{\bullet}(\overline{M})$ denotes the set of all C^{∞} forms with compact support on $\overline{M}$.

For any complex 1-form α denote by $i(\alpha)$ the substitution operator $i(v)$ of the (complex) vector field v corresponding to the form α with the use of the given hermitian metric, so $i(\alpha) : \Lambda^p(M) \to \Lambda^{p-1}(M)$. In particular we shall use $i(\partial\rho)$ where $\partial : \Lambda^{p,q} \to \Lambda^{p+1,q}$ appears in the standard decomposition $d = \partial + \bar{\partial}$ for the de Rham differential d.

The following Lemma gives a description of the operators $\bar{\partial}^*$, $\Box$ (as well as their domains $D(\bar{\partial}^*)$, $D(\Box)$).

LEMMA 1.1. (i) *The operator $\bar{\partial}^*$ can be obtained as the closure from the initial domain*

$$(1.1) \qquad D_0(\bar{\partial}^*) = \{\omega \mid \omega \in \Lambda_c^{\bullet}(\overline{M}), \ i(\partial\rho)\omega = 0 \ \text{on } bM\}.$$

(ii) *The operator $\Box = \Box_{p,q}$ can be obtained as the closure from the initial domain*

$$(1.2) \qquad D_0(\Box) = \{\omega \mid \omega \in \Lambda_c^{\bullet}(\overline{M}), \ i(\partial\rho)\omega = 0 \ \text{and } i(\partial\rho)\bar{\partial}\omega = 0 \ \text{on } bM\}.$$

PROOF. The proof is a simple combination of the one given by Hörmander [12] in the compact case and Gaffney's cut-off trick (see [8], [10]). $\quad\square$

REMARK. The conditions on ω in (1.2) are called the $\bar{\partial}$-*Neumann conditions*.

PROPOSITION 1.2. *The domain $D(\Box_{p,q})$, $q > 0$, is included into $W^{1/2}$ and there exists a constant $C > 0$ such that the following estimate is true*

$$\|\omega\|_{1/2}^2 \le C((\Box_{p,q}\omega, \omega) + \|\omega\|_0^2), \qquad \omega \in D(\Box_{p,q}), \ q > 0.$$

PROOF. Summing up the corresponding local Morrey estimates from [13], [14] with the use of a Γ-invariant partition of unity we obtain the following fundamental inequality:

$$\|\omega\|_{L^2(bM)}^2 \le C((\Box_{p,q}\omega, \omega) + \|\omega\|_0^2), \qquad \omega \in \Lambda_c^{p,q}(\overline{M}), \ q > 0.$$

This inequality, the Kohn-type estimate (see [7])

$$\|u\|_{1/2}^2 \le C(\|u\|_{L^2(bM)}^2 + \|\bar{\partial}u\|_{L^2(M)}^2), \qquad u \in C^{\infty}(\overline{M}),$$

and Lemma 1.1 imply the necessary statement.

COROLLARY 1.3. *There exists $C > 0$ such that*

$$\|\omega\|_{1/2} \leq C\|\omega\|_0, \qquad \omega \in \operatorname{Ker}\square_{p,q}, \ q > 0 \,,$$

where $\operatorname{Ker}\square = \{\omega \mid \omega \in D(\square), \square\omega = 0\}$.

2. Let us formulate the necessary version of the Hodge-Kodaira decomposition (see e.g. [7]):

PROPOSITION 1.4. *The following orthogonal decompositions hold:*

(1.3) $$L^2\Lambda^\bullet(M) = \overline{\operatorname{Im}\bar\partial} \oplus \operatorname{Ker}\square \oplus \overline{\operatorname{Im}\bar\partial^*}$$

and

$$\operatorname{Ker}\bar\partial = \overline{\operatorname{Im}\bar\partial} \oplus \operatorname{Ker}\square.$$

In particular we have an isomorphism of Hilbert Γ-modules

$$L^2\bar H^{p,q}(M) = \operatorname{Ker}\square_{p,q}.$$

We shall use the following rather general

LEMMA 1.5. *Let L be a closed Γ-invariant subspace in $L^2(M, E)$, $L \subset W^\varepsilon$ for some $\varepsilon > 0$ and there exists $C > 0$ such that*

(1.4) $$\|u\|_\varepsilon \leq C\|u\|_0, \qquad u \in L \,.$$

Then $\dim_\Gamma L < \infty$.

To prove this Lemma we need the following simple Lemma about estimates of Sobolev norms on compact manifolds with boundary.

LEMMA 1.6. *Let X be a compact Riemannian manifold, possibly with a boundary. Let E be a (complex) vector bundle with an hermitian metric over $\overline X$. Denote by $(\cdot, \cdot)$ the induced hermitian inner product in the Hilbert space $L^2(X, E)$ of square-integrable sections of E over X. Denote by $W^s = W^s(X, E)$ the corresponding Sobolev space of sections of E over X, $\|\cdot\|_s$ the norm in this space. Let us choose a complete orthonormal system $\{\psi_j;\ j = 1, 2, \dots\}$ in $L^2(X, E)$. Then for all $\varepsilon > 0$ and $\delta > 0$ there exists an integer $N > 0$ such that*

$$\|u\|_0 \leq \delta\|u\|_\varepsilon \text{ provided } u \in W^\varepsilon \text{ and } (u, \psi_j) = 0, \qquad j = 1, \dots, N.$$

PROOF. Assuming the opposite we conclude that there exist $\varepsilon > 0$ and $\delta > 0$ such that for every $N > 0$ there exists $u_N \in W^\varepsilon$ with $(u_N, \psi_j) = 0$, $j = 1, \dots, N$ satisfying the estimate $\|u_N\|_\varepsilon \leq \delta^{-1}\|u_N\|_0$. Normalizing u_N we can assume that $\|u_N\|_0 = 1$, so the previous estimate gives $\|u_N\|_\varepsilon \leq \delta^{-1}$ for all N. It follows from the Sobolev compactness theorem that the set $\{u_N \mid N = 1, 2, \dots\}$ is compact in $L^2 = L^2(X, E)$. On the other hand obviously $u_N \to 0$ weakly in L^2 as $N \to \infty$. Therefore $\|u_N\|_0 \to 0$ as $N \to \infty$ which contradicts to the chosen normalization.

PROOF OF LEMMA 1.5. Let us choose a Γ-invariant covering of M by balls γB_k, $k = 1, \ldots, m$, $\gamma \in \Gamma$, so that all the balls have smooth boundary (e.g. have sufficiently small radii). Let us choose a complete orthonormal system $\{\psi_j^{(k)} \ j = 1, 2, \ldots\}$ in $L^2(B_k, E)$ for every $k = 1, \ldots, m$. Then $\{(\gamma^{-1})^* \psi_j^{(k)}, \ j = 1, 2, \ldots\}$ will be an orthonormal system in γB_k (here we identify the element γ with the corresponding transformation of M).

Given the subspace L satisfying the conditions in the Lemma let us define a map

$$P_N : L \longrightarrow L^2\Gamma \otimes \mathbb{C}^{mN}$$

$$u \mapsto \{(u, (\gamma^{-1})^* \psi_j^{(k)}), \qquad j = 1, 2, \ldots, N; \ k = 1, \ldots, m; \ \gamma \in \Gamma\}.$$

Since $\dim_\Gamma L^2\Gamma \otimes \mathbb{C}^{mN} = mN < \infty$ the desired result will follow if we prove that P_N is injective for large N. Assume that $u \in L$ and $P_N u = 0$. Using Lemma 1.6 we get then

$$\|u\|_{0,\gamma B_k}^2 \leq \delta_N^2 \|u\|_{\varepsilon, \gamma B_k}^2, \qquad k = 1, \ldots, m; \ \gamma \in \Gamma,$$

where $\delta_N \to 0$ as $N \to \infty$ and $\|u\|_{s,\gamma B_k}$ means the norm in the Sobolev space W^s over the ball γB_k. Summing over all k and γ we get

$$\|u\|_0^2 \leq C_1^2 \delta_N^2 \|u\|_\varepsilon^2,$$

where $C_1 > 0$ does not depend on N. This clearly contradicts (1.4) unless $u = 0$.

REMARK 1.7. It is not necessary to require that L is closed in L^2 in Lemma 1.5. For any L satisfying (1.4) we can consider its closure $\bar{L}$ in L^2. Then obviously $\bar{L} \subset W^\varepsilon$ and Lemma 1.5 implies that $\dim_\Gamma \bar{L} < \infty$.

PROOF OF THEOREM 0.1. Propositions 1.2, 1.4 and Lemma 1.5 immediately imply Theorem 0.1.

2. L^2 holomorphic functions

1. We shall use some simple linear algebra and Γ-Fredholm operators in Hilbert Γ-modules. Necessary background and similar arguments can be found in [3] and [19].

LEMMA 2.1. *Let L be a Hilbert Γ-module, L_1, L_2 its Hilbert Γ-submodules such that $\dim_\Gamma L_1 > \operatorname{codim}_\Gamma L_2$ where $\operatorname{codim}_\Gamma L_2$ means the Γ-dimension of the orthogonal complement of L_2 in L. Then $L_1 \cap L_2 \neq \{0\}$. Moreover*

$$(2.1) \qquad \dim_\Gamma L_1 \cap L_2 \geq \dim_\Gamma L_1 - \operatorname{codim}_\Gamma L_2.$$

PROOF. Denote by $L_1 \ominus L_2$ the orthogonal complement of $L_1 \cap L_2$ in L_1. Clearly $\dim_\Gamma L_1 \ominus L_2 \leq \operatorname{codim}_\Gamma L_2$. Therefore if (2.1) is not true, then we get

$$\dim_\Gamma L_1 = \dim_\Gamma L_1 \cap L_2 + \dim_\Gamma L_1 \ominus L_2 \leq \dim_\Gamma L_1 \cap L_2 + \operatorname{codim}_\Gamma L_2 < \dim_\Gamma L_1$$

which is a contradiction.

We will use unbounded Γ-Fredholm operators. The corresponding definition slightly extends the corresponding definition for bounded operators given by M. Breuer [3] (see also [19]).

DEFINITION 2.2. Let L_1, L_2 be Hilbert Γ-modules, $A : L_1 \to L_2$ a closed densely defined linear operator (with the domain $D(A)$) which commutes with the action of Γ in L_1 and L_2. The operator A is called Γ-*Fredholm* if the following conditions are satisfied:

(i) $\dim_\Gamma \operatorname{Ker} A < \infty$;

(ii) there exists a closed Γ-invariant subspace $Q \subset L_2$ such that $Q \subset \operatorname{Im} A$ and $\operatorname{codim}_\Gamma Q (= \dim_\Gamma(L_2 \ominus Q)) < \infty$.

Let us also recall the following definition from [19]:

DEFINITION 2.3. Let L be a Hilbert Γ-module, $Q \subset L$ is a Γ-invariant subspace (not necessarily closed). Then

(i) Q is called Γ-*dense* in L if for every $\varepsilon > 0$ there exists a Γ-invariant subspace $Q_\varepsilon \subset Q$ such that Q_ε is closed in L and $\operatorname{codim}_\Gamma Q_\varepsilon < \varepsilon$ in L.

(ii) Q is called *almost closed* if Q is Γ-dense in its closure $\bar{Q}$.

If Q is Γ-dense in L then it is also dense in L in the usual sense i.e. $\bar{Q} = L$ (see Lemma 1.8 in [19]). Note also that if Γ is trivial (or finite) then Q is Γ-dense in L if and only if $Q = L$ (in particular in this case Q is almost closed if and only if it is closed).

LEMMA 2.4. *If* $A : L_1 \to L_2$ *is a* Γ-*Fredholm operator then* $\operatorname{Im} A$ *is almost closed.*

PROOF. This statement can be reduced to the case when A is bounded by replacing L_1 by $D(A)$ with the graph norm. Then the statement is due to M. Breuer [3] (see also Lemma 1.15 in [19]).

LEMMA 2.5. ([19]) *Let* L *be a Hilbert* Γ-*module,* $L_1 \subset L$ *and* $Q \subset L$ *its* Γ-*invariant subspaces in* L *such that* L_1 *is closed and* Q *is* Γ-*dense in* L. *Then* $Q \cap L_1$ *is* Γ-*dense in* L_1. *More generally, if* Q *is almost closed then* $Q \cap L_1$ *is almost closed and its closure equals* $\bar{Q} \cap L_1$.

COROLLARY 2.6. *Let* $A : L_1 \to L_2$ *be a* Γ-*Fredholm operator,* $L_3 \subset L_2$ *is a closed* Γ-*invariant subspace such that* $L_3 \subset \overline{\operatorname{Im} A}$. *Then* $L_3 \cap \operatorname{Im} A$ *is* Γ-*dense in* L_3.

Now let us return to the analytic situation described above.

PROPOSITION 2.7. *Let* A *be a self-adjoint linear operator in* $L^2(M, E)$ *such that* A *commutes with the action of* Γ, $D(A) \subset W^\varepsilon$ *where* $\varepsilon > 0$ *and*

$$(2.2) \qquad \|u\|_\varepsilon^2 \le C(\|Au\|^2 + \|u\|_0^2), \quad u \in D(A).$$

Then A *is* Γ-*Fredholm.*

PROOF. It follows from (2.2) that the estimate (1.4) is satisfied on $L = \operatorname{Ker} A$. Therefore Lemma 1.5 implies that $\dim_\Gamma \operatorname{Ker} A < \infty$.

Let $\tilde{E}_\delta$ be the spectral projection of A corresponding to the interval $(-\delta, \delta)$. Then again $\dim_\Gamma \operatorname{Im} \tilde{E}_\delta < \infty$ by Lemma 1.5. On the other hand

$$\operatorname{Im}(I - \tilde{E}_\delta) = (\operatorname{Im} \tilde{E}_\delta)^\perp \subset \operatorname{Im} A,$$

which immediately implies that A is Γ-Fredholm.

2. Now using Theorem 0.1 we will be able to provide the complete proof of Theorem 0.2. We shall start with the following elementary

LEMMA 2.8. *Let U be an arbitrary set, $g : U \to \mathbb{C}$ an unbounded function. Then for any integer $N > 0$ the functions $g, g^2, \ldots, g^N$ are linearly independent modulo bounded functions i.e. if $B(U)$ is the space of all bounded functions on U and*

(2.3) $$c_1 g + c_2 g^2 + \cdots + c_N g^N \in B(U),$$

then $c_1 = \ldots c_N = 0$.

PROOF. Assuming that (2.3) is fulfilled consider the polynomial

$$p(t) = c_1 t + c_2 t^2 + \ldots c_N t^N, \qquad t \in \mathbb{C}.$$

Then (2.3) implies that this polynomial is bounded along an unbounded sequence of complex values of t. Clearly this is only possible if the polynomial p is identically 0.

PROOF OF THEOREM 0.2. We shall use the notations from the introduction to this paper.

Let us choose a defining function ρ of the manifold M (see (0.1)) so that the Levi form (0.3) is positive for all $w \in \mathbb{C}^n - \{0\}$ (and not only for $w \in T_z^c(bM) - \{0\}$) at all points $z \in bM$. Using (0.4) we see that $\mathrm{Re}\, f(x, z) < 0$ if $x \in bM$ and $z \in M$ is sufficiently close to x. It follows that we can choose a branch of $\log f(x, z)$ so that $g_x(z) = \log f(x, z)$ is a holomorphic function in $z \in M \cap U_x$ where U_x is a sufficiently small neighbourhood of x in $\overline{M}$. Note that we can (and will) choose $U_{\gamma x} = \gamma U_x$.

Let us fix an arbitrary point $x \in bM$. Clearly $g_x^m \in L^2(\overline{M} \cap U)$ for all $m = 1, 2, \ldots$ and all functions g_x^m have a peak point at x. Besides all these functions are linearly independent modulo bounded functions by Lemma 2.8.

Let us choose a cut-off function $\chi \in C_c^\infty(U)$ where U is a sufficiently small neighbourhood of x, so that $\chi = 1$ in a neighbourhood of x. We shall identify χ with its extension by 0 to $\overline{M}$, so it becomes a function from $C_c^\infty(\overline{M})$. The translation of χ by $\gamma \in \Gamma$ is a function $\gamma^* \chi$ which is supported in a small neighbourhood of γx: $\gamma^* \chi(z) = \chi(\gamma^{-1} z)$.

Denote by L the closed Γ-invariant subspace in $L^2(M)$ generated by all functions χg_x^m; $m = 1, \ldots, N$. Clearly

(2.4) $$L = \left\{ f \mid f = \sum_{\gamma \in \Gamma} \sum_{m=1}^N c_{m,\gamma} \gamma^*(\chi g_x^m); \sum_{m,\gamma} |c_{m,\gamma}|^2 < \infty \right\},$$

where $c_{m,\gamma}$ are complex constants. It follows that L has the form $L^2\Gamma \otimes \mathbb{C}^N$, hence $\dim_\Gamma L = N$.

Let us consider the set of (0,1)-forms (which are smooth on $\overline{M}$ and have compact support):

(2.5) $$\bar{\partial}(\chi g_x^m); \qquad m = 1, 2, \ldots, N.$$

They are linearly independent for any integer $N > 0$. Indeed, assuming that

$$c_1 \bar{\partial}(\chi g_x) + c_2 \bar{\partial}(\chi g_x^2) + c_N \bar{\partial}(\chi g_x^N) = 0$$

with some complex constants $c_1, \ldots, c_N$, we see that

$$c_1 \chi g_x + c_2 \chi g_x^2 + \cdots + c_N \chi g_x^N$$

is holomorphic on M and has a compact support, hence it is identically 0, which implies that $c_1 = \cdots = c_N = 0$ due to Lemma 2.8.

Let L_1 be a closed Γ-invariant subspace in $L^2 \Lambda^{0,1}(M)$ generated by the set of forms (2.5). Then again

$$L_1 = \left\{ \omega \mid \omega = \sum_{\gamma \in \Gamma} \sum_{m=1}^{N} c_{m,\gamma} \bar{\partial}(\gamma^*(\chi g_x^m)), \quad \sum_{m,\gamma} |c_{m,\gamma}|^2 < \infty \right\},$$

where $c_{m,\gamma}$ are complex constants, and $\dim_\Gamma L_1 = N$. Clearly $L_1 \subset C^\infty \Lambda^{0,1}(\overline{M})$ i.e. all elements of L_1 are C^∞ forms of type (0,1) on $\overline{M}$. Also $L_1 \subset \mathrm{Im}\,\bar{\partial}$, hence $L_1 \subset \overline{\mathrm{Im}\,\square}$ due to the orthogonal decomposition (1.3).

Now we can apply Corollary 2.6 to conclude that $\mathrm{Im}\,\square \cap L_1$ is Γ-dense in L_1. Hence for any $\delta > 0$ there exists a closed Γ-invariant subspace $Q_1 \subset L_1$ such that $Q_1 \subset \mathrm{Im}\,\square$ and $\dim_\Gamma Q_1 > N - \delta$. Solving the equation $\square \omega = \alpha$ with $\alpha \in Q_1$ we can assume that $\omega \perp \mathrm{Ker}\,\square$ and in this case the solution ω will be unique. Denote the space of all such solutions by K. Then $\dim_\Gamma K = \dim_\Gamma Q_1 > N - \delta$.

Applying $\bar{\partial}$ to both sides of the equation $\square \omega = \alpha$ we see that $\bar{\partial}\bar{\partial}^*\bar{\partial}\omega = 0$, hence $\bar{\partial}^*\bar{\partial}\omega = 0$ and $\bar{\partial}\omega = 0$. Therefore $\bar{\partial}\bar{\partial}^*\omega = \alpha$. Also $\omega \in \Lambda^{0,1}(\overline{M})$ (i.e. $\omega \in C^\infty$ on $\overline{M}$) for any such solution ω due to the local regularity theorem for the $\bar{\partial}$ Neumann problem (see [7]).

Now denote

$$Q = \{f \mid f \in L, \bar{\partial}f = \alpha \in Q_1\}.$$

As we have seen earlier $\bar{\partial}$ is injective on L, hence $\dim_\Gamma Q = \dim_\Gamma Q_1 > N - \delta$. If $f \in Q$ then we can find a (unique) solution $\omega \in K$ of the equation $\square \omega = \alpha = \bar{\partial}f$ and then $h = f - \bar{\partial}^*\omega \in L^2 \mathcal{O}(M)$. All these functions h form a closed Γ-invariant subspace $H \subset L^2 \mathcal{O}(M)$ with $\dim_\Gamma H > N - \delta$. Hence $\dim_\Gamma L^2 \mathcal{O}(M) = \infty$. Besides using the Γ-invariance of H we see that we can always find a function $h \in H$ such that one of the coefficients $c_{m,e}; m = 1, \ldots, N$, in the expansion (2.4) (for the corresponding function f) does not vanish. The point x will be a local peak point for this function. This completes the proof of Theorem 0.2.

PROOF OF THEOREM 0.3. We should modify the proof of Theorem 0.2 by another choice of locally given holomorphic functions with singularities at a point $x \in bM$. Namely, if f is a holomorphic polynomial from (0.4), then we should use $\{f^{-k}, f^{-2k}, \ldots, f^{-kN}\}$ with sufficiently large integer $k > 0$ instead of $\{\log f, \ldots, (\log f)^N\}$ as we did in the proof of Theorem 0.2. It is easy to check that all functions χf^{-k} are in appropriate Sobolev spaces. Then we should apply Lemma 2.1 to evaluate the Γ-dimension of the intersection $L_1 \cap L_2$ where L_1 is the Γ-invariant subspace generated by all forms $\bar{\partial}(\chi f^{-km})$, $m = 1, \ldots, N$, and $L_2 = \overline{\mathrm{Im}\,\bar{\partial}}$ in $L^2 \Lambda^{0,1}(M)$.

All other arguments are similar to the ones used in the proof of Theorem 0.2.

REMARK. An interesting feature of Theorem 0.3 is that its proof does not use the regularity results for the $\bar{\partial}$-Neumann problem and so this theorem can be extended to a number of less regular situations.

PROOF OF THEOREM 0.4. We should apply the arguments given in the proof of Theorem 0.3 to a strongly pseudoconvex Γ-invariant neighbourhood $\hat{M}$ of $\overline{M}$, find a sufficiently large space H of holomorphic functions on $\hat{M}$ with singularities on the boundary of $\hat{M}$ and then take the space L of restrictions of all functions from H to M. Since the restriction operator is injective the closure of L will have the same Γ-dimension as H.

3. Open questions

Here we give a list of open questions of various difficulty. It is assumed in all questions that we are in the situation of Theorems 0.1–0.4.

1. Does there exist a finite number of functions in $L^2\mathcal{O}(M) \cap C(\overline{M})$ which separate all points in bM?

2. Does there exist $f \in L^2(M) \cap C(\overline{M})$ such that $f(x) \neq 0$ for all $x \in bM$?

3. Is it true that for every CR-function $f \in L^2(bM) \cap C(bM)$ $(\bar{\partial}_b f = 0)$ there exists $F \in L^2\mathcal{O}(M) \cap C(\overline{M})$ such that $F|_{bM} = f$?

References

[1]   M. Atiyah, *Elliptic operators, discrete groups and von Neumann algebras*, Astérisque **32–33** (1976), 43–72.

[2]   A. Andreotti and E. Vesentini, *Carleman estimates for the Laplace-Beltrami equation on complex manifolds*, Publications Math. IHES **25** (1965), 81–130.

[3]   M. Breuer, *Fredholm theories in von Neumann algebras. I, II*, Math. Annalen **178** (1968), 243–254; **180** (1969), 313–325.

[4]   J. M. Cohen, *Von Neumann dimension and the homology of covering spaces*, Quart. J. Math. Oxford, Ser. (2) **30** (1974), 133–142.

[5]   A. Connes, *Sur la théorie non commutative de l'integration*, Lecture Notes Math. **725** (1979), 19–143.

[6]   J. Dixmier, *Von Neumann algebras*, North Holland Publishing Co., Amsterdam, 1981.

[7]   G. B. Folland and J. J. Kohn, *The Neumann problem for the Cauchy-Riemann complex*, Annals of Mathematics Studies, 75, Princeton Univ. Press, Princeton, 1972.

[8]   M. Gaffney, *A special Stokes theorem for Riemannian manifolds*, Annals of Math. **60** (1954), 140–145.

[9]   H. Grauert, *On Levi's problem and the imbedding of real-analytic manifolds*, Annals of Math. **68** (1958), 460–472.

[10] M. Gromov, *Kähler hyperbolicity and L_2-Hodge theory*, J. Diff. Geom. **33** (1991), 263–292.

[11] G. Henkin, *The Lewy equation and analysis on pseudoconvex manifolds*, Russian Math. Surveys **32** (1977), 59–130.

[12] L. Hörmander, *L^2-estimates and existence theorem for the $\bar{\partial}$-operator*, Acta Math. **113** (1965), 89–152.

[13] C. B. Morrey, Jr., *The analytic embedding of abstract real analytic manifolds*, Annals of Math. **68** (1958), 159–201.

[14] C. B. Morrey, Jr., *The $\bar{\partial}$-Neumann problem on strongly pseudo-convex manifolds*, Differential Analysis, Bombay Colloq., Oxford University Press, London, 1964, pp. 81–133.

[15] M. A. Naimark, *Normed rings*, Noordhoff, Groningen, 1964.

[16] T. Napier, *Convexity properties of coverings of smooth projective varieties*, Math. Annalen **286** (1990), 433–479.

[17] J. Roe, *Coarse cohomology and index theory on complete Riemannian manifolds*, Mem. Amer. Math. Soc., 104, no. 497, AMS, Providence, RI, 1993.

[18] M. A. Shubin, *Spectral theory of elliptic operators on non-compact manifolds*, Astérisque **207** (1992), 35–108.

[19] M. A. Shubin, *L^2 Riemann-Roch theorem for elliptic operators*, Geometric Analysis and Functional Analysis **5** (1995), 482–527.

[20] K. Stein, *Überlagungen holomorph-vollstandiger komplexe Räume*, Arch. Math. **7** (1956), 354–361.

[21] M. Takesaki, *Theory of operator algebras, I*, Springer-Verlag, 1979.

IHES, BURES-SUR-YVETTE, FRANCE
E-mail address: gromov@ihes.fr

UNIVERSITÉ PARIS-VI, FRANCE

NORTHEASTERN UNIVERSITY, BOSTON, USA
E-mail address: shubin@neu.edu

Contemporary Mathematics
Volume **212**, 1998

On Some Operators in Clifford Analysis

Klaus Gürlebeck

ABSTRACT. The aim of this paper is to consider properties of some operators arising in Clifford analysis. We study the properties of integral operators related to the generalized Cauchy-Riemann operator and the Laplacian, respectively. Particularly, projections defined by the Cauchy integral operator F and the Bergman operator B will be considered. Preparing further investigations on generalized analytic functions and hypercomplex Beltrami equations we study a generalized Π-operator. As a main result we show connections between orthogonal decompositions of $\mathcal{L}_2$ and the invertibility of Π. At the end we investigate some relations between the fundamental solution of the generalized Cauchy-Riemann operator, the fundamental solution of the Laplacian, classical Green's function of the Laplacian and the Bergman kernel.

1. Introduction

Methods of Clifford analysis became a powerful tool for the treatment of boundary value problems in higher dimensions (see [1], [6], [8] and further references therein). Many of the obtained representation formulas are restricted to problems connected with the Dirac operator or related to certain generalized Cauchy-Riemann operators [8]. In this paper we will study properties of integral operators related to the usual generalized Cauchy-Riemann operator [1]. Furthermore, it is our aim to study connections between regular functions with respect to the Cauchy-Riemann operator or its conjugate operator.

Introducing a single layer potential and a Newtonian potential we study a link to the classical potential theory of the Laplacian. We prove an orthogonal decomposition of $\mathcal{L}_2$ generated by the Laplacian. These studies make it possible to compare the unbounded projection $(I - F)$ and the orthoprojection $(I - B)$, where F stands for the Cauchy integral operator and B denotes the Bergman projection. While the relation $\ker(I - F) = \ker(I - B)$ is trivial the question $\operatorname{im}(I - F) \cap \operatorname{im}(I - B) = \emptyset$ is stated in [11] as an open problem.

Having in mind Vekua's theory of generalized analytic functions and the solution of hypercomplex Beltrami equations we look for a spatial generalization of the

---

1991 *Mathematics Subject Classification.* Primary 30G35; Secondary 45P05, 47G10.
Key words and phrases. Clifford analysis, boundary value problems, projection operators, kernel functions.

complex Π-operator. Starting from a definition of a generalized Π-operator proposed by Sprößig ([14]) we prove a representation formula, continuity in Sobolev spaces, and a norm estimate. The main result in this part is the conservation of orthogonal decompositions by Π_Ω. This will be the basis to prove a result on the invertibility of Π_Ω in the case of bounded domains Ω.

At the end we investigate some relations between the fundamental solution of the generalized Cauchy-Riemann operator, the fundamental solution of the Laplacian, classical Green's function of the Laplacian and the Bergman kernel.

2. Preliminaries

Let $\{e_1, e_2, \ldots, e_n\}$ be an orthonormal basis in $\mathbb{R}^n$. Consider the 2^n-dimensional real Clifford algebra $C\ell_{0,n}$ generated by $e_1, \ldots, e_n$ according to the multiplication rules $e_i e_j + e_j e_i = -2\delta_{ij} e_0$, where e_0 is the identity of $C\ell_{0,n}$. The elements $e_A : A \subseteq \{1, \ldots, n\}$ define a basis of $C\ell_{0,n}$, where $e_A = e_{h_1} \cdots e_{h_k} = e_{h_1 \cdots h_k}$, $1 \leq h_1 < \cdots < h_k \leq n$, and $e_\emptyset = e_0$. Recall that for the product of general elements in $C\ell_{0,n}$ we have the following multiplication rule

$$e_A e_B = (-1)^{p(A \cap B)} (-1)^{p(A,B)} e_{A \triangle B},$$

where $p(A) = \mathrm{card}(A)$, $p(A,B) = \sum_{j \in B} p(A,j)$ and $p(A,j) = \mathrm{card}\,\{i \in A : i > j\}$. The sets A, B and $A \triangle B$ are ordered in the above mentioned way. Let $u = \sum_A u_A e_A$ be an arbitrary element from $C\ell_{0,n}$. We will use the involution, called *conjugation*, defined by

$$\overline{e}_A = (-1)^{p(A)(p(A)-1)/2} e_A.$$

We suppose $\Omega \subset \mathbb{R}^{n+1}$ to be a domain with a smooth boundary Γ and $x = x_0 e_0 + \underline{x}$ with $\underline{x} = \sum_{i=1}^n x_i e_i$. For each $x \in \mathbb{R} + \mathbb{R}^n$ we have $x\overline{x} = x_0^2 + x_1^2 + \cdots + x_n^2 = |x|^2$. Then functions f defined in Ω with values in $C\ell_{0,n}$ are considered. These functions may be written as

$$f(x) = \sum_{k=0}^n e_k f_k(x), \qquad x \in \Omega.$$

Properties such as continuity, differentiability, integrability, and so on, which are ascribed to f have to be possessed by all components $f_k(x), k = 0, \ldots, n$. In this way the usual Banach spaces of these functions are denoted by C^α, $\mathcal{L}_p$ and $\mathcal{W}_p^k$. In the case of $p = 2$ we introduce in $\mathcal{L}_2(\Omega)$ the $C\ell_{0,n}$-valued inner product

$$(1) \qquad\qquad (u,v) = \int_\Omega \overline{u}(\xi) v(\xi)\, d\Omega_\xi.$$

We now define the generalized Cauchy-Riemann operator by

$$D = \sum_{k=0}^n e_k \frac{\partial}{\partial x_k}.$$

For this operator we have that

$$(2) \qquad\qquad D\overline{D} = \overline{D}D = \Delta,$$

where Δ is the Laplacian and $\overline{D} = \sum_{k=0}^n \overline{e}_k (\partial/\partial x_k)$ is the conjugate Cauchy-Riemann operator. We consider in Ω the equation

$$(Du)(x) = 0,$$

and look for its solutions which are called *left-monogenic* functions in Ω.

Now we define the Cauchy kernel in $\mathbb{R}^{n+1}$ by

$$e(x) = |S_n|^{-1}\overline{x}/|x|^{n+1}, \quad x \neq 0,$$

where $|S_n|$ denotes the surface area of the unit sphere in $\mathbb{R}^{n+1}$. It is well known that $e(x)$ is monogenic. Using the function $e(x)$ we introduce the following integral operators:

$$(T_\Omega u)(x) := \int_\Omega e(x-y)u(y)dy, \quad x \in \mathbb{R}^{n+1} \qquad \text{(Teodorescu transform)}$$

$$(\overline{T}_\Omega u)(x) := \int_\Omega \overline{e(x-y)}u(y)dy, \quad x \in \mathbb{R}^{n+1}$$

$$(F_\Gamma u)(x) := -\int_\Gamma e(x-y)n(y)u(y)d\Gamma_y, \quad x \notin \Gamma \quad \text{(Cauchy type operator)}$$

$$(\overline{F}_\Gamma u)(x) := -\int_\Gamma \overline{e(x-y)}\ \overline{n(y)}u(y)d\Gamma_y, \quad x \notin \Gamma$$

$$(S_\Gamma u)(x) := -\int_\Gamma 2e(x-y)n(y)u(y)d\Gamma_y, \ x \in \Gamma \quad \text{(singular integral operator)}$$

$$(\overline{S}_\Gamma u)(x) := -\int_\Gamma 2\overline{e(x-y)}\ \overline{n(y)}u(y)d\Gamma_y, \ x \in \Gamma$$

where $n(y) = \sum_{i=0}^n \mathbf{e}_i n_i(y)$ is the outward pointing normal (unit) vector to Γ at the point y. The integrals which define the operators S_Γ and $\overline{S}_\Gamma$ are to be taken in the sense of Cauchy's principal value. The operator $P_\Gamma := (1/2)(I + S_\Gamma)$ denotes the projection onto the space of all $C\ell_{0,n}$-valued functions which may be left monogenic extended into the domain Ω. $Q_\Gamma := (1/2)(I - S_\Gamma)$ denotes the projection onto the space of all $C\ell_{0,n}$-valued functions which may be left monogenic extended into the domain $\mathbb{R}^{n+1} \setminus \overline{\Omega}$ and vanish at infinity. The operators $\overline{P}_\Gamma$ and $\overline{Q}_\Gamma$ have analogous properties with respect to the operator $\overline{D}$.

Immediately from [6] we get the following statements.

LEMMA 1. *Let* $u \in C^1(\Omega, C\ell_{0,n}) \cap C(\overline{\Omega}, C\ell_{0,n})$. *Then we have*
(i)

$$(F_\Gamma u)(x) + (T_\Omega D)u(x) = \begin{cases} u(x), & x \in \Omega \\ 0, & x \in \mathbb{R}^{n+1} \setminus \overline{\Omega} \end{cases} \qquad \text{(Borel-Pompeiu's formula)}$$

(ii)

$$(DT_\Omega u)(x) = \begin{cases} u(x), & x \in \Omega \\ 0, & x \in \mathbb{R}^{n+1} \setminus \overline{\Omega} \end{cases}$$

(iii)

$$(DF_\Gamma)u(x) = 0 \qquad in \ \Omega \cup (\mathbb{R}^{n+1} \setminus \overline{\Omega}).$$

LEMMA 2. (Plemelj-Sokhotzkij's formulas). *Let* $u \in C^{0,\alpha}(\Omega, C\ell_{0,n})$, $0 < \alpha < 1$. *Then we have*

(i)
$$\lim_{\substack{x \to \xi \in \Gamma \\ x \in \Omega}} (F_\Gamma u)(x) = P_\Gamma u(\xi)$$

(ii)
$$\lim_{\substack{x \to \xi \in \Gamma \\ x \in \mathbb{R}^{n+1} \setminus \overline{\Omega}}} (F_\Gamma u)(x) = -Q_\Gamma u(\xi)$$

for any $\xi \in \Gamma$.

COROLLARY 1. *Let* $u \in C^{0,\alpha}(\Gamma, C\ell_{0,n})$. *Then the equations*
(i) $(S_\Gamma^2 u)(\xi) = u(\xi)$
(ii) $(F_\Gamma P_\Gamma u)(\xi) = (F_\Gamma u)(\xi)$
(iii) $(P_\Gamma^2 u)(\xi) = (P_\Gamma u)(\xi)$
(iv) $(Q_\Gamma^2 u)(\xi) = (Q_\Gamma u)(\xi)$
are valid for any $\xi \in \Gamma$.

We remark that the operators $F_\Gamma, S_\Gamma, P_\Gamma$, and Q_Γ allow an extension to $\mathcal{L}_2(\Gamma, C\ell_{0,n})$. The restriction of a $C\ell_{0,n}$-valued function u to a function defined on the boundary Γ is expressed by $tr\, u$. For simplicity we will omit here the exact definition of the considered operators in the scale of Sobolev spaces. Note that in this case the formulas above are to be understood in the generalized sense.

THEOREM 1. *The Hilbert space* $\mathcal{L}_2(\Omega, C\ell_{0,n})$ *allows the orthogonal decompositions*

$$\mathcal{L}_2(\Omega, C\ell_{0,n}) = [\ker D \cap \mathcal{L}_2(\Omega, C\ell_{0,n})] \oplus \overline{D}[\overset{\circ}{\mathcal{W}}{}_2^1(\Omega, C\ell_{0,n})]$$

$$\mathcal{L}_2(\Omega, C\ell_{0,n}) = [\ker \overline{D} \cap \mathcal{L}_2(\Omega, C\ell_{0,n})] \oplus D[\overset{\circ}{\mathcal{W}}{}_2^1(\Omega, C\ell_{0,n})].$$

For the proof we refer to [5].
We shall abbreviate in the following by im A the range of an operator A.

COROLLARY 2. *There exist four orthoprojections* $\mathbf{P}_D$, $\mathbf{Q}_D$, $\mathbf{P}_{\overline{D}}$, *and* $\mathbf{Q}_{\overline{D}}$

$$\mathbf{P}_D : \mathcal{L}_2(\Omega, C\ell_{0,n}) \to \ker D \cap \mathcal{L}_2(\Omega, C\ell_{0,n}),$$

$$\mathbf{Q}_D : \mathcal{L}_2(\Omega, C\ell_{0,n}) \to \overline{D}[\overset{\circ}{\mathcal{W}}{}_2^1(\Omega, C\ell_{0,n})], \quad \mathbf{Q}_D = I - \mathbf{P}_D.$$

$$\mathbf{P}_{\overline{D}} : \mathcal{L}_2(\Omega, C\ell_{0,n}) \to \ker \overline{D} \cap \mathcal{L}_2(\Omega, C\ell_{0,n}),$$

$$\mathbf{Q}_{\overline{D}} : \mathcal{L}_2(\Omega, C\ell_{0,n}) \to D[\overset{\circ}{\mathcal{W}}{}_2^1(\Omega, C\ell_{0,n})], \quad \mathbf{Q}_{\overline{D}} = I - \mathbf{P}_{\overline{D}}.$$

COROLLARY 3. *Let* $f \in \mathcal{L}_2(\Omega, C\ell_{0,n})$, $(\overline{T}_\Omega f)(x) = 0 \; \forall x \in \mathbb{R}^{n+1} \setminus \overline{\Omega} \Longleftrightarrow f \in$ *im* $\mathbf{Q}_D$.

PROOF. ($\Rightarrow$) We use the representation $f = \overline{D} g$ with $g = \overline{T}_\Omega f$. From the assumption it follows that $tr\, g = 0$ and hence, $f = \overline{D} g \in \overline{D}[\overset{\circ}{\mathcal{W}}{}_2^1(\Omega, C\ell_{0,n})]$.

($\Leftarrow$) From $f \in$ im $\mathbf{Q}_D$ we immediately obtain $tr\, Tf = 0$. Then, using the identity theorem for monogenic functions we conclude $(\overline{T}_\Omega f)(x) = 0 \; \forall x \in \mathbb{R}^{n+1} \setminus \overline{\Omega}$.

This Corollary shows that there exist functions with finite support in im T. Note that the condition $(\overline{T}_\Omega f)(x) = 0 \; \forall x \in \mathbb{R}^{n+1} \setminus \overline{\Omega}$ can be replaced by $tr T_\Omega f = 0$.

3. Projection onto harmonic functions

In addition to the above defined operators we introduce a generalized single layer potential by

$$(V_{\overline{\alpha}}f)(x) = \int\limits_{\Gamma} E(x-y)\overline{n(y)}f(y)\,d\Gamma_y$$

and a Newtonian potential in the usual way

$$(Kf)(x) = \int\limits_{\Omega} -E(x-y)f(y)\,d\Omega_y.$$

Here, E stands for the fundamental solution of $\triangle$. Then we can prove the following properties:

(3) $DK = -\overline{T}, \quad \overline{D}K = -T, \quad DV_{\overline{\alpha}} = \overline{F}, \quad \overline{D}V_{\alpha} = F, \quad V_{\overline{\alpha}}f = Tf + K\overline{D}f.$

These equations are valid in $\mathcal{C}^{\alpha}$ and in Sobolev spaces $\mathcal{W}_p^k$. We will omit the detailed proofs here. The equations (3) connect operators from the classical potential theory with operators arising in Clifford analysis. Using integration by parts we can prove

(4) $(Du, v) = \int\limits_{\Gamma} \overline{u}\,\overline{n}\,vd\Gamma - (u, \overline{D}v) \quad \text{and} \quad (\overline{D}u, v) = \int\limits_{\Gamma} \overline{u}\,n\,vd\Gamma - (u, Dv)$

By the help of $D\overline{D} = \overline{D}D = \triangle$ we obtain from (4)

(5) $(\triangle u, v) = (u, \triangle v) - \int\limits_{\Gamma} \overline{u}nDvd\Gamma + \int\limits_{\Gamma} \overline{Du}nvd\Gamma$

The formula (5) shows that the subspaces $\ker \triangle \cap \mathcal{L}_2$ and $\triangle(\mathcal{W}_2^2(\Omega)\cap\ker tr\cap\ker trD)$ are orthogonal subspaces. If we assume that $(u, \phi) = 0 \ \forall \phi \in \ker \triangle(\Omega)$ and $(u, \psi) = 0 \ \forall \psi \in \triangle(\mathcal{W}_2^2(\Omega) \cap \ker tr \cap \ker trD)$ then we have that $(u, \psi) = 0 \ \forall \psi \in \triangle(\overset{\circ}{\mathcal{W}}_2^2 (\Omega))$. Using (5) we immediately obtain $(\triangle u, v) = 0 \ \forall u \in \overset{\circ}{\mathcal{W}}_2^2 (\Omega) \Rightarrow \triangle u = 0 \Rightarrow u = 0$. This consideration proves

THEOREM 2. *The space* $\mathcal{L}_2(\Omega, Cl_{0,n})$ *allows the orthogonal decomposition*

(6)
$$\mathcal{L}_2(\Omega, Cl_{0,n}) = \ker \triangle \cap \mathcal{L}_2(\Omega, Cl_{0,n}) \oplus \triangle(\overset{\circ}{\mathcal{W}}_2^2 (\Omega, Cl_{0,n}))$$
$$= \ker \triangle \cap \mathcal{L}_2(\Omega, Cl_{0,n}) \oplus \triangle(\mathcal{W}_2^2(\Omega, Cl_{0,n}) \cap \ker tr \cap \ker trD).$$

We denote the corresponding orthoprojections by

$$\mathbf{P}_{\triangle} : \mathcal{L}_2(\Omega, Cl_{0,n}) \to \ker \triangle(\Omega, Cl_{0,n}) \cap \mathcal{L}_2(\Omega, Cl_{0,n})$$
$$\mathbf{Q}_{\triangle} : \mathcal{L}_2(\Omega, Cl_{0,n}) \to \triangle(\mathcal{W}_2^2(\Omega, Cl_{0,n}) \cap \ker tr \cap \ker trD).$$

4. The projections $I - F$ and $I - B$

Now, we are able to study the projections $I - F$ and $I - B$. Porter, Shapiro, and Vasilevski [11] stated the problem to describe the intersection $\mathrm{im}(I-F)\cap\mathrm{im}(I-B)$. It is not clear whether this intersection is empty or not. We will describe in this section a subspace of $\mathrm{im}(I - F) \cap \mathrm{im}(I - B)$.

Let us begin with $V_{\overline{\alpha}}f = Tf + K\overline{D}f$. Using $f = Dg$, Borel-Pompeiu's formula, and (3) we obtain $V_{\overline{\alpha}}Dg + \overline{D}V_{\alpha}g = g + K \triangle g$. If we assume that $g \in \overset{\circ}{\mathcal{W}}{}_2^2\,(\Omega, C\ell_{0,n})$ then it follows that $g = -K \triangle g$. Hence, we have $\triangle g \in \mathrm{im}\ \mathbf{Q}_\triangle$. If we substitute $\triangle g = u$ we get $-Ku = g$. Now, we solve the boundary value problem

$$\triangle g = u \quad \text{in } \Omega$$
$$g = 0 \quad \text{on } \Gamma.$$

By the help of [6] we obtain

$$g = \overline{T}\mathbf{Q}_D Tu = -Ku$$

Applying $\overline{D}$ we get $\mathbf{Q}_D Tu = Tu \in \mathrm{im}\ \mathbf{Q}_D$. Now, we have that on the one hand

$$(I - F)Tu = Tu \qquad \text{because} \quad trTu \in \mathrm{im}\ Q_\Gamma \quad \forall u \in \mathcal{L}_2(\Omega, C\ell_{0,n})$$

and, on the other hand

$$(I - \mathbf{P}_D)Tu = (I - B)Tu = Tu \qquad \text{because} \quad Tu \in \mathrm{im}\ \mathbf{Q}_D$$

for the above considered u. Therefore, we have the following theorem.

THEOREM 3. $Tu \in \mathrm{im}(I - F) \cap \mathrm{im}(I - \mathbf{P}_D) \quad \forall u \in \mathrm{im}\mathbf{Q}_\triangle$

This theorem enables us to describe functions u with the property that $tr\overline{T}Tu = 0$. As a consequence we proved sufficient conditions which ensure that $\overline{T}Tu$ will be a function with finite support. The set of functions Tu defined in Theorem 3 forms a subspace of $\mathrm{im}\ (I - F) \cap \mathrm{im}\ (I - \mathbf{P}_D) \cap \mathcal{L}_2(\Omega, C\ell_{0,n})$. In this we have got a partial answer to the problem mentioned above.

5. Generalization of the complex Π-operator

Starting with his generalization of the complex T-operator in [14] Sprößig proposed the following definition of a hypercomplex analogy of the complex Π-operator.

DEFINITION 1. The operator Π, defined by

$$\Pi f = \overline{D}Tf$$

is called *generalized Π-operator*.

This operator acts from $\mathcal{C}^\alpha(\Omega, C\ell_{0,n})$ to $\mathcal{C}^\alpha(\Omega, C\ell_{0,n})$, $0 < \alpha \leq 1$ [14]. Applying the definition of the T- and the $\overline{D}$-operator we get an integral representation formula for the Π-operator.

THEOREM 4. *Assume that* $f \in \mathcal{C}^\alpha(\Omega, C\ell_{0,n})$, $0 < \alpha \leq 1$. *Then,*

$$(7) \qquad (\Pi f)(x) = -\frac{1}{|S_n|} \int_\Omega \frac{(n-1) + (n+1)\dfrac{\overline{(y-x)}^2}{|y-x|^2}}{|y-x|^{n+1}} f(y)\, d\Omega_y + \frac{1-n}{1+n} f(x)$$

holds.

PROOF. Using [10] Chapter IX §7 we get for $f \in C^\alpha(\Omega, C\ell_{0,n})$, $0 < \alpha \leq 1$, $k = 0, \ldots, n$, the equation

$$\frac{\partial}{\partial x_k} \int_\Omega \frac{\overline{(y-x)}}{|y-x|^{n+1}} f(y) \, d\Omega_y$$

$$= \int_\Omega \frac{-\bar{\mathbf{e}}_k + (n+1)(y_k - x_k)\dfrac{\overline{(y-x)}}{|y-x|^2}}{|y-x|^{n+1}} f(y) \, d\Omega_y - |S_n| \frac{\bar{\mathbf{e}}_k}{n+1} f(x)$$

because

$$\frac{\partial}{\partial x_k} \left[\frac{\overline{(y-x)}}{|y-x|^{n+1}} \right] = \frac{-\bar{\mathbf{e}}_k + (n+1)(y_k - x_k)\dfrac{\overline{(y-x)}}{|y-x|^2}}{|y-x|^{n+1}}$$

and

$$\int_S \frac{\overline{(y-x)}}{|y-x|} \cos(r, x_k) \, dS = |S_n| \frac{\bar{\mathbf{e}}_k}{n+1}.$$

From this we obtain our representation formula by summation over k.

REMARK 1. Obviously we get from this representation formula that Π is a strongly singular operator of Calderon-Zygmund type.

REMARK 2. In the case of $n = 1$ this hypercomplex Π-operator coincides with the usual complex Π-operator

$$\Pi_\Omega h(z) = \frac{\partial}{\partial z} T_\Omega h(z) = -\frac{1}{\pi} \int\!\!\int_\Omega \frac{h(\zeta)}{(\zeta - z)^2} \, d\xi \, d\eta$$

up to the factor two. Note that in [9] Malonek and Müller defined another generalization of the complex Π-operator. We will skip these results here because this Π-operator acts between other function spaces. Also in the papers [11] and [12] operators are considered which may be called generalized Π-operators. But, these operators are not studied as generalizations of the complex Π-operator and therefore, other properties as in the following are investigated.

THEOREM 5. *Under the same assumptions as in Theorem 4 we have for the conjugate operator* $\overline{\Pi}$

$$\overline{\Pi}f = D\overline{T}f.$$

PROOF. Using our representation formula of the Π-operator and the definition of the conjugate operator we get

$$(\overline{\Pi}f)(x) = -\frac{1}{|S_n|} \int_\Omega \frac{(n-1) + (n+1)\dfrac{(y-x)^2}{|y-x|^2}}{|y-x|^{n+1}} f(y) \, d\Omega_y + \frac{1-n}{1+n} f(x),$$

with $f \in C^\alpha(\Omega, C\ell_{0,n}), 0 < \alpha \le 1$. Calculating $D\overline{T}f$ we have

$$
\begin{aligned}
D\overline{T}f &= \sum_{k=0}^{n} \mathbf{e}_k \left(\frac{-1}{|S_n|}\right) \int_\Omega \frac{-\mathbf{e}_k + (n+1)(y_k - x_k)\dfrac{(y-x)}{|y-x|^2}}{|y-x|^{n+1}} f(y)\, d\Omega_y \\
&\quad + \sum_{k=0}^{n} \frac{\mathbf{e}_k^2}{n+1} f(x) \\
&= -\frac{1}{|S_n|} \int_\Omega \frac{(n-1) + (n+1)\dfrac{(y-x)^2}{|y-x|^2}}{|y-x|^{n+1}} f(y)\, d\Omega_y + \frac{1-n}{1+n} f(x).
\end{aligned}
$$

From this we conclude $\overline{\Pi}f = D\overline{T}f$.

REMARK 3. In general the equation

$$
\overline{\Pi}f = \overline{T}Df
$$

is not true.

If we look for applications of the generalized Π-operator then we need its mapping properties within Sobolev spaces. Because we have that the Π-operator is an operator of Calderon-Zygmund type we can apply the theory of Calderon and Zygmund. This means that first we have to study the symbol of the Π-operator. For the sake of brevity we refer to the textbook [10] and the papers [5] and [7] for more information and the details. In our case we obtain

THEOREM 6. *Suppose that* $1 < p < \infty$, $k \in \mathbb{N} \cup \{0\}$. *Then, we have*

$$
\Pi : \mathcal{W}_p^k(\Omega, C\ell_{0,n}) \mapsto \mathcal{W}_p^k(\Omega, C\ell_{0,n}).
$$

By the help of the theory of Calderon and Zygmund one can also estimate the norm of the generalized Π-operator in the following form

$$
\|\Pi\|_{[\mathcal{L}_2(\Omega, C\ell_{0,n}), \mathcal{L}_2(\Omega, C\ell_{0,n})]} \le \frac{n-1}{1+n} + \frac{2n\sqrt{c_4}}{|S_n|^{\frac{1}{4}}},
$$

where c_4 is a constant which is exactly computed in [5]. For the proof of this estimate and further details see [7]. Notice that Theorem 6 remains true in the case of $\Omega = \mathbb{R}^{n+1}$.

THEOREM 7. *Suppose that* $f \in \mathcal{W}_p^1(\Omega, C\ell_{0,n})$, $1 < p < \infty$, *then we have*

(8) $$D\Pi f = \overline{D}f$$
(9) $$\Pi D f = \overline{D}f - \overline{D}\, F_\Gamma f$$
(10) $$F_\Gamma \Pi f = (\overline{D}T - T\overline{D})f$$
(11) $$(D\Pi - \Pi D)f = \overline{D}F_\Gamma f$$
(12) $$F_\Gamma \Pi \overline{F}_\Gamma f = \Pi \overline{F}_\Gamma f$$

PROOF.

of (8): $D\Pi f = D\overline{D}Tf = \overline{D}DTf = \overline{D}f$.

of (9): $\Pi Df = \overline{D}TDf = \overline{D}(I - F_\Gamma)f = \overline{D}f - \overline{D}F_\Gamma f$

of (10): $F_\Gamma \Pi f = (I - TD)\overline{D}Tf = \overline{D}Tf - TD\overline{D}Tf = (\overline{D}T - T\overline{D})f$

of (11): $(D\Pi - \Pi D)f = (\overline{D} - \overline{D} + \overline{D}F_\Gamma)f = \overline{D}F_\Gamma f$

of (12): $F_\Gamma \Pi \overline{F}_\Gamma f = (\overline{D}T - T\overline{D})(I - \overline{T}\,\overline{D})f = \overline{D}T(I - \overline{T}\,\overline{D})f = \Pi \overline{F}_\Gamma f$

From equation (8) we obtain the following important mapping property of the generalized Π-operator:

PROPOSITION 1. *We have*

$$\Pi : \ker \overline{D}(\Omega, C\ell_{0,n}) \cap \mathcal{L}_2(\Omega, C\ell_{0,n}) \mapsto \ker D(\Omega, C\ell_{0,n}) \cap \mathcal{L}_2(\Omega, C\ell_{0,n}).$$

In terms of the orthoprojections we obtained

$$\Pi : \operatorname{im} \mathbf{P}_{\overline{D}} \mapsto \operatorname{im} \mathbf{P}_D.$$

From equation (9) and $F_\Gamma u = 0$ for any function $u \in \overset{\circ}{\mathcal{W}}{}_2^1 (\Omega, C\ell_{0,n})$ we get:

PROPOSITION 2. *We have that*

$$\Pi : D(\overset{\circ}{\mathcal{W}}{}_2^1 (\Omega, C\ell_{0,n})) \mapsto \overline{D}(\overset{\circ}{\mathcal{W}}{}_2^1 (\Omega, C\ell_{0,n})).$$

Proposition 2 means that

$$\Pi : \operatorname{im} \mathbf{Q}_{\overline{D}} \mapsto \operatorname{im} \mathbf{Q}_D.$$

Therefore, the Bergman operator preserves the orthogonal decompositions of $\mathcal{L}_2(\Omega, C\ell_{0,n})$ in a certain sense. More exactly, decompositions generated by D are transformed into those generated by $\overline{D}$. In [11] the authors study the $\bar{\partial}$-problem for quaternionic-valued functions. They prove the existence and a representation formula of the solution using the subspaces $\operatorname{im} \mathbf{P}_D$ and $\ker \mathbf{P}_D$, respectively. In [6] the orthoprojection $\mathbf{P}_D$ is used to solve second order boundary value problems of Dirichlet's type. Therefore, from the present point of view it seems to be advantageous to preserve the mentioned invariance properties also in the class of problems connected with applications of Π.

Using the same ideas we can obtain similar results for $\overline{\Pi}$:

REMARK 4. Again investigating the above mapping properties for the conjugate operator $\overline{\Pi}$ we get

$$\overline{\Pi} : \operatorname{im} \mathbf{P}_D \mapsto \operatorname{im} \mathbf{P}_{\overline{D}},$$

$$\overline{\Pi} : \operatorname{im} \mathbf{Q}_D \mapsto \operatorname{im} \mathbf{Q}_{\overline{D}}.$$

The complex Π-operator is an unitary operator for special domains, see e.g. [16]. Investigating the hypercomplex Π-operator we get the following connection between the Π-operator and its conjugate operator $\overline{\Pi}$:

THEOREM 8. *Suppose* $f \in \mathcal{W}_p^k(\Omega, C\ell_{0,n})$, $1 < p < \infty, k \in \mathbb{N}$, *then we have*

$$\overline{\Pi}\Pi f = D\overline{T}\ \overline{D}Tf = D(I - \overline{F}_\Gamma)Tf = f - D\overline{F}_\Gamma Tf$$

$$\Pi\overline{\Pi} f = \overline{D}TD\overline{T}f = \overline{D}(I - F_\Gamma)\overline{T}f = f - \overline{D}F_\Gamma \overline{T}f$$

The first part of this theorem was proved by Sprößig ([14]) in the case of Hölder-continuous functions. This result was used in [14] to investigate $\overline{\Pi}\Pi$ in $C_0^\infty(\mathbb{R}^{n+1})$ by asymptotic estimates of the remainder term $D\overline{F}_\Gamma Tf$. These results coincide with the corresponding properties of the complex Π-operator. The case of bounded domains is also very important. Using the connections between Π and the orthoprojections studied above we get from Theorem 8 the following properties of Π and $\overline{\Pi}$, respectively.

COROLLARY 4. *We have*

(13) $\overline{\Pi}\Pi f = f \quad \forall f \in \text{im } \mathbf{Q}_{\overline{D}}$

$\Pi\overline{\Pi} f = f \quad \forall f \in \text{im } \mathbf{Q}_D.$

PROOF. From $f \in \text{im } \mathbf{Q}_{\overline{D}}$ we get $tr\ Tf = 0$ and $f \in \text{im } \mathbf{Q}_D$ implies $tr\ \overline{T}f = 0$.

If we try to combine these results we have to look for the intersection of $\text{im } \mathbf{Q}_D$ and $\text{im } \overline{\mathbf{Q}}_D$. Using the results of the previous section we can prove that

$$\text{im } \mathbf{Q}_\triangle \subset \text{im } \mathbf{Q}_D \cap \text{im } \overline{\mathbf{Q}}_D.$$

From Corollary 4 we now deduce the following theorem on the invertibility of Π.

THEOREM 9. $\overline{\Pi}\Pi f = \Pi\overline{\Pi} f \quad \forall f \in \text{im } \mathbf{Q}_\triangle$.

Now let Ω be the whole $\mathbb{R}^{n+1}$. Then the following theorem was proved in [14]:

THEOREM 10. *Let* $f \in C_0^\infty(\mathbb{R}^{n+1})$, *then we have*

$$\overline{\Pi}\Pi u = \Pi\overline{\Pi} u = u.$$

Using this theorem and (7) we get a result for the invertibility of Π defined on a class of functions containing essentially more than in the above theorem.

PROPOSITION 3. *Suppose* $u \in \mathcal{L}_2(\mathbb{R}^{n+1})$ *then we have*

$$\overline{\Pi}\Pi u = \Pi\overline{\Pi} u = u.$$

PROOF. The proof follows from the fact that $C_0^\infty(\mathbb{R}^{n+1})$ is dense in $\mathcal{L}_2(\mathbb{R}^{n+1})$ and from the boundedness of Π as an operator from $\mathcal{L}_2(\mathbb{R}^{n+1})$ to $\mathcal{L}_2(\mathbb{R}^{n+1})$.

Now, let us discuss the case of Sobolev spaces $\overset{\circ}{\mathcal{W}}_2^k (\Omega, C\ell_{0,n})$, $k \in \mathbb{N}$. Moreover, we denote by A^* the to a given operator A $\mathcal{L}_2$-adjoint operator with respect to our scalar product (1) introduced above.

THEOREM 11. *Assume* $u, w \in \overset{\circ}{\mathcal{W}}_2^k (\Omega, C\ell_{0,n})$, $k \in \mathbb{N}$, *then*

$$(\Pi u, v) = (u, \overline{T}Dv)$$

holds.

For the proof we make use of $T^* = -\overline{T}$ and $D^* = -\overline{D}$. Taking this theorem we can put a representation of the $\mathcal{L}_2$-adjoint operator Π^* in a special case.

PROPOSITION 4. *Let $f \in \overset{\circ}{\mathcal{W}_2^k} (\Omega, C\ell_{0,n})$, $k \in \mathbb{N}$, then it holds*

$$\Pi^* f = \overline{T} D f.$$

Applying this result we obtain that in this special case Π^* is a left inverse of the generalized Π-operator.

PROPOSITION 5. *We have*

$$\Pi^* \Pi u = u \quad \forall u \in \overset{\circ}{\mathcal{W}_2^k} (\Omega, C\ell_{0,n}), k \in \mathbb{N}.$$

PROOF. Let $u, v \in \overset{\circ}{\mathcal{W}_2^k} (\Omega, C\ell_{0,n})$, $k \in \mathbb{N}$, v arbitrary. Then, $(\Pi^* \Pi u, v) = (\overline{T}\,\overline{D}u, v) = (u, DTv) = (u, v) \ \forall v \in \overset{\circ}{\mathcal{W}_2^k} (\Omega, C\ell_{0,n}) \Longrightarrow \Pi^* \Pi u = u$.

Considering the whole space $\mathbb{R}^{n+1}$ and using the fact that the spaces $\overset{\circ}{\mathcal{W}_2^k}$ ($\mathbb{R}^{n+1}$), $k \in \mathbb{N}$, are dense in $\mathcal{L}_2(\mathbb{R}^{n+1})$ we can derive from Proposition 3 an expression of the $\mathcal{L}_2$-adjoint operator in the case of $\mathcal{L}_2(\mathbb{R}^{n+1})$.

PROPOSITION 6. *Suppose $u \in \mathcal{L}_2(\mathbb{R}^{n+1})$ then we have*

$$\Pi^* u = \overline{\Pi} u.$$

We see from the Propositions 3 and 6 that $\Pi : \mathcal{L}_2(\mathbb{R}^{n+1}) \longrightarrow \mathcal{L}_2(\mathbb{R}^{n+1})$ is an unitary operator. Therefore, we get from this a proposition for applications of the hypercomplex Π-operator important property.

PROPOSITION 7. *Let $u, v \in \mathcal{L}_2(\mathbb{R}^{n+1})$; then it holds*

$$(\Pi u, \Pi v) = (u, v),$$

$$\|\Pi u\|_{\mathcal{L}_2(\mathbb{R}^{n+1})} = \|u\|_{\mathcal{L}_2(\mathbb{R}^{n+1})}.$$

6. Green's function, Bergman kernel, and fundamental solutions

In [12], [13] the authors have considered some connections between Green's function and the Bergman kernel. We will study these connections again. Furthermore, we look for representations of the Green's function by the help of the Bergman projection and the fundamental solution of the Cauchy Riemann operator. The Green function of $\{\triangle, tr\}$ is defined by

$$G(x, y) = E(x, y) + h(x, y),$$

where $(x, y) \in \Omega \times \Omega \setminus \{(x, x) \ x \in \Omega\}$, E is the fundamental solution of $\triangle$ and h is the solution of the boundary value problem

$$\triangle_y h(x, y) = 0 \qquad\qquad \text{in } \Omega$$
$$tr\, h(x, y) = -tr\, E(x, y) \qquad \text{on } \Gamma$$

Then, using ideas from [6], we have

$$h(x, y) = -\overline{F}_\Gamma tr E_x - \overline{T} F_\Gamma (tr \overline{T} F_\Gamma)^{-1} \overline{Q_\Gamma} E_x$$

and

$$G(x,y) = E_x(y) + h_x(y) = E - \overline{F}_\Gamma E - \overline{T}F_\Gamma (tr\overline{T}F)^{-1} tr(E - \overline{F}_\Gamma E)$$
$$= \overline{T}\,\overline{D}E - \overline{T}P_D\overline{D}E = (\overline{T}\mathbf{Q}_D e_x)(y).$$

Now, we have a description of the Green function G by the help of $\mathbf{Q}_D$ and $\overline{T}$. Note that this representation of G is advantageous in cases where norm estimates of G are required. Obviously, $\|\mathbf{Q}_D\|_{[\mathcal{L}_2(\Omega,C\ell_{0,n}),\mathcal{L}_2(\Omega,C\ell_{0,n})]} = 1$ and the norm of $\overline{T}$: $\mathcal{L}_2 \cap \operatorname{im}\mathbf{Q}_D \mapsto \overset{\circ}{\mathcal{W}}{}^1_2 (\Omega, C\ell_{0,n})$ is exactly known using the first eigenvalue of $\{\triangle, tr\}$ (see e.g. [6]).

In the following we study a similar boundary value problem.

$$\triangle_y h_1(x,y) = 0 \qquad \text{in } \Omega$$
$$tr\, h_1(x,y) = -tr\overline{e_x}(y) \quad \text{on } \Gamma$$

Using again representation formulas from [6] we get

$$h_1(x,y) = -\overline{F}\,\overline{e_x}(y) - \overline{T}\,F(tr\overline{T}\,F)^{-1}\overline{Q_\Gamma e_x}(y) = \overline{T}\,F(tr\overline{T}\,F)^{-1}\overline{Q_\Gamma e_x}(y).$$

Applying the conjugate Cauchy-Riemann operator we obtain

$$b(x,y) := \overline{D}_y h_1(x,y) = -F(tr\overline{T}\,F)^{-1}\overline{Q_\Gamma e_x}(y)$$

It is easy to see that $b(x,y) \in \ker D_y \cap \mathcal{L}_2(\Omega, C\ell_{0,n})$. Let $\phi \in \ker D \cap \mathcal{W}^1_2(\Omega, C\ell_{0,n})$. Then we have

$$(\overline{D_y h_1}, \phi) = -(h_1, D\phi) + \int_\Gamma \overline{h_1}(x,y)n(y)\phi(y)\,d\Gamma_y = F\phi = \phi.$$

If we assume that $\phi \in \operatorname{im}\mathbf{Q}_D, \phi = \overline{D}u, u \in \overset{\circ}{\mathcal{W}}{}^1_2 (\Omega, C\ell_{0,n})$ then we immediately obtain

$$(\overline{D}_y h_1, \phi) = (\overline{D}_y h_1, \overline{D}u) = (\triangle h_1, u) = 0.$$

These considerations prove that $b(x,y)$ has the reproducing property in the subspace of monogenic functions. Together with the projection property we obtain that $b(x,y)$ is the Bergman kernel function.

References

[1] F. Brackx, R. Delanghe, and F. Sommen, *Clifford Analysis,* Research Notes in Mathematics **76**, Pitman Advanced Publishing Program, 1982.

[2] J. Cnops, *Hurwitz Pairs and Applications of Möbius Transformations,* Thesis, Universiteit Gent, 1994.

[3] J. Cnops and H. Malonek, *Introduction to Clifford Analysis,* Lecture Notes, 1. European Intensive Course "Complex Analysis and it's generalizations", preprint.

[4] R. Delanghe, F. Sommen, and V. Souček, *Clifford Algebra and Spinor-Valued Functions,* Kluwer, 1992.

[5] K. Gürlebeck and U. Kähler, *On a spatial generalization of the complex Π-operator,* J. Anal. Appl. **15** (1996), 283–298.

[7] K. Gürlebeck and W. Sprößig, *Quaternionic Analysis and Elliptic Boundary Value Problems,* ISNM 89, Birkhäuser Verlag, Basel, 1990.

[8] U. Kähler, *Über einige räumliche Verallgemeinerungen eines komplexen singulären Integraloperators,* Diplomarbeit, TU Chemnitz-Zwickau, 1995.

[9] V. V. Kravchenko and M. V. Shapiro, *Integral representations for spatial models of mathematical physics,* Pitman Research Notes in Math. Series 351, 1996.

[10] H. Malonek and B. Müller, *Definition and properties of a hypercomplex singular integral operator*, Results in Mathematics **22** (1992), 713–724.

[11] S. G. Michlin und S. Prößdorf, *Singuläre Integraloperatoren*, Akademie-Verlag, Berlin, 1980.

[12] R. M. Porter, M. V. Shapiro, and N. L. Vasilevski, *Quaternionic Differential and Integral Operators and the $\bar{\partial}$-problem*, Journal of Natural Geometry **6** (1994), 101–124.

[13] E. Ramirez de Arellano, M. V. Shapiro and N. L. Vasilevski, *The hyperholomorphic Bergman projector and its properties*, Reporte Interno No. 140, Departamento de Mathemáticas, CINVESTAV del I. P. N., Mexico City, 1993.

[14] M. V. Shapiro and N. L. Vasilevski, *On the Bergman kernel function in hyperholomorphic analysis*, Acta Applicandae Mathematica, to appear.

[15] W. Sprößig, *Über eine mehrdimensionale Operatorrechnung über beschränkten Gebieten des $\mathbb{R}^n$*, Thesis, TH Karl-Marx-Stadt (1979).

[16] W. Sprößig, *Räumliches Analogon zum komplexen T-Operator*, Beiträge zur Analysis 12, Verlag der Wiss., Berlin, 1978, pp. 127–138.

[17] I. N. Vekua, *Generalized analytic functions*, Reading, 1962.

TECHNISCHE UNIVERSITÄT CHEMNITZ-ZWICKAU, FAKULTÄT FÜR MATHEMATIK, D-09107 CHEMNITZ, GERMANY

Current address: Bauhaus-Universität Weimar, Fakultät Bauingenieurwesen, Professur Analysis, D-99423 Weimar, Germany

E-mail address: k.guerlebeck@infotmatik.uni-weimar.de

Contemporary Mathematics
Volume **212**, 1998

Toeplitz C*-Algebras Over Non-Convex Cones and Pseudo-Symmetric Spaces

U. Hagenbach and H. Upmeier

0. Introduction

Toeplitz operators and Toeplitz C*-algebras for multi-variable domains of holomorphy have been studied extensively in the literature, notably for tube domains [12], symmetric domains [18] and Reinhardt domains [15], [2]. In [19], a systematic framework for studying Toeplitz C*-algebras for non-smooth domains has been developed based on the notion of *polar decomposition with respect to a group action*. The above mentioned cases arise for euclidean vector groups (tube domains), compact tori (Reinhardt domains) and non-commutative compact Lie groups (symmetric domains). However, due to advances in representation theory (discrete series of semi-simple Lie groups, non-Riemannian symmetric spaces, Gelfand-Gindikin program [7]) it is important to allow more general groups (neither compact nor commutative) in the polar decomposition and also to treat non-pseudoconvex domains for which the relevant "Hardy spaces" consist of $\bar{\partial}$-closed cohomology classes [8].

The purpose of this paper is to provide a general framework for studying Toeplitz C*-algebras over "S-bicircular domains" Ω where S is a Lie group and Ω may be pseudoconcave. The basic idea is to relate the structure theory for the associated Toeplitz C*-algebra $\mathcal{T}(S)$ to the fundamental results of non-commutative C*-algebra duality (coactions and cocrossed products [11]). As an application, we then determine the ideal structure of $\mathcal{T}(S)$ explicitly for an important example of a non-convex tube domain, namely the symmetric 2×2-matrices of negative determinant (this is the exterior of the double light cone in 3 dimensions). In a subsequent paper this analysis will be generalized to cover arbitrary pseudoconcave domains defined for semi-simple Jordan algebras, and also for non-commutative groups such as $SL(2, \mathbf{R})$. In view of this we will formulate many Propositions for a general Lie group S, although in this paper only the case of euclidean vector groups is considered in depth.

1991 *Mathematics Subject Classification.* Primary 47B35, 47C15; Secondary 32M15, 17C20.

1. Tube domains and quasi-symmetric embedding

Let $\underline{S}$ be the Lie algebra and exp the exponential map of a Lie group S. For $s \in S$ let $Int(s)$ be the mapping $x \longmapsto sxs^{-1}$ and put $Ad(s) := T_e\,(Int\,(s))$. A cone $V \subset i\underline{S}$ is called invariant if for all $s \in S$

$$Ad(s)(V) = V.$$

By general Lie theory it follows that

(1.1) $$Int(s)(\exp(V)) = \exp(V)$$

for all $s \in S$. The *tube domain* $S \odot V$ corresponding to S and V is defined by

(1.2) $$S \odot V := S \cdot \exp(V).$$

This is a subset of the complexification of S. In the commutative *vector group* case the Lie group is denoted by

(1.3) $$S = \underline{S} = iX,$$

where X is a euclidean space $\mathbf{R}^n$. Then the invariance condition is fulfilled for every cone $V \subset i\underline{S} = X$, and we have for such cones

(1.4) $$S \odot V = iX \oplus V.$$

EXAMPLE 1.1. For the Jordan algebra

$$X := Sym(n, \mathbf{R}) := \left\{ A \in M(n, \mathbf{R}) \mid A^T = A \right\}$$

of symmetric matrices there exist for $k = 0, \ldots, n$ the open cones

$$V_k := \{A \in Sym(n, \mathbf{R}) \text{ invertible} \mid A \text{ has exactly } k \text{ positive eigenvalues} \}.$$

These cones satisfy

(1.5) $$\bigcup_{k=0,\ldots,n} V_k \underset{dense}{\subset} Sym(n, \mathbf{R}).$$

One sees at once that only V_0 and V_n are convex cones. By [1] and [10] we have the following:

LEMMA 1.1. *The cones in* $Sym(n, \mathbf{R})$ *are (right) homogeneous spaces*

(1.6)
$$\begin{aligned}
V_0 &= SO(0, n) \setminus Gl^+\,(n, \mathbf{R})\,, \\
V_k &= SO(k, n-k) \setminus Gl^+\,(n, \mathbf{R})\,, \quad k = 1, \ldots, n-1, \\
V_n &= SO(n, 0) \setminus Gl^+\,(n, \mathbf{R})
\end{aligned}$$

and for the corresponding tube domains we have

(1.7)
$$\begin{aligned}
iX \oplus V_0 &= U(0, n) \setminus Sp(2n, \mathbf{R}), \\
iX \oplus V_k &\underset{dense}{\subset} U(k, n-k) \setminus Sp(2n, \mathbf{R}), \quad k = 1, \ldots, n-1, \\
iX \oplus V_n &= U(n, 0) \setminus Sp(2n, \mathbf{R}).
\end{aligned}$$

Via suitable involutions the spaces $U(k, n-k) \setminus Sp(2n, \mathbf{R})$ are quasi-symmetric and even symmetric in case $k = 0, n$. Thus we obtain the *quasi-symmetric embedding* of $iX \oplus V_k$ for $k \neq 0, n$ [1].

Later we will treat in detail the cone given by the exterior of the *double light cone*, which arises when $n = 2$ and $k = 1$. So we summarize the results for that special case:

EXAMPLE 1.2. For $n = 2$ we have:

(1.8)
$$i\mathbf{R}^3 \oplus V_{--} = U(0,2) \setminus Sp(4,\mathbf{R}); \qquad V_{--} = SO(0,2) \setminus Gl^+(2,\mathbf{R}),$$
$$i\mathbf{R}^3 \oplus V_{+-} \underset{dense}{\subset} U(1,1) \setminus Sp(4,\mathbf{R}); \qquad V_{+-} = SO(1,1) \setminus Gl^+(2,\mathbf{R}),$$
$$i\mathbf{R}^3 \oplus V_{++} = U(2,0) \setminus Sp(4,\mathbf{R}); \qquad V_{++} = SO(2,0) \setminus Gl^+(2,\mathbf{R}).$$

2. Hardy spaces and Gindikin cohomology

Usually one defines for a locally compact unimodular group S with a Haar measure λ and a convex invariant cone V the *Hardy space* of the tube domain $S \odot V$ as

$$H^2(S \odot V) := \left\{ h : S \odot V \to \mathbf{C} \text{ holomorphic} \mid \|h\|^2 := \sup_{z \in S \odot V} \int_S |h(zs)|^2 ds < \infty \right\}.$$

For a nonconvex cone in $\mathbf{R}^n$, Gindikin modified this concept in [8] to obtain the cohomology space $H_2^{(d)}(S \odot V)$, where d is a concavity index for V. To obtain the Fourier transform of this space, define the *unitary dual* $\hat{S}$ of S as the set of all irreducible continuous representations of S. In the vector group case $S = iX$ we have $\widehat{(iX)} = X^\#$ (linear dual) and the dual cone is to define in the sense of duality introduced by [8]. By Fourier theory [16] we get an isometrical isomorphism

(2.1)
$$\mathcal{F} : L^2(iX) \xrightarrow{\approx} L^2(X^\#),$$

given by

(2.2)
$$(\mathcal{F}h)(\xi) = \hat{h}(\xi) := c \int_X e^{i\xi x} h(ix) dx$$

for all $h \in L^2(iX)$ and $\xi \in X^\#$, where c is a constant. For the closed cone

$$\Lambda := \hat{V}$$

the multiplication operator χ_Λ on $L^2(X^\#)$ maps onto the closed subspace $L^2(\Lambda)$. Since $\|\chi_\Lambda(\phi)\| \leq \|\phi\|$, $(\chi_\Lambda)^* = \chi_\Lambda$ and $\chi_\Lambda \chi_\Lambda = \chi_\Lambda$, χ_Λ is a projection of $L^2(X^\#)$ onto $L^2(\Lambda)$. Because of the isometrical property of $\mathcal{F}$ the space

$$H_\Lambda^2(iX) := \mathcal{F}^{-1}\left(L^2(\Lambda)\right) = \left\{ h \in L^2(iX) \mid \hat{h}(\xi) = 0 \text{ for almost all } \xi \in X^\# \setminus \Lambda \right\}$$

is a closed subspace of $L^2(iX)$. Putting $E = \widehat{\chi_\Lambda}$ the following diagram commutes

(2.3)
$$\begin{array}{ccc} L^2(iX) & \xrightarrow{\mathcal{F}} & L^2(X^\#) \\ {\scriptstyle l_E} \downarrow & & \downarrow {\scriptstyle \chi_\Lambda} \\ H_\Lambda^2(iX) & \xrightarrow{\mathcal{F}} & L^2(\Lambda), \end{array}$$

where l_E is left convolution.

For a convex cone it is well known that taking the limit function h defined by

$$(2.4) \qquad h(s) := \lim_{S \odot V \ni z \to e} h(s \cdot z)$$

gives an isomorphism of $H^2(S \odot \Lambda)$ to $H^2_\Lambda(S)$ [14]. For nonconvex cones $V \subset X = \mathbf{R}^n$ we have by [8, Theorem 1] the following description of the Hardy space

$$(2.5) \qquad H_2^{(d)}(iX \oplus V) \cong H^2_\Lambda(iX)$$

using the Fourier transform (2.1) and a suitable boundary evaluation.

3. C^*-algebras, coactions and cocrossed products

We will now apply the methods of "non-commutative C^*-duality" [11] to the construction of representations of Toeplitz algebras. We will use the symbol f not only for a function in $L^\infty(S)$ but also for the multiplication operator in $\mathcal{L}\left(L^2(S)\right)$ (bounded operators) defined by f.

DEFINITION 3.1. The (spatial) group C^*-algebra of S is defined as

$$C^*(S) := C^*\left(l_f \mid f \in L^1(S)\right) \subset \mathcal{L}\left(L^2(S)\right),$$

where l_f is the left convolution operator defined by $f \in L^1(S)$. For an element $u \in M^1(S)$, the algebra of bounded complex measures on S (the linear dual of $\mathcal{C}_0(S)$ [4, 1.38]), we define the left convolution operator on $L^2(S)$ by

$$l_u(h)(s) := \int_S u(dx)h(x^{-1}s), \quad s \in S.$$

Then the group W^*-algebra of S is

$$W^*(S) := W^*\left(l_u \mid u \in M^1(S)\right) \subset \mathcal{L}\left(L^2(S)\right).$$

As a bounded left convolution operator, l_E commutes with all right translations, denoted by γ_s, so [17, p. 261] yields

$$(3.1) \qquad l_E \in W^*(S).$$

For any Banach algebra A there is the following contractive action of A on its linear dual space $A^\#$:

$$(3.2) \qquad A^\# \otimes A \longrightarrow A^\#, \ (\alpha, a) \longmapsto \alpha \cdot a$$

defined for all $b \in A$ by

$$\langle \alpha \cdot a, b \rangle := \langle \alpha, ab \rangle.$$

An easy consequence is the continuous action of the *Fourier algebra* $A(S)$ on $W^*(S)$ in the above sense, since by [5] $A(S)$ is the predual of $W^*(S)$, i.e., $A(S)^\# = W^*(S)$, via the pairing

$$(3.3) \qquad \langle l_u, f \rangle := \int_S u(ds)f(s).$$

In that case we have for $f \in A(S)$ and $u \in M^1(S)$

$$(3.4) \qquad l_u \cdot f = l_{uf}$$

and obtain by contractivity

$$\|l_u \cdot f\|_{W^*(S)} = \|l_{uf}\|_{W^*(S)} \leq \|f\|_{A(S)} \cdot \|l_u\|.$$

Define the C^*-algebra

(3.5)
$$\begin{aligned} C_\Lambda^*(S) &:= C^* \left(l_E \cdot f \mid f \in A(S) \right) \\ &= C^* \left(l_{fE} \mid f \in A(S) \right) \subset W^*(S). \end{aligned}$$

In case S is not compact, the constant function $\mathbf{1}$ is not contained in $A(S) \subset \mathcal{C}_0(S)$ and $l_E \notin C_\Lambda^*(S)$. Specializing to the commutative case $S = iX$, we have [5, p. 12]

(3.6)
$$A(iX) = \left\{ \hat\phi \mid \phi \in L^1(X^\#) \right\},$$

and

$$\|\hat\phi\|_{A(iX)} = \|\phi\|_{L^1(X^\#)}.$$

Via the calculation

$$\begin{aligned} Ad(\mathcal{F})(C_\Lambda^*(iX)) &= Ad(\mathcal{F})C^* \left(l_{fE} \mid f \in A(iX) \right) \\ &= C^* \left(l_{\hat{f}}(\hat{E}) \mid f \in A(iX) \right) \overset{(14)}{=} C^* \left(l_\phi(\hat{E}) \mid \phi \in L^1(X^\#) \right) \\ &= C^* \left(\phi \star \chi_\Lambda \mid \phi \in L^1(X^\#) \right) \;=:\; \mathcal{C}_0^\Lambda(X^\#) \end{aligned}$$

we can pass to the dual picture, and obtain

LEMMA 3.2.

$$\mathcal{C}_0^\Lambda(X^\#) \subset \mathcal{U}_b(X^\#) := \left\{ \phi : X^\# \to \mathbf{C} \mid \phi \ \textit{uniformly continuous and bounded} \right\}.$$

PROOF. Let $\psi = \phi \star \chi_\Lambda \in \mathcal{C}_0^\Lambda(X^\#)$ and let (ϕ_n) be a sequence in $\mathcal{C}_c(X^\#)$ with $\phi_n \to \phi$ in the L^1-norm. Then Young's inequality gives

(3.7) $$\|\phi_n \star \chi_\Lambda - \phi \star \chi_\Lambda\|_\infty \leq \|\phi_n - \phi\|_1 \cdot \|\chi_\Lambda\|_\infty = \|\phi_n - \phi\|_1 \to 0,$$

which proves that

(3.8) $$\phi_n \star \chi_\Lambda =: \psi_n \to \psi \quad \text{in the } L^\infty\text{-norm}.$$

Now we calculate for $\xi, \eta \in X^\#$:

(3.9) $$(\phi_n \star \chi_\Lambda)(\xi) = \int_{X^\#} \phi_n(\xi - \tau)\chi_\Lambda(\tau)d\tau = \int_\Lambda \phi_n(\xi - \tau)d\tau = \int_{\xi-\Lambda} \phi_n(\tau)d\tau,$$

hence

$$|\psi_n(\xi) - \psi_n(\eta)| = \left| \int_{\xi-\Lambda} \phi_n(\tau)d\tau - \int_{\eta-\Lambda} \phi_n(\tau)d\tau \right| = \left| \int_{(\xi-\Lambda)\triangle(\eta-\Lambda)} \phi(\tau)d\tau \right|$$

$$\leq \|\phi_n\|_\infty \cdot |K_n \cap ((\xi - \Lambda) \triangle (\eta - \Lambda))|,$$

where K_n is the compact support of ϕ_n, $\triangle$ denotes the symmetric difference and $|\cdot|$ denotes the measure. Since

$$K_n \cap ((\xi - \Lambda) \triangle (\eta - \Lambda)) \to \emptyset \quad \text{for } \|\xi - \eta\| \to 0$$

the ψ_n are uniformly continuous. The boundedness is clear by Young's inequality. Since $\mathcal{U}_b(X^\#)$ is closed under uniform convergence we have proved the assertion.

As a consequence there is a continuous action of $X^{\#}$ on $\mathcal{C}_0^{\Lambda}(X^{\#})$ by right translations given by

$$(3.10) \qquad X^{\#} \times \mathcal{C}_0^{\Lambda}(X^{\#}) \longrightarrow \mathcal{C}_0^{\Lambda}(X^{\#}), \quad \xi \propto \psi(\eta) := \psi(\eta + \xi).$$

This action can be described by the embedding

$$(3.11) \qquad \rho : \mathcal{C}_0^{\Lambda}(X^{\#}) \longrightarrow \tilde{M}(\mathcal{C}_0^{\Lambda}(X^{\#}) \otimes \mathcal{C}_0(X^{\#})) = \mathcal{C}_b\left(X^{\#}, \mathcal{C}_0^{\Lambda}(X^{\#})\right)$$

where

$$\rho(\psi)(\xi) = \xi \propto \psi.$$

Here $\otimes$ is the spatial C^*-tensor product and $\tilde{M}$ denotes the (restricted) multiplier algebra [11]. Because

$$Ad(\mathcal{F}^*) : \mathcal{C}_0^{\Lambda}(X^{\#}) \cong C_{\Lambda}^*(iX)$$

there is also an action

$$\hat{\rho} : C_{\Lambda}^*(iX) \longrightarrow \tilde{M}(C_{\Lambda}^*(iX) \otimes \mathcal{C}_0(X^{\#})).$$

By [11, 2.2.(4)] this is equivalent to the existence of a *coaction* of iX on $C_{\Lambda}^*(iX)$, given as the mapping

$$\delta : C_{\Lambda}^*(iX) \longrightarrow \tilde{M}(C_{\Lambda}^*(iX) \otimes C^*(iX))$$

defined by

$$(3.12) \qquad \delta(a) = (1 \otimes \mathcal{F}^*)\hat{\rho}(a)(1 \otimes \mathcal{F}), \; a \in C_{\Lambda}^*(iX).$$

This coaction can also be defined in the non-commutative case as the restriction of the standard W^*-coaction $\delta(l_s) = l_s \otimes l_s$ ($s \in S$), where l_s denotes left translation. Thus we can define the corresponding *cocrossed product*

$$(3.13) \qquad A \otimes_{\delta} \mathcal{C}_0(S) := C^*\left(\delta(a)(1 \otimes f) \mid a \in A, \; f \in \mathcal{C}_0(S)\right),$$

where $A := C_{\Lambda}^*(S)$.

4. Representations of $C_{\Lambda}^*(S) \otimes_{\delta} \mathcal{C}_0(S)$

First recall the following fundamental result [13, Theorem A.1(b)]:

THEOREM 4.1. *There is a bijection between the nondegenerate representations μ of the Fourier algebra $A(S)$ on a Hilbert space H and the unitaries $W \in \mathcal{B}(H) \bar{\otimes} W^*(S)$, fulfilling the condition*

$$(4.1) \qquad (W \otimes 1)(W \otimes 1)^{\sigma} = (1 \otimes W_S)(W \otimes 1)(1 \otimes W_S^*)$$

on $H \otimes L^2(S \times S)$. It is given by

$$(4.2) \qquad \mu(f) = S_f(W) \quad for \; f \in A(S).$$

Here, for an operator W on $L^2(S \times S) = L^2(S) \otimes L^2(S)$ we put $W^{\sigma} := \Sigma W \Sigma^*$, with $(\Sigma \Phi)(s,t) := \Phi(t,s)$, $\Phi \in L^2(S \times S)$, W_S is the unitary operator on $L^2(S \times S)$, given by

$$(4.3) \qquad (W_S \Phi)(s,t) := \Phi(s, s^{-1}t)$$

and the *slice map* S_f is the mapping

$$(4.4) \qquad \left(\sum_{i=1}^{n} a_i \otimes b_i\right) \longmapsto \sum_{i=1}^{n} a_i f(b_i),$$

extended to the multiplier algebra $M(A \otimes B)$ for an arbitrary $f \in B^*$ [11, 1.5]. Note that for $f \in A(S)$ the multiplication operator $M_f = S_f(W_S)$.

DEFINITION 4.2. Let δ be a coaction of a Lie group S on a C^*-algebra $\mathcal{A}$. A pair (π, W) consisting of a nondegenerate representation π of A on a Hilbert space H and a unitary $W \in \mathcal{B}(H) \bar{\otimes} W^*(S)$, corresponding by Theorem 4.1 to a nondegenerate representation μ of $A(S)$ on H, is called a *covariant pair* for the system (A, S, δ), if for all $a \in A$ the condition

$$(4.5) \qquad \overline{(\pi \otimes i)}\,(\delta(a)) = W\,(\pi(a) \otimes 1)\,W^*$$

holds. Here the bar denotes strict extension. In this case one obtains a nondegenerate representation $\Pi(\pi, \mu)$ of the cocrossed product $A \otimes_\delta \mathcal{C}_0(S)$ satisfying

$$(4.6) \qquad \delta(a)(1 \otimes f) \longmapsto \pi(a) \cdot \mu(f),$$

for all $a \in A$, $f \in \mathcal{C}_0(S)$. The mapping

$$(4.7) \qquad (\pi, \mu) \longmapsto \Pi(\pi, \mu)$$

is a bijection between covariant pairs of representations of (A, S, δ) and nondegenerate representations of $A \otimes_\delta \mathcal{C}_0(S)$ [11, Theorem 3.7].

Our aim is to define for every face of $\Lambda = \hat{V}$ a covariant pair of representations of $(C^*_\Lambda(S), S, \delta)$ in the vector group case $S = iX$. In general we will deal with cones $\Lambda = T \cdot \Lambda_0$, where Λ_0 is a *polyhedral* cone (in a lower dimensional space $X_0 \subset X$) and T is a compact group acting by linear transformations on X and hence on $X^\#$.

LEMMA 4.3. *There is a strongly continuous right action of T on $\mathcal{C}_0^\Lambda(X^\#)$ defined by*

$$(4.8) \qquad (\psi \cdot t)(\xi) := \psi(t\xi)$$

for all $t \in T$ and $\xi \in X^\#$.

PROOF. It suffices to show continuity for generators $\psi = \phi \star \chi_\Lambda$, $\phi \in L^1(X^\#)$ at $t = e$. Since $\mathcal{C}_c^\infty(X^\#)$ is a dense subset of $L^1(X^\#)$ we may assume $\phi \in \mathcal{C}_c^\infty(X^\#)$ and $K = B_R(0) \subset X^\#$ contains the support of ϕ. Now

$$(4.9) \qquad |\psi(t\xi) - \psi(\xi)| \leq \|(\phi \star \chi_\Lambda) \cdot t - \phi \star \chi_\Lambda\|_\infty \leq \|\phi \cdot t - \phi\|_1 \|\chi_\Lambda\|_\infty.$$

The mean value theorem implies $\phi \cdot t \to \phi$ uniformly as $t \to e$. Therefore $\|\phi \cdot t - \phi\|_1 \to 0$ and the assertion follows.

Define for the faces F of Λ_0 the "annihilator"

$$(4.10) \qquad F^\perp := \{x \in iX_0 \;;\; \langle x, F \rangle = 0\},$$

let $T_F \subset T$ be the stabilizer group of F in T and

$$(4.11) \qquad S_F := T_F \cdot F^\perp.$$

The faces of Λ are all faces F of Λ_0 rotated by the action of T/T_F and have the form tF, for some $t \in T/T_F$.

In our case where $\Lambda = \widehat{V_{+-}}$, a 3-dimensional cone of convex rank 2 (c.f. Example 1.2), we have $T = \mathbf{T}$ (1-torus) and $\Lambda_0 = (\mathbf{R}^+)^2$, with faces $\Lambda'_+ = \mathbf{R}^+ \times 0$ (1-dimensional face of rank 1), $\Lambda'_- = 0 \times \mathbf{R}^+$ (perpendicular face of rank 1) and $\Lambda'' = (0)$ is the vertex (rank 0) (c.f. figure 4.1). Note that the space of all 1-dimensional faces of Λ is parametrized by $\mathbf{T} \mathbin{\dot\cup} \mathbf{T}$ (disjoint union), in particular this space is not connected. Such a phenomenon does not occur for convex cones.

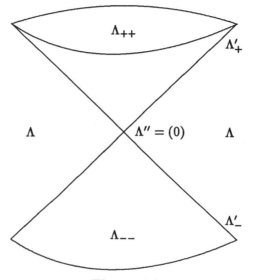

Figure 4.1

We define a nondegenerate representation μ_{tF} of $A(iX)$ on $L^2(tS_F)$ corresponding to a face tF of Λ by the restriction

$$\mu_{tF} : f \longmapsto f\,|_{tS_F} =: \bar{f}.$$

It is easy to show that μ_{tF} is nondegenerate, because $A(tS_F)$ is a dense subset of $\mathcal{C}_0(tS_F)$ by [5, Prop. 3.7].

The construction of the associated representation π_{tF} (4.6) requires more work, and in fact constitutes the main technical part of this paper, carried out in the vector group setting. Because $Ad(\mathcal{F}) : C^*_\Lambda(iX) \to \mathcal{C}_0^\Lambda(X^{\#})$ and $\mathcal{F}\,|_{L^2(tS_F)} : L^2(tS_F) \to L^2((tS_F)^{\#})$ are isomorphisms we need to find a nondegenerate representation $\hat{\pi}_{tF}$ of $\mathcal{C}_0^\Lambda(X^{\#})$ on $L^2((tS_F)^{\#})$.

THEOREM 4.4. *Let* $(\xi_n) \subset tX_0^{\#}$ *be a sequence with*
1. $\xi_n\,|_{tS_F} \to \bar{\xi} \in (tS_F)^{\#}$,
2. $pr_{\langle tF \rangle}(\xi_n) \to +\infty$ *relative to the order determined by* tF. *Here* $\langle tF \rangle$ *is the linear span of* tF.

For short we write $\xi_n \overset{tF}{\to} \bar{\xi} \in (tS_F)^{\#}$. *Then the limit*

(4.12) $$\bar{\psi}(\bar{\xi}) = \lim_{\xi_n \overset{tF}{\to} \bar{\xi}} \psi(\xi_n)$$

exists for all $\psi \in C_0^\Lambda(X^\#)$.

In case $F = \Lambda'' = (0)$, we have $S_F = S$ and π_F is the natural representation. Now consider the 1-dimensional case $F = \Lambda'_+ = (\mathbf{R}^+ \times 0)$ with perpendicular face $\Lambda'_- = (0 \times \mathbf{R}^+) =: (\Lambda'_+)_\perp$ (similarly for Λ'_-). Then $(S_{\Lambda'})^\# = \langle \Lambda'_\perp \rangle$.

PROPOSITION 4.5. *Let (ξ_n) be a sequence in Λ with $\xi_n \overset{t\Lambda'}{\to} \xi'$. Suppose $\xi' \in \widehat{V_{++}}$. Then $\chi_{\xi_n + \Lambda}$ converges almost everywhere to the characteristic function of*

(4.13) $$\xi' + (X^\# \setminus (\langle t\Lambda' \rangle + \widehat{V_{++}})) \cup C$$

where C is a subset of the tangential plane on Λ through $t\Lambda'$. A similar conclusion holds for $\widehat{V_{--}}$.

PROOF. Since C has measure zero it suffices to show

$$\lim_{n \to \infty} \chi_{\xi_n + \Lambda}(\eta) = \chi_{H(\xi')}(\eta)$$

for all η not in the hyperplane through ξ' and orthogonal to $t\Lambda'_\perp$. Here

(4.14) $$H(\xi') := \xi' + (X^\# \setminus (\langle t\Lambda' \rangle + \widehat{V_{++}}))$$

is the closed halfspace determined by the conditions $\xi' \in \partial H(\xi')$, $\partial H(\xi')$ is orthogonal to $t\Lambda'_\perp$ and $\xi' + \widehat{V_{++}} \not\subset H(\xi')$. Since $\xi_n \overset{t\Lambda'}{\to} \xi'$ we introduce cartesian coordinates according to $t\Lambda'$ and $t\Lambda'_\perp$ and write $\xi_n = (\xi_n^1, \xi_n^2, \xi_n^3)$, with $t\Lambda'_\perp \ni \xi_n^1 \to \xi'$, $t\Lambda' \ni \xi_n^2 \to \infty$ and $\xi_n^3 = 0$. For $\eta \in X^\#$, $\xi_n - \eta = (\xi_n^1 - \eta^1, \xi_n^2 - \eta^2, -\eta^3)$ converges to $(\xi' - \eta^1, \infty, -\eta^3)$. Thus we get for sufficiently large n

$$\xi_n - \eta \in \Lambda \quad \text{if } \eta^1 < \xi'$$
$$\xi_n - \eta \notin \Lambda \quad \text{if } \eta^1 > \xi'$$

implying

$$\chi_{\xi_n - \Lambda}(\eta) = \begin{cases} 1 & \text{if } \eta^1 < \xi' \\ 0 & \text{if } \eta^1 > \xi' \end{cases}$$

for all n large enough.

THEOREM 4.6. *For each 1-dimensional face $tF = t\Lambda'$, there is a nondegenerate representation $\hat{\pi}_{tF}$ of $C_0^\Lambda(X^\#)$ on $L^2((tS_F)^\#)$ determined by*

$$\hat{\pi}_{tF} : \psi = \phi \star \chi_\Lambda \longmapsto \psi', \quad \psi'(\xi') := \int_{H(\xi')} \phi(\xi) d\xi.$$

PROOF. Consider the set

$$\mathcal{A} := \left\{ \psi \in \mathcal{U}_b(X^\#) \mid \lim_{\xi_n \overset{t\Lambda'}{\to} \xi'} \psi(\xi_n) =: \psi'(\xi') \text{ exists for all } \xi' \in \langle t\Lambda'_\perp \rangle \right\}.$$

Clearly $\mathcal{A}$ is a *-algebra and $\psi \mapsto \psi'$ is a contractive *-homomorphism.

To show that $\mathcal{A}$ is a closed algebra, consider a sequence ψ_m in $\mathcal{A}$ converging uniformly to a function ψ. Since ψ_m is a uniform Cauchy sequence it is bounded by c and we get $|\psi(\xi)| = \lim_{n \to \infty} |\psi_n(\xi)| \leq c$ for all $\xi \in X^\#$. Therefore a subsequence $(\psi(\xi_{m_k}))$ converges to a M in $\mathbf{C}$. For arbitrary $\epsilon > 0$, choose m_0 so that $\|\psi - \psi_{m_0}\| <$

$\epsilon/6$, an l so that $|M - \psi(\xi_{n_l})| < \epsilon/6$ and $|\psi'_{m_0}(\xi') - \psi_{m_0}(\xi_{n_l})| < \epsilon/6$, and n_0 so that $|\psi'_{m_0}(\xi') - \psi_{m_0}(\xi_n)| < \epsilon/6$ for all $n > n_0$. Then for $n > n_0$

$$|M - \psi(\xi_n)| = |M - \psi(\xi_{n_l})| + |\psi(\xi_{n_l}) - \psi_{m_0}(\xi_{n_l})|$$
$$+ |\psi_{m_0}(\xi_{n_l})| - \psi'_{m_0}(\xi')| + |\psi'_{m_0}(\xi') - \psi_{m_0}(\xi_n)| + |\psi_{m_0}(\xi_n) - \psi(\xi_n)|$$
$$< \epsilon.$$

Therefore $M = \lim_{\xi_n \overset{t\Lambda'}{\to} \xi'} \psi(\xi_n)$ exists. By Proposition 1 and dominated convergence we get for all $\psi = \phi \star \chi_\Lambda$, $\phi \in L^1(X^\#)$:

$$(4.15) \qquad \psi'(\xi') := \lim_{\xi_n \overset{t\Lambda'}{\to} \xi'} \psi(\xi_n) = \int_{H(\xi')} \phi(t)dt.$$

Hence the generators of $C_0^\Lambda(X^\#)$ are contained in $\mathcal{A}$. Since $\mathcal{A}$ is closed, $C_0^\Lambda(X^\#) \subset \mathcal{A}$ and, by contractivity, the limit function $\psi' \in \mathcal{U}_b(\langle t\Lambda'_\perp \rangle)$ for all $\psi \in C_0^\Lambda(X^\#)$. To see that the representation on $L^2(\langle t\Lambda'\perp \rangle) = L^2((t\Lambda'^\perp)^\#)$ is nondegenerate, consider a function $\phi \in L^1(X^\#)$ with the properties $\int_{H(0)} \phi(\xi)d\xi = 1$ and $\phi(\xi) = 0$ for all ξ in the exterior of the half-space $H(0)$. Then $\hat{\pi}(\phi \star \chi_\Lambda)(\vartheta) = 0$ implies $\vartheta = 0 \in L^2(\langle t\Lambda'_\perp \rangle)$.

The corresponding representation π_{tF} of $C_\Lambda^*(iX)$ is given by applying $Ad(\mathcal{F})$ and $Ad(\mathcal{F}|_{\langle tS_F \rangle})$ and has the explicit form:

COROLLARY 4.7. *For a 1-dimensional face $tF = t\Lambda'$ there exists a representation π_{tF} of $C_\Lambda^*(iX)$ on $L^2(tS_F)$ determined by*

$$(4.16) \qquad \pi_{tF}(l_{fE}) = l_{\bar{f}E_{tF}},$$

where $\bar{f} := f|_{tS_F}$ and $E_{tF} := (\widehat{\chi_{\Lambda \cap tF_\perp}})$. Here $tF_\perp$ is the perpendicular face.

PROOF. Since $\mathcal{C}_c(X^\#)$ is dense in $L^1(X^\#)$, we may assume $\phi \in \mathcal{C}_c(X^\#) \subset L^2(X^\#)$. Since $L^2(iX) = L^2(\mathbf{R}^2) \otimes L^2(tS_F)$, we may assume $\hat{\phi} = g \otimes h$, $g \in \mathcal{C}_c(\mathbf{R}^2)$, $h \in \mathcal{C}_c(tS_F)$. Since $|e^{ix_0\beta}| \le 1$ for all $x_0 \in \mathbf{R}^2, \beta \in X^\#$, the function $g_\beta : x_0 \longmapsto e^{ix_0\beta}g(x_0)$ is an element of $L^2(\mathbf{R}^2)$ and

$$(4.17) \qquad g_\beta(x) = \int_{R^2} e^{-i\xi_0 x}\hat{g}_\beta(\xi_0)d\xi_0 = \int_{R^2} e^{-i\xi_0 x} \int_{R^2} e^{i\xi_0 x_0} g_\beta(x_0)dx_0 d\xi_0,$$

hence

$$(4.18) \qquad g(0) = e^{i0\beta}g(0) \overset{(Def)}{=} g_\beta(0) = \int_{R^2} 1 \int_{R^2} e^{i\xi_0 x_0} g_\beta(x_0)dx_0 d\xi_0$$

and we get, using (4.15)

$$\psi'(\xi') = g(0) \int_{\xi_1 \le \xi', \xi_1 \in \langle tF_\perp \rangle} \hat{h}(\xi_1)d\xi_1$$
$$= g(0) \int_{\xi' - tF_\perp} \hat{h}(\xi_1)d\xi_1 = g(0) \int_{\langle tF_\perp \rangle} \chi_{tF_\perp}(\xi' - \xi_1)h(\xi_1)d\xi_1$$
$$= g(0) \cdot (\chi_{tF_\perp} \star \hat{h})(\xi'),$$

with the convolution relative $\langle tF_\perp \rangle$. Applying $Ad(\mathcal{F})$ we get

$$(4.19) \qquad Ad(\mathcal{F})(\psi') = g(0) \cdot l_{h\hat{\chi}_{tF_\perp}} = l_{g(0)h\hat{\chi}_{tF_\perp}}$$

and the assertion follows.

Finally, let $F = \Lambda_0 = (\mathbf{R}^+)^2$.

PROPOSITION 4.8. *Let (ξ_n) be a sequence in Λ with $\xi_n \xrightarrow{\Lambda_0} 0$. Then*

$$\phi \star \chi_\Lambda(\xi_n) \to \hat{\phi}(0)$$

for all $\phi \in L^1(X^\#)$.

PROOF. Let $\eta \in X^\#$. Then $\xi_n - \eta = (\xi_n^1 - \eta_1, \xi_n^2 - \eta_2, -\eta_3)$ converges to infinity in the first two components. Therefore $\chi_{\xi_n - \Lambda}$ converges pointwise to the constant function 1 on $X^\#$. Dominated convergence implies

$$(4.20) \qquad \phi \star \chi_\Lambda(\xi_n) = \int_{X^\#} \phi(\tau)\chi_{\xi_n - \Lambda}(\tau)d\tau \to \int_{X^\#} \phi(\tau)d\tau = \hat{\phi}(0), \quad \phi \in L^1(X^\#).$$

REMARK 4.9. By the symmetric properties of Λ one sees, that $\chi_{\xi_n - \Lambda}$ converges to 1 for *every* sequence $\xi_n \xrightarrow{t\Lambda_0} 0$ with $t \in T$.

THEOREM 4.10. *There exists a character, denoted by π_Λ $(= \pi_{\Lambda_0})$, of $C_\Lambda^*(iX)$ on $L^2(0) = \mathbf{C}$, determined by*

$$(4.21) \qquad \pi_\Lambda : l_{fE} \longmapsto f(0).$$

PROOF. The same techniques as in the proof of Theorem 4.6 together with (4.20) show that

$$\hat{\pi}_\Lambda : \psi = \phi \star \chi_\Lambda \longmapsto \hat{\phi}(0)$$

is a representation of $\mathcal{C}_0^\Lambda(X^\#)$ on $\mathbf{C}$. Applying $Ad(\mathcal{F})$ to the equation $\widehat{fE} = \phi \star \chi_\Lambda$ yields

$$l_{fE} \mapsto f(0) \in \mathbf{C}.$$

Since $f(0) \neq 0$ for some $f \in A(S)$, the character (4.21) is non-zero.

Let $\bar{S}$ be the space tS_F and put $l_{\bar{f}\bar{E}} = \pi_{tF}(l_{fE})$. We will now show the covariance condition of Definition 4.2 for the representation π_{tF} associated with any face tF of Λ.

THEOREM 4.11. *For all g in $A(S)$ we have*

$$(\pi_{tF} \otimes i)^-(\delta(l_{gE})) = W_{\bar{S}}(\pi_{tF}(l_{gE}) \otimes 1)W_{\bar{S}}^*,$$

where

$$W_{\bar{S}}\Phi(s,r) = \Phi(s, r - s) \ (additive\ notation)$$

for all $s \in \bar{S}, r \in S$.

PROOF. Because $(\pi_{tF} \otimes i)^-(\delta(l_{gE}))$ and $W_{\bar{S}}(\pi_{tF}(l_{gE}) \otimes 1)W_{\bar{S}}^*$ commute with right translations $\gamma_s \otimes \gamma_r$, $(s,r) \in \bar{S} \times S$ both are elements of $W^*(\bar{S})\bar{\otimes}W^*(S) = W^*(\bar{S} \times S)$. Therefore it's enough to show, that the Fourier transform of both define the same multiplication operator on $L^2(\bar{S} \times S)$. First we compute

$$W_{\bar{S}}(l_{\bar{g}\bar{E}} \otimes 1)W_{\bar{S}}^*(\phi \otimes \psi)(s,r) = (l_{\bar{g}\bar{E}} \otimes 1)W_{\bar{S}}^*(\phi \otimes \psi)(s,r-s)$$

$$= \int_{\bar{S}} \bar{g}\bar{E}(x)\, W_{\bar{S}}^*(\phi \otimes \psi)(s-x,r-s)\, dx$$

$$= \int_{\bar{S}} \bar{g}\bar{E}(x)\, \phi(s-x)\psi(r-x)\, dx.$$

Hence, putting $s - x =: y$, $r - x =: z$ and $\bar{\eta} := \eta\,|_{\bar{S}}$,

$$\mathcal{F}(W_{\bar{S}}(l_{\bar{g}\bar{E}} \otimes 1)W_{\bar{S}}^*(\phi \otimes \psi))(\bar{\xi},\eta)$$

(4.22)
$$= \int_{\bar{S}} \int_S \int_{\bar{S}} \bar{g}\bar{E}(x)\phi(s-x)\psi(r-x)e^{-i\bar{\xi}(s)}e^{-i\eta(s)}\,dx\,ds\,dr$$

$$= \int_{\bar{S}} \int_S \phi(y)\psi(z)e^{-i\bar{\xi}(y)}e^{-i\eta(z)} \int_{\bar{S}} \bar{g}\bar{E}(x)e^{-i\bar{\xi}(x)}e^{-i\bar{\eta}(x)}\,dx,$$

$$= \mathcal{F}(\phi \otimes \psi)(\bar{\xi},\eta) \cdot \mathcal{F}(\bar{g}\bar{E})(\bar{\xi}+\bar{\eta}).$$

On the other hand we get $\delta(l_{gE}) = \lim_i \sum l_{u_i} \otimes l_{v_i} \in (C_\Lambda^*(S) \otimes C^*(S))^-$ (strict closure), hence $(\pi_{tF} \otimes i)^-(\delta(l_{gE})) = \lim_i \sum \pi_{tF}(l_{u_i}) \otimes l_{v_i}$ and

$$Ad(\mathcal{F})((\pi_{tF} \otimes i)(\delta(l_{gE})))(\bar{\xi},\eta) = \lim_i \sum Ad(\mathcal{F})(\pi_{tF}(l_{u_i}))(\bar{\xi}) \otimes \mathcal{F}(v_i)(\eta)$$

$$= \lim_{\xi \xrightarrow{tF} \bar{\xi}} \lim_i \sum \mathcal{F}(u_i)(\xi) \otimes \mathcal{F}(v_i)(\eta)$$

(4.23)
$$= \lim_{\xi \xrightarrow{tF} \bar{\xi}} Ad(\mathcal{F})(\delta(l_{gE}))(\xi,\eta)$$

$$= \lim_{\xi \xrightarrow{tF} \bar{\xi}} \mathcal{F}(gE)(\xi+\eta)$$

$$= \mathcal{F}(\bar{g}\bar{E})(\bar{\xi}+\bar{\eta}),$$

since for all tF we have $\xi+\eta \xrightarrow{tF} \bar{\xi}+\bar{\eta}$ if $\xi \xrightarrow{tF} \bar{\xi}$. Viewed as a multiplication operator on $L^2(\bar{S} \times S)$, (4.22) and (4.23) agree.

All together we have proved the following

THEOREM 4.12. *For every face tF of Λ we have a nondegenerate representation Π_{tF} of the cocrossed product C^*-algebra $C_\Lambda^*(S) \otimes_\delta C_0(S)$ on $L^2(\bar{S})$, determined for all $a \in C_\Lambda^*(S)$, $g \in A(S)$ by*

(4.24)
$$\delta(a)(1 \otimes g) \longmapsto \pi_{tF}(a)\mu_{tF}(g).$$

5. Structure theory of Toeplitz C*-algebras

Consider the *Hardy-Toeplitz algebra*

$$\mathcal{T}(S) := C^*\left(T_S(f) \mid f \in C_0(S)\right),$$

where l_E is the projection of $L^2(S)$ on the Hardy space $H^2_\Lambda(S)$ and

$$T_S(f) := l_E \, f \, l_E$$

is the Hardy-Toeplitz operator corresponding to $f \in C_0(S)$. Our aim is to find all irreducible representations of $\mathcal{T}(S)$.

Applying Katayama's isomorphism Φ [11, p. 768] we get for every representation Π_{tF} of $C^*_\Lambda(S) \otimes_\delta C_0(S)$ defined by (4.24) a nondegenerate representation ν_{tF} of

(5.1) $$\mathcal{K}_\Lambda(L^2(S)) := \Phi(C^*_\Lambda(S) \otimes_\delta C_0(S)) = C^*(l_{fE}g \,; f \in A(S),\ g \in C_0(S))$$

on $L^2(\bar{S})$, given by $\nu_{tF}(\Phi(x)) = \Pi_{tF}(x)$ for all x in $C^*_\Lambda(S) \otimes_\delta C_0(S)$. Therefore

(5.2) $$\nu_{tF}(l_{fE}g) = \pi_{tF}(l_{fE})\mu_{tF}(g).$$

LEMMA 5.1. *Let* $f_1, \dots, f_n, g_1, \dots, g_n, h_0, \dots, h_n \in C_c^\infty(S)$ *fulfill the condition*

(5.3) $$\prod_{j=1}^{n} f_j(t_{j-1}t_j^{-1})g_j(t_j) = \prod_{i=0}^{n} h_i(t_i)$$

for all $t_0, \dots, t_n \in S$. *Then we have the operator identity (on* $L^2(S)$*)*

(5.4) $$h_0 l_E h_1 l_E \cdots l_E h_n = l_{f_1 E} g_1 \cdots l_{f_n E} g_n.$$

PROOF. For any $k \in C_c^\infty(S)$ we calculate the inner product (antilinear in the first variable)

$$\langle k \mid h_0 l_E \cdots l_E h_n k \rangle = \int_S \cdots \int_S \overline{k(t_0)} h_0(t_0) h_1(s_1^{-1}t_0) \cdots h_n(s_n^{-1} \cdots s_1^{-1}t_0) \cdot$$
$$\cdot k(s_n^{-1} \cdots s_1^{-1}t_0) E(ds_1) \cdots E(ds_n) dt_0$$
$$= \int_S \cdots \int_S \overline{k(t_0)} g_1(s_1^{-1}t_0) g_2(s_2^{-1}s_1^{-1}t_0) \cdots g_n(s_n^{-1} \cdots s_1^{-1}t_0) \cdot$$
$$\cdot k(s_n^{-1} \cdots s_1^{-1}t_0) f_1(s_1) \cdots f_n(s_n) E(ds_1) \cdots E(ds_n) dt_0$$
$$= \langle k \mid l_{f_1 E} g_1 \cdots l_{f_n E} g_n k \rangle.$$

Note that because $l_E \in W^*(S)$ the integral $\int h(st^{-1})E(dt)$ is well defined for all $h \in A(S)$, and therefore for all $h \in C_c^\infty(S)$. Analogously for the other integrals.

PROPOSITION 5.2.

(5.5) $$l_E \, \mathcal{K}_\Lambda(L^2(S)) \, l_E = \mathcal{T}(S).$$

PROOF. To show the first inclusion it suffices (by density) to show

(5.6) $$l_E \left(\prod_j l_{f_j E} \, g_j \right) l_E \in \mathcal{T}(S)$$

for all $f_1, \ldots, f_n \in A(S)$, $g_1, \ldots, g_n \in \mathcal{C}_c^\infty(S)$. Since $\mathcal{C}_c^\infty(S)$ is dense in $A(S)$ [5, Proposition 3.26], we may assume $f_1, \ldots, f_n \in \mathcal{C}_c^\infty(S)$. To see that

$$(5.7) \qquad F(t_0, \ldots, t_n) := \prod_{j=1}^n f_j(t_{j-1}t_j^{-1})g_j(t_j)$$

belongs to $\mathcal{C}_c^\infty(S \times \ldots \times S) = \mathcal{C}_c^\infty(S) \otimes \ldots \otimes \mathcal{C}_c^\infty(S)$ let K_{g_j}, K_{f_j} be the compact support of f_j, g_j and put

$$(5.8) \qquad K := (K_{f_1} \cdot K_{g_1}) \times K_{g_1} \times \cdots \times K_{g_n}.$$

If $(t_0, \ldots, t_n) \notin K$ then there are three cases. If $t_0 \notin K_{f_1} \cdot K_{g_1}$ and $t_1 \in K_{g_1}$, then $t_0 t_1^{-1} \notin K_{f_1}$ implying $f_1(t_0 t_1^{-1}) = 0$, hence $F(t_0, \ldots, t_n) = 0$. Now assume $t_0 t_1^{-1} \notin K_{f_1}$ and $t_1 \notin K_{g_1}$. Then $g_1(t_1) = 0$ and the assertion holds. Finally let $t_0 \in K_{f_1} \cdot K_{g_1}$. Then there is a $j \geq 1$ with $t_j \notin K_{g_j}$ implying $g_j(t_j) = 0$ and hence $F(t_0, \ldots, t_n) = 0$. Thus $\mathrm{supp}(F) \subset K$ is compact. Hence F is a uniform limit of finite sums of the form $\prod_{i=0}^n h_i(t_i)$ where $h_i \in \mathcal{C}_c^\infty(S)$. Therefore Lemma 5.1 implies (5.6). Conversely, let $A \in \mathcal{T}(S)$. Since $\mathcal{T}(S) \cdot \mathcal{T}(S)$ is dense in $\mathcal{T}(S)$, we may assume

$$(5.9) \qquad A = \prod_{i=0}^n l_E h_i l_E,$$

where $h_i \in \mathcal{C}_0(S)$ and $n \geq 1$. Since $\mathcal{C}_0(S) \cdot \mathcal{C}_0(S)$ is dense in $\mathcal{C}_0(S)$ we may assume $h_0, \ldots, h_n \in \mathcal{C}_c^\infty(S)$ and $h_i = h_i^1 \cdot h_i^2$ for $i = 1, \ldots, n-1$, where $h_i^1, h_i^2 \in \mathcal{C}_c^\infty(S)$. Then

$$(5.10) \qquad G(s_1, \ldots, s_n, t_1, \ldots, t_n) := h_0(s_1 t_1)\left(\prod_{i=1}^{n-1} h_i^1(t_i)h_i^2(s_{i+1}t_{i+1})\right)h_n(t_n)$$

belongs to $\mathcal{C}_c^\infty(S \times \ldots \times S)$ ($2n$ times). Hence G is a uniform limit of finite sums of the form $\prod_{j=1}^n f_j(s_j)g_j(t_j)$ where f_j, $g_j \in \mathcal{C}_c^\infty(S)$. By putting $s_i := t_{i-1}t_i^{-1}$, $i = 1, \ldots, n$ we get $\prod_{i=0}^n h_i(t_i)$ as a uniform limit of finite sums of the form $\prod_{j=1}^n f_j(t_{j-1}t_j^{-1})g_j(t_j)$. Applying Lemma 5.1 again, we obtain $A \in l_E \mathcal{K}_\Lambda(L^2(S)) l_E$.

LEMMA 5.3. *For all* $a, b \in \mathcal{K}_\Lambda(L^2(S))$ *we have* $a l_E b \in \mathcal{K}_\Lambda(L^2(S))$ *and*

$$(5.11) \qquad \nu_{tF}(a l_E b) = \nu_{tF}(a) l_{\bar{E}} \nu_{tF}(b).$$

PROOF. Let $f, h \in A(S)$ and $g, k \in \mathcal{C}_0(S)$. As in Proposition 5.2 and Lemma 5.1 one shows that $l_{fE} g \cdot l_E \cdot l_{hE} k$ is a limit of finite sums of the form $(h_0 l_E h_1)l_E \cdot (h_2 l_E h_3)$, $h_i \in \mathcal{C}_c^\infty(S)$, and $(h_0 l_E h_1)l_E(h_2 l_E h_3)$ is a limit of finite sums of the form $l_{f_1 E} g_1 \cdot l_{f_2 E} g_2 \cdot l_{f_3 E} g_3$, $f_i, g_i \in \mathcal{C}_c^\infty(S)$. Therefore $l_{fE} g \cdot l_E \cdot l_{hE} k \in \mathcal{K}_\Lambda(L^2(S))$. Since

$$\nu_{tF}(l_{f_1 E}g_1 \cdot l_{f_2 E}g_2 \cdot l_{f_3 E}g_3) = \nu_{tF}(l_{f_1 E}\, g_1)\,\nu_{tF}(l_{f_2 E}\, g_2)\,\nu_{tF}(l_{f_3 E}\, g_3)$$
$$= l_{\bar{f}_1 \bar{E}}\, \bar{g}_1 \cdot l_{\bar{f}_2 \bar{E}}\, \bar{g}_2 \cdot l_{\bar{f}_3 \bar{E}}\, \bar{g}_3,$$

we see by the same calculations as above, only in a lower dimension, that $\nu_{tF}(l_{fE}\, g \cdot l_E \cdot l_{hE}\, k) = l_{\bar{f}\bar{E}}\, \bar{g} \cdot l_{\bar{E}} \cdot l_{\bar{h}\bar{E}}\, \bar{k} = \nu_{tF}(l_{fE}\, g)\, l_{\bar{E}}\, \nu_{tF}(l_{hE}\, k)$.

THEOREM 5.4. *For each face tF of Λ there exists a representation ρ_{tF} of $\mathcal{T}(S) = l_E \, \mathcal{K}_\Lambda(L^2(S)) \, l_E$ such that $\rho_{tF}(l_E \, a \, l_E) := l_{\bar{E}} \, \nu_{tF}(a) \, l_{\bar{E}}$ and*

$$(5.12) \qquad \nu_{tF}(b \, l_E \, a \, l_E \, c) = \nu_{tF}(b) \rho_{tF}(l_E \, a \, l_E) \nu_{tF}(c)$$

for all a, b, c in $\mathcal{K}_\Lambda(L^2(S))$.

PROOF. To show (5.12) consider a sequence of sums of products $\sum_i \prod_j (l_{f_{ij}E} g_{ij})$ of typical generators of $\mathcal{K}_\Lambda(L^2(S))$ converging to a. For each of them we have by Lemma 5.3

$$b \, l_E \sum_i \prod_j (l_{f_{ij}E} \, g_{ij}) \, l_E \, c = \sum_i b \, l_E \prod_p (l_{f_{ip}E} \, g_{ip}) \prod_q (l_{f_{iq}E} \, g_{iq}) \, l_E \, c \in \mathcal{K}_\Lambda(L^2(S)).$$

Here the j-indices are split into $p's$ and $q's$, with both products non-empty. Hence

$$\nu_{tF}(b \, l_E \, a \, l_E \, c) = \lim \sum_i \nu_{tF}(b \, l_E \prod_j (l_{f_{ij}E} \, g_{ij}) \, l_E \, c)$$

$$= \lim \sum_i \nu_{tF}(b \, l_E \prod_p (l_{f_{ip}E} \, g_{ip})) \nu_{tF}(\prod_q (l_{f_{iq}E} \, g_{iq}) \, l_E \, c)$$

$$\stackrel{(5.11)}{=} \lim \sum_i \nu_{tF}(b) \, l_{\bar{E}} \, \nu_{tF}(\prod_p (l_{f_{ip}E} \, g_{ip})) \nu_{tF}(\prod_q (l_{f_{iq}E} \, g_{iq})) \, l_{\bar{E}} \, \nu_{tF}(c)$$

$$= \lim \sum_i \nu_{tF}(b) \, l_{\bar{E}} \, \nu_{tF}(\prod_j (l_{f_{ij}E} \, g_{ij})) \, l_{\bar{E}} \, \nu_{tF}(c)$$

$$= \lim \sum_i \nu_{tF}(b) \rho_{tF}(l_E \prod_j (l_{f_{ij}E} \, g_{ij}) \, l_E) \nu_{tF}(c)$$

$$= \nu_{tF}(b) \rho_{tF}(l_E \, a \, l_E) \nu_{tF}(c).$$

This implies that the map is well defined, because for an a with $l_E \, a \, l_E = 0$ we get for all b, c in $\mathcal{K}_\Lambda(L^2(S))$

$$(5.13) \qquad 0 = \nu_{tF}(b \, l_E \, a \, l_E \, c) = \nu_{tF}(b) \rho_{tF}(l_E \, a \, l_E) \nu_{tF}(c)$$

implying $\rho_{tF}(l_E \, a \, l_E) = 0$ by the irreducibility of ν_{tF} (we will show later that the compact operators $\mathcal{K}(L^2(S)) \subset \mathcal{K}_\Lambda(L^2(S))$). To see the homomorphism property calculate for a, b in $\mathcal{K}_\Lambda(L^2(S))$:

$$\rho_{tF}(l_E \, a \, l_E \cdot l_E \, b \, l_E) = \rho_{tF}(l_E \, a \, l_E \, b \, l_E), \quad \text{because } l_E \text{ is a projection,}$$

$$= l_{\bar{E}} \, \nu_{tF}(a \, l_E \, b) \, l_{\bar{E}} \stackrel{(60)}{=} l_{\bar{E}} \, \nu_{tF}(a) \, l_{\bar{E}} \, \nu_{tF}(b) \, l_{\bar{E}}$$

$$= l_{\bar{E}} \, \nu_{tF}(a) \, l_{\bar{E}} \cdot l_{\bar{E}} \, \nu_{tF}(b) \, l_{\bar{E}} = \rho_{tF}(l_E \, a \, l_E) \cdot \rho_{tF}(l_E \, b \, l_E).$$

The representations ρ_{tF} just defined give rise to "translated" representations. For $s \in S$ the right translation γ_s maps $f \in \mathcal{C}_0(S)$ to f_s defined by $f_s(t) = f(ts)$. The translated representation ρ_{tF}^s corresponding to s and ρ_{tF} is defined by the following diagram

$$(5.14) \qquad \begin{array}{ccc} \mathcal{T}(S) & \xrightarrow{\rho_{tF}} & \mathcal{L}(l_{\bar{E}} \, L^2(\bar{S})) \\ {\scriptstyle Ad(\gamma_s)} \downarrow & & \downarrow {\scriptstyle Id} \\ \mathcal{T}(S) & \xrightarrow{\rho_{tF}^s} & \mathcal{L}(l_{\bar{E}} \, L^2(\bar{S})). \end{array}$$

The fact that $\mathcal{T}(S) = l_E \, \mathcal{K}_\Lambda(S) \, l_E$ is $Ad(\gamma_s)$-invariant follows from the formula

(5.15) $\gamma_s \, l_E \, l_{fE} \, g \, l_E \, \gamma_{-s} = l_E \, l_{fE} \, \gamma_s g \gamma_{-s} \, l_E = l_E \, l_{fE} \, g_s \, l_E$

applied to a typical generator $l_E \, l_{fE} \, g \, l_E$ of the Toeplitz algebra, since $g_s \in \mathcal{C}_0(S)$.

LEMMA 5.5. *Two representations $\rho_{t_1 F_1}^{s_1}$ and $\rho_{t_2 F_2}^{s_2}$ are pairwise inequivalent, whenever $F_1 \neq F_2$ or $F_1 = F_2 = F$ and $t_1 t_2^{-1} \notin T_F$ or $s_1 - s_2 \notin S_F$. All representations are irreducible.*

PROOF. For the trivial face $F = \Lambda_0$ we have only $\rho_{tF}^s = \rho_{tF}$, the identity representation of $\mathcal{T}(S)$ on $H_\Lambda^2(S)$. It is irreducible as a consequence of Corollary 5.7 below. For a 1-dimensional face tF, the classical Toeplitz extension

(5.16) $0 \to \mathcal{K}(H_{tF}^2(t S_F)) \to \mathcal{T}(t S_F) \to \mathcal{C}_0(t S_F) \to 0$

on the half-line is an exact sequence, so that the representation ρ_{tF} is irreducible and

(5.17) $\|T_{t S_F}(\tilde{\varphi})\|_{\mathcal{T}/\mathcal{K}} = \|\tilde{\varphi}\|_\infty$

for $\tilde{\varphi} \in \mathcal{C}_0(t S_F)$. Now consider distinct $\rho_{t_1 F_1}^{s_1}, \rho_{t_2 F_2}^{s_2}$, where $\dim(F_i) = 1$. Then we have to discuss two cases. First let $s_1 + t_1 S_{F_1} \neq s_2 + t_2 S_{F_2}$. Then there exists $\varphi \in \mathcal{C}_0(S)$ such that $\varphi_1 := \gamma_{s_1^{-1}} \varphi \mid_{t_1 S_{F_1}} \neq 0$ and $\varphi_2 := \gamma_{s_2^{-1}} \varphi \mid_{t_2 S_{F_2}} = 0$. Then $\rho_{t_1 F_1}^{s_1}(l_E \, \varphi \, l_E) = l_{E_{t_1 F_1}} \varphi_1 \, l_{E_{t_1 F_1}} = T_{t_1 S_{F_1}}(\varphi_1) \notin \mathcal{K}(H_{t_1 F_1}^2(t_1 S_{F_1}))$ since $\|\varphi_1\| \neq 0$, hence $T_{t_1 S_{F_1}}(\varphi_1) \neq 0$, but $\rho_{t_2 F_2}^{s_2}(l_E \, \varphi \, l_E) = l_{E_{t_2 F_2}} \varphi_2 \, l_{E_{t_2 F_2}} = 0$. Now assume $s_1 + t_1 S_{F_1} = s_1 + t_2 S_{F_2}$. Then $F_1 \neq F_2$ and we can assume $t_1 = 1$ and $t_2 = -1$. Then there exists a ψ in $\mathcal{C}_0^\Lambda(X^\#)$ in $Ker \hat{\pi}_{F_1} \setminus Ker \hat{\pi}_{F_2}$ (e.g. constant in the direction of F_2 and converging to 0 in the direction of F_1) such that by taking a suitable $f \in A(S)$ not in $Ker \mu_{t_2 F_2}^{s_2}$ we have

(5.18) $l_{E_{t_2 F_2}} \pi_{t_2 F_2}(\hat{\psi}) \mu_{t_2 F_2}^{s_2}(f) l_{E_{t_2 F_2}} \neq 0.$

Therefore $Ker \, \rho_{t_1 F_1}^{s_1} \neq Ker \, \rho_{t_2 F_2}^{s_2}$ and the representations are inequivalent. The same reasoning shows that all translated vertex representations are inequivalent. For pairs F_1 and F_2 of different dimension the assertion is clear.

LEMMA 5.6.
$$C^*(S) \subset C_\Lambda^*(S).$$

PROOF. Let Λ_{++} (resp. Λ_{--}) denote the forward (resp. backward) lightcone (c.f. Example 1.2). Then

(5.19) $\hat{\chi}_{\Lambda_{++}}(ix) = (Det(ix + 0))^{-3/2} := \lim_{y \searrow 0} (Det(ix + y))^{-3/2}$

and

(5.20) $\hat{\chi}_{\Lambda_{--}}(ix) = (Det(-ix + 0))^{-3/2} := \lim_{y \searrow 0} (Det(-ix + y))^{-3/2}$

as follows from the general theory of euclidean Jordan algebras [6]. Since $Det(ix+y)$ and $Det(-ix + y)$ are not in $\mathbf{R}^+ \cup \{0\}$ for $y > 0$ we can define (5.19) and (5.20) using the same branch of the logarithm. Now

(5.21) $\begin{aligned} E(ix) &= \mathcal{F}^{-1}(1 - \chi_{\Lambda_{--}} - \chi_{\Lambda_{++}})(ix) \\ &= c\delta_0(ix) - (Det(ix + 0))^{-3/2} - (Det(-ix + 0))^{-3/2}, \end{aligned}$

where c is a constant. Let $f(ix) := -(Det(ix + 0))^{3/2} \cdot (Det(-ix + 0))^{3/2} \in C(iX)$, $u \in \mathcal{C}_c(iX)$. Thus $ufE = cuf\delta_0 + u\tilde{f}$, where $\tilde{f}(ix) := (Det(ix + 0))^{3/2} + (Det(-ix + 0))^{3/2}$ is a continuous function on iX. Since $l_{uf\delta_0}(g) = u(0)f(0)g = 0$ for all g in $L^2(iX)$ we get for all $u \in \mathcal{C}_c(iX)$

$$(5.22) \qquad\qquad l_{ufE} = l_{u\tilde{f}}.$$

Since $N := \{ix \in iX; \; \tilde{f}(ix) = 0\}$ has measure 0, one may choose functions $\varphi_\epsilon \in \mathcal{C}_c^\infty(iX)$, $\epsilon > 0$, with $\varphi_\epsilon = 0$ in the $\epsilon/2$-neighborhood of N and $\varphi_\epsilon = 1$ outside the ϵ-neighborhood of N. For each element $v \in \mathcal{C}_c^\infty(iX)$ the function

$$(5.23) \qquad\qquad u_\epsilon := \begin{cases} v\varphi_\epsilon/\tilde{f} & \text{on } iX \setminus N \\ 0 & \text{on } N \end{cases}$$

is well defined and we get

$$(5.24) \qquad\qquad u_\epsilon \cdot \tilde{f} = v \cdot \varphi_\epsilon \in \mathcal{C}_c^\infty(iX).$$

Since $v \cdot \varphi_\epsilon$ converges to v in the L^1-norm and $\mathcal{C}_c^\infty(iX) \subset A(iX)$ by [5, Proposition 3.26] the proof is complete.

COROLLARY 5.7.
$$\mathcal{K}(H_\Lambda^2(S)) \subset \mathcal{T}(S).$$

PROOF. The C^*-algebra $\mathcal{K}(L^2(S))$ of compact operators on $L^2(S)$ is given by the image of $C^*(S) \otimes_\delta C_0(S)$ under Katayama's isomorphism Φ [11, p. 768]. By Lemma 5.13 $\mathcal{K}(L^2(S))$ is contained in $\Phi(C_\Lambda^*(S) \otimes_\delta C_0(S)) = \mathcal{K}_\Lambda(L^2(S))$. Since every $x \in \mathcal{K}(H_\Lambda^2(S))$ can be extended trivially to a compact operator on $L^2(S)$, we get $\mathcal{K}(H_\Lambda^2(S)) = l_E \mathcal{K}(L^2(S)) l_E \subset l_E \mathcal{K}_\Lambda(L^2(S)) l_E = \mathcal{T}(S)$.

As a consequence of Lemma 5.6 the Fourier transform $\hat{f}$ of every $f \in L^1(iX)$ is contained in $\mathcal{C}_0^\Lambda(X^\#)$. Thus $\hat{\pi}_{tF}(\hat{f})(\bar{\xi}) = \lim_{\xi_n \xrightarrow{tF} \bar{\xi}} \hat{f}(\xi_n) = 0$ for all $\bar{\xi}$, implying $\pi_{tF}(f) = 0$ whenever $tF \neq (0)$. It follows that $\nu_{tF}(l_f g) = 0$ for all generators of $\mathcal{K}(L^2(S))$, and so

$$(5.25) \qquad\qquad \nu_{tF}\left(\mathcal{K}(L^2(S))\right) = 0.$$

For the translated representations one has the dual action $Ad(\gamma_s)(\mathcal{K}(L^2(S))) = \mathcal{K}(L^2(S))$ and so for all $s \in S$ $\nu_{tF}^s(\mathcal{K}(L^2(S))) = \nu_{tF}(Ad(\gamma_s))(\mathcal{K}(L^2(S))) = (0)$ and hence

$$(5.26) \qquad\qquad \mathcal{K}(L^2(S)) \subset \bigcap_{\dim(F)\geq 1} \bigcap_{t\in T/T_F} \bigcap_{s\in S/S_F} Ker\, \nu_{tF}^s$$

and (5.26) implies

$$(5.27) \qquad\qquad \mathcal{K}(H_\Lambda^2(S)) \subset \bigcap_{\dim(F)\geq 1} \bigcap_{t\in T/T_F} \bigcap_{s\in S/S_F} Ker\, \rho_{tF}^s.$$

Our aim is to prove equality, i.e.

THEOREM 5.8.

(5.28)
$$\mathcal{K}(H^2_\Lambda(S)) = \bigcap_{\dim(F)\geq 1} \bigcap_{t\in T/T_F} \bigcap_{s\in S/S_F} Ker\, \rho^s_{tF}.$$

The proof uses techniques from C*-duality [11]. Let π, μ and W denote the direct sum of π_{tF}, μ_{tF} and $W_{tF} = W_{\bar{S}}$ (respectively), where $tF = t\Lambda'_\pm$ or $tF = \Lambda_0$. Then (π, W) defines a covariant pair of representations of $(C^*_\Lambda(S), S, \delta)$.

PROPOSITION 5.9. *For all* $x \in C^*_\Lambda(S) \otimes_\delta C_0(S)$,

$$h = (h_{tF}) \in \sum_{tF=t\Lambda'_\pm,\Lambda_0} L^2(tS_F),$$

$$\phi = (\phi_{tF}) \in \sum_{tF=t\Lambda'_\pm,\Lambda_0} L^2(S),$$

$$s = (s_{tF}) \in \prod_{tF=t\Lambda'_\pm,\Lambda_0} tS_F$$

and

$$u = (u_{tF}) \in \prod_{tF=t\Lambda'_\pm,\Lambda_0} S$$

we have

(5.29)
$$(\pi \otimes I)^-(x)(h\otimes\phi)(s,u) = (\pi\otimes\mu^{s^{-1}u})(x)(m(h\otimes\phi^{s^{-1}u}))(s),$$

where

$$s^{-1}u = (s^{-1}_{tF}u_{tF}) \in \prod_{tF=t\Lambda'_\pm,\Lambda_0} S,\ \mu^{s^{-1}u} = \sum_{tF=t\Lambda'_\pm,\Lambda_0} \mu^{s^{-1}_{tF}u_{tF}}_{tF},\ I = \sum_{tF=t\Lambda'_\pm,\Lambda_0} i$$

and

$$m(h\otimes\phi)(s) = (m(h_{tF}\otimes\phi_{tF})(s_{tF})) = (h_{tF}(s_{tF})\cdot\phi_{tF}(s_{tF})).$$

PROOF. First we show (5.29) for generators $x = \delta(l_{fE})(1\otimes g)$ of $C^*_\Lambda(S)\otimes_\delta C_0(S)$, and then prove that it respects the C*-structure. With the above notations

$$\begin{aligned}(\pi\otimes I)(x)(h\otimes\phi)(s,u) &= (\pi\otimes I)(\delta(l_{fE})(1\otimes g))(h\otimes\phi)(s,u)\\ &= W(\pi(l_{fE})\otimes 1)W^*(h\otimes g\,\phi)(s,u)\\ &= (\pi(l_{fE})\otimes 1)W^*(h\otimes g\,\phi)(s,s^{-1}u)\\ &= \pi(l_{fE})\mu^{s^{-1}u}(g)(m(h\otimes\phi^{s^{-1}u}))(s)\\ &= \pi\otimes\mu^{s^{-1}u}(x)(m(h\otimes\phi^{s^{-1}u}))(s).\end{aligned}$$

To see that (5.29) is multiplicative for $x, y \in C^*_\Lambda(S)\otimes_\delta C_0(S)$, we know that $(\pi\otimes I)(y)(h\otimes\phi)$ is an element of $\sum_{tF=t\Lambda'_\pm,\Lambda_0} L^2(tS_F)\otimes\sum_{tF=t\Lambda'_\pm,\Lambda_0} L^2(S)$, thus can be approximated by functions $f_1\otimes f_2$. Hence

$$\begin{aligned}(\pi\otimes I)(x\cdot y)(h\otimes\phi)(s,u) &= (\pi\otimes I)(f_1\otimes f_2)(s,u)\\ &= (\pi\otimes\mu^{s^{-1}u})(x)(m(f_1\otimes f_2^{s^{-1}u}))(s)\end{aligned}$$

$$= (\pi \otimes \mu^{s^{-1}u})(x)(u \longmapsto (\pi \otimes \mu^{r^{-1}rs^{-1}u})(y)$$
$$\cdot (m(h \otimes \phi^{r^{-1}rs^{-1}u})))(s)$$
$$= (\pi \otimes \mu^{s^{-1}u})(x)(\pi \otimes \mu^{s^{-1}u})(y)(m(h \otimes \phi^{s^{-1}u}))(s)$$
$$= (\pi \otimes \mu^{s^{-1}u})(x \cdot y)(m(h \otimes \phi^{s^{-1}u}))(s).$$

Similar arguments give the *-property:

$$(\pi \otimes I)(x^*)(h \otimes \phi)(s,u) = (\pi \otimes I)(1 \otimes g^*)(\pi \otimes I)(\delta(l_{fE}^*))(h \otimes \phi)(s,u)$$
$$= g^*(t)(\pi(l_{fE}^*)(m(h \otimes \phi^{s^{-1}u})))(s)$$
$$= (g^*)^{s^{-1}u}(s)\pi(l_{fE}^*)(m(h \otimes \phi^{s^{-1}u}))(s)$$
$$= (\pi \otimes \mu^{s^{-1}u})(x^*)(m(h \otimes \phi^{s^{-1}u}))(s).$$

COROLLARY 5.10.
$$\bigcap_{tF=t\Lambda'_\pm,\Lambda_0} \bigcap_{r \in S} Ker(\pi_{tF} \otimes \mu_{tF}^r) = Ker(\pi \otimes I)^- \ \text{(strict extension)}.$$

PROOF. Let $x \in \bigcap_{tF=t\Lambda'_\pm,\Lambda_0} \bigcap_{r \in S} Ker(\pi_{tF} \otimes \mu_{tF}^r)$, then

$$(\pi_{tF} \otimes \mu_{tF}^{s_{tF}^{-1}u_{tF}})(x) = 0 \in \mathcal{L}(L^2(tS_F))$$

for all $(s_{tF}, u_{tF}) \in tS_F \times S$, hence

$$(\pi \otimes \mu^{s^{-1}u})(x) = 0 \text{ on } \sum_{tF=t\Lambda'_\pm,\Lambda_0} L^2(tS_F).$$

Therefore $(\pi \otimes \mu^{s^{-1}u})(x)(m(h \otimes \phi^{s^{-1}u})) = 0$ for all h and ϕ, thus for all s, u

$$(\pi \otimes I)^-(x)(h \otimes \phi)(s,u) = (\pi \otimes \mu^{s^{-1}u})(x)(m(h \otimes \phi^{s^{-1}u})) = 0,$$

implying $(\pi \otimes I)^-(x) = 0$ and we have proved

(5.30) $$\bigcap_{tF=t\Lambda'_\pm,\Lambda_0} \bigcap_{r \in S} Ker(\pi_{tF} \otimes \mu_{tF}^r) \subset Ker(\pi \otimes I)^-.$$

The converse inclusion is trivial.

PROPOSITION 5.11. *The induced coaction δ on $Ker\pi$ satisfies*

$$Ker(\pi \otimes I)^- \subset Ker\pi \otimes_\delta C_0(S).$$

PROOF. By [11, Proposition 4.9] $Ker\pi$ is an ideal invariant under δ [11, Definition 4.1] and $\delta^{Ker\pi}$, given by [11, Lemma 4.6], defines a coaction of S on $C_\Lambda^*(S)/Ker\pi$. Define the quotient map $\Theta : C_\Lambda^*(S) \to C_\Lambda^*(S)/Ker\pi$ and let $(\Theta \otimes_\delta i)$ be the surjection of $C_\Lambda^*(S) \otimes_\delta C_0(S)$ onto $C_\Lambda^*(S)/Ker\pi \otimes_{\delta^{Ker\pi}} C_0(S)$ induced by $(\Theta \otimes i)^-$ on the multiplier algebra $M(C_\Lambda^*(S) \otimes \mathcal{K}(L^2(S)))$. The proofs of [11, Lemma 4.6, Proposition 4.9] imply

(5.31) $$Ker(\Theta \otimes i) = Ker\Theta \otimes C^*(S) = Ker\pi \otimes C^*(S) = Ker(\pi \otimes i),$$

so we get by [11, Lemma 1.1] that $Ker(\pi \otimes i)^- = Ker(\Theta \otimes i)^-$. Therefore the kernel of $(\pi \otimes i)^-$ restricted to the cocrossed product $C^*_\Lambda(S) \otimes_\delta C_0(S)$ agrees with $Ker(\Theta \otimes_\delta i)$. The proof of [11, Proposition 4.8] based on properties of the quotient coaction $\delta^{Ker\pi}$ and Katayama's Theorem, yields the equation

$$(5.32) \qquad Ker\pi \otimes_\delta C_0(S) = Ker(\Theta \otimes_\delta i),$$

which completes the proof.

LEMMA 5.12.

$$(5.33) \qquad \hat{a} \mid_\Lambda \in C_0(\Lambda) \text{ for all } a \in Ker\pi.$$

PROOF. For $a \in Ker\pi$ we get $\hat{a} \in Ker\hat{\pi}_{tF}$ whenever $tF \neq (0)$. Suppose $\xi_n = t_n \cdot (\xi_n^1, \xi_n^2)$, $(\xi_n^1, \xi_n^2) \in X_0$ is a sequence in Λ converging to infinity with the property $|\hat{a}(\xi_n)| > \epsilon$ for all n. Since $\mathbf{T}$ is compact we find a subsequence (t_{n_l}) converging to a $t_0 \in \mathbf{T}$. Write l for n_l.

1. Case: One of the sequences $\text{pr}_{\langle t_0 \Lambda'_\pm \rangle}(\xi_l)$ is bounded.

Without loss of generality $\text{pr}_{\langle t_0 \Lambda'_+ \rangle}(\xi_l)$ is bounded, hence we find a subsequence (ξ_{l_m}) such that $\text{pr}_{\langle t_0 \Lambda'_+ \rangle}(\xi_{l_m}) \to \xi' \in \langle t_0(\Lambda'_+) \rangle$. Now $t_{l_m}(\xi_{l_m})$ is admissible for $\hat{\pi}_{t_0 \Lambda'_-}$ and the definition of $\hat{\pi}_{t_0 \Lambda'_-}$ yields $0 = |\hat{\pi}_{t_0 \Lambda'_-}(\hat{a})(\bar{\xi})| = \lim_{m \to \infty} |\hat{a}(\xi_{l_m})|$. Since $\mathbf{T}$ acts continuously on $C_0^\Lambda(X^\#)$ we get the contradiction

$$(5.34) \qquad \epsilon < |\hat{a}(t_{l_m}\xi_{l_m})| \le |\hat{a}(t_{l_m}\xi_{l_m}) - \hat{a}(\xi_{l_m})| + |\hat{a}(\xi_{l_m})| \to 0.$$

2. Case: None of the sequences $\text{pr}_{\langle t_0 \Lambda'_\pm \rangle}(\xi_l)$ is bounded.

Now we find a subsequence (ξ_{l_m}) admissible for $\hat{\pi}_\Lambda$ and get in view of Remark 4.9 the contradiction

$$(5.35) \qquad 0 = |\bar{\psi}(0)| = \lim_{m \to \infty} |\psi(\xi_{l_m})| = \lim_{m \to \infty} |\psi(t_{l_m}\xi_{l_m})| > \epsilon.$$

LEMMA 5.13.

$$l_E \left(\bigcap_{tF=t\Lambda'_\pm, \Lambda_0} \bigcap_{s \in S} Ker\, \nu^s_{tF} \right) l_E \subset \mathcal{K}(H^2_\Lambda(S)).$$

PROOF. Corollary 5.10 and Proposition 5.11 yield together with Katayama's Theorem

$$(5.36) \qquad \bigcap_{tF=t\Lambda'_\pm, \Lambda_0} \bigcap_{s \in S} Ker\, \nu^s_{tF} \subset \Phi\left(Ker\pi \otimes_\delta C_0(S)\right).$$

Each element a of the right hand side is a limit of sums of products $(\sum_i \prod_j l_{a_{ij}} g_{ij})$, with $a_{ij} \in Ker\pi$, $g_{ij} \dot{\in} C_0(S)$. Let $u_i \in C_0(X^\#)$ be a function with $u_i \mid_\Lambda = \widehat{a_{i1}}$ and $(u_{i_m}) \subset A(S)$ be a sequence converging uniformly to u_i. Thus $\lim_{m \to \infty} \chi_\Lambda u_{i_m} = \chi_\Lambda u_i = \chi_\Lambda \widehat{a_{i1}}$ and applying $Ad(\mathcal{F})$ yields

$$(5.37) \qquad \lim_{m \to \infty} l_E\, l_{\overline{u_{i_m}}}\, g_{i1} \prod_{i>1} a_{ij} g_{ij} = l_E\, a_{i1} g_{i1} \prod_{i>1} a_{ij} g_{ij}.$$

Since $l_{\overline{u_{i_m}}} g_{i1} \in \mathcal{K}(L^2(S))$ and this is an ideal we see that $l_E\, l_{\overline{u_{i_m}}} g_{ij} \prod_{i>1} a_{ij}\, g_{ij}$ is an element of $l_E \mathcal{K}(L^2(S))$ for all m. Therefore

$$(5.38) \qquad\qquad l_E \prod_j a_{ij}\, g_{ij} \in l_E\, \mathcal{K}(L^2(S))$$

for all of the above products and hence $l_E\, a \in l_E\, \mathcal{K}(L^2(S))$, thus $l_E\, a\, l_E \in l_E\, \mathcal{K}(L^2(S))\, l_E = \mathcal{K}(H^2_\Lambda(S))$.

In order to complete the proof of Theorem 5.8 let $l_E\, a\, l_E$ be an element of the right side of (5.28) and suppose without loss of generality $a^* = a$. As in Theorem 5.4 one shows for all $tF = t\Lambda'_\pm$

$$(5.39) \qquad \nu^s_{tF}(a\, l_E\, a\, l_E\, a) = \nu^s_{tF}(a)\rho^s_{tF}(l_E\, a\, l_E)\nu^s_{tF}(a) = 0,$$

implying by Lemma 5.13 that

$$(5.40) \qquad l_E\, a\, l_E\, a\, l_E\, a\, l_E = (l_E\, a\, l_E)^3 \in \mathcal{K}(H^2_\Lambda(S)).$$

Let $\mathcal{L}(H^2_\Lambda(S))/\mathcal{K}(H^2_\Lambda(S))$ be the Calkin C^*-algebra, then the Gelfand-Naimark Theorem yields

$$(5.41) \qquad \|l_E\, a\, l_E\|^3_{\mathcal{L}/\mathcal{K}} = \|(l_E\, a\, l_E)^3\|_{\mathcal{L}/\mathcal{K}} = 0$$

and Theorem 5.8 is proved.

REMARK 5.14. Applying the same reasoning to each 1-dimensional face $t\Lambda'_\pm$ one shows

$$(5.42) \qquad l_{E_{tF}}\, \mathcal{K}(L^2(tS_F))\, l_{E_{tF}} = \bigcap_{\sigma \in tS_F} Ker\rho^\sigma_{tF}$$

where ρ^σ_{tF} is the vertex representation of $\mathcal{L}(L^2(tS_F))$ (translated by σ) and E_{tF} is the corresponding Szegö projection on the half-line.

The next Theorem is our main result:

THEOREM 5.15. *For the Toeplitz algebra* $\mathcal{T}(S)$ *(over the exterior of the double lightcone) define a filtration*

$$(5.43) \qquad I_0 = (0) \lhd I_1 = \mathcal{K}(H^2_\Lambda(S)) \lhd I_2 \lhd I_3 = \mathcal{T}(S)$$

by

$$(5.44) \qquad I_k = \bigcap_{\dim(F)=k} \bigcap_{s \in S/S_F} \bigcap_{t \in T/T_F} Ker\rho^s_{tF}.$$

Then the graduation is as follows

$$(5.45) \qquad I_3/I_2 \cong \mathcal{C}_0(S)$$

$$(5.46) \qquad I_2/I_1 \cong \overset{\oplus}{\int_{\dim(F)=1}} \overset{\oplus}{\int_{s \in S/S_F}} \overset{\oplus}{\int_{t \in T/T_F}} \mathcal{K}(l_{E_{tF}}\, L^2(tS_F))$$

$$\text{(continuous field of compact operators)}$$

$$(5.47) \qquad I_1/I_0 \cong \mathcal{K}(H^2_\Lambda(S)).$$

PROOF. Define

$$(5.48) \qquad \rho_k := \int_{\dim(F)=k}^{\oplus} \int_{s \in S/S_F}^{\oplus} \int_{t \in T/T_F}^{\oplus} \rho_{tF}^s.$$

Then $Ker \rho_k = I_k$ by definition and we have to show that

$$(5.49) \qquad \rho_k(I_{k+1}) = \int_{\dim(F)=k}^{\oplus} \int_{s \in S/S_F}^{\oplus} \int_{t \in T/T_F}^{\oplus} \mathcal{K}(l_{E_{tF}} L^2(tS_F)).$$

For $k = 0$ we have $\rho_0(I_1) = I_1 = \mathcal{K}(H_\Lambda^2(S))$, as shown in Theorem 5.8, since $F = (0), T_F = T$ and $S_F = S$. For $k = 1$ let $x \in I_2 = Ker \rho_2$ (given by the translated vertex representations). Then $\rho_{tF}^s(x)$ is compact for all $t \in T/T_F = T$ and $s \in S/S_F$, since for each $\sigma \in tS_F$ we have $\rho_{tF}^\sigma(\rho_{tF}^s(x)) = \rho_\Lambda^{\sigma+s}(x) = 0$, where ρ_{tF}^σ is as in Remark 5.14. It is easy to see that $\rho_{tF}^s(x)$ depends continuously on $t \in T$ and $s \in S_F$ (recall $F = \Lambda'_\pm$). Since

$$(5.50) \qquad \rho_{tF}^s(x)(l_E (l_{gE} f)l_E) = l_{E_{tF}} (l_{\bar{g}E_{tF}} \bar{f}_s)l_{E_{tF}}$$

(with the bar denoting restriction to tS_F) it follows that (5.50) vanishes as $s \to \infty$ in S/S_F since $f \in \mathcal{C}_0(S)$. Thus $\rho_1(x)$ belongs to the continuous field algebra (5.49). Also, (5.50) and Lemma 5.6, applied to the half-line case tS_F, imply that $\rho_{tF}^s(I_2)$ gives all compact operators. Now suppose $(F_1, s_1, t_1), (F_2, s_2, t_2)$ are distinct in the sense that either $F_1 \neq F_2$ or $F_1 = F_2$ and $s_1 - s_2 \notin S_F$ or $t_1 \neq t_2$. If $s_1 + t_1 S_{F_1} \neq s_2 + t_2 S_{F_2}$ there exists $x = l_E f l_E$ with $f_{s_1} |_{t_1 F_1} = 0, f_{s_2} |_{t_2 F_2} \neq 0$. Then for any $y = l_E g l_E$ we have $z := xy - yx \in I_2$ and $\rho_{t_1 F_1}^{s_1}(z) = 0$ but $\rho_{t_2 F_2}^{s_2}(z) \neq 0$ since g is still arbitrary. If $s_1 + t_1 S_{F_1} = s_2 + t_2 S_{F_2}$ we construct z as in Lemma 5.5 by taking a suitable function ψ, now even fulfilling $\lim \psi(\xi) = 0$ if $\xi \to \infty$ relative Λ to ensure that $z \in I_2$. Again $\rho_{t_1 F_1}^{s_1}(z) = 0, \rho_{t_2 F_2}^{s_2}(z) \neq 0$, so that $\rho_{t_2 F_2}^{s_2}(\mathcal{T}(S) z \mathcal{T}(S)) = \mathcal{K}(l_{E_{t_2 F_2}} L^2(t_2 S_{F_2}))$. The Stone-Weierstrass Theorem for operator fields [3, 11.5.3] completes the proof.

For $k = 2$, the same reasoning shows the assertion using the scalar Stone-Weierstrass Theorem.

References

[1] J. E. D'Atri and S. Gindikin, *Siegel domain realization of pseudo-Hermitian symmetric manifolds*, Geometrica Dedicata **46(1)** (1993), 91–125.

[2] R. Curto, P. S. Muhly, *C*-algebras of multiplication operators on Bergmann spaces*, J. Funct. Anal. **64** (1985), 315–329.

[3] J. Dixmier, *C*-Algebras*, North-Holland Publishing Company, Amsterdam-New York-Oxford, 1969.

[4] R. Douglas, *Banach Algebra Techniques in Operator Theory*, Academic Press, New York-London, 1972.

[5] P. Eymard, *L'algèbre de Fourier d'un groupe localement compact*, Bull. Soc. Math. France **92** (1964), 181–236.

[6] J. Faraut and A. Korányi, *Analysis on Symmetric Cones*, Clarendon Press, Oxford, 1994.

[7] I. Gelfand and S. Gindikin, *Complex manifolds whose skeletons are semisimple real Lie Groups, and analytic discrete series of representations*, Funkt. Anal. Prilozh **11(4)** (1977), 20–28.

[8] S. Gindikin, *Fourier transform and Hardy spaces of $\bar{\partial}$-cohomology in tube domains*, C. R. Acad. Sci. Paris **315(11)** (1992), 1139–1143.

[9] J. Hilgert and K.-H. Neeb, *Grouppoid C-*-algebras of order compactified symmetric spaces*, Japan. J. Math. **21(1)** (1995), 117–188.

[10] S. Kaneyuki, *Pseudo-Hermitian symmetric spaces and symmetric domains over non-degenerate cones*, Hokkaido Math. J. **20** (1991), 213–239.

[11] M. B. Landstad, J. Phillips, I. Raeburn, and C. E. Sutherland, *Representations of crossed products by coactions and principal bundles*, Trans. Amer. Math. Soc. **299** (1987), 747–784.

[12] P. S. Muhly, J. N. Renault, *C*-algebras of multivariable Wiener-Hopf operators*, Trans. Amer. Math. Soc. **1** (1982), 1–44.

[13] Y. Nakagami and M. Takesaki, *Duality for crossed products of von Neumann algebras*, Lecture Notes in Math. 731, Springer Verlag, Heidelberg-New York, 1979.

[14] G. I. Olshanski, *Complex Lie semigroups, Hardy spaces and the Gelfand Gindikin program*, Diff. Geom. Appl. **1** (1991), 235–246.

[15] N. Salinas, A. Sheu, and H. Upmeier, *Toeplitz operators on pseudoconvex domains and foliation algebras*, Ann. Math. **130** (1989), 531–565.

[16] E. M. Stein and D. Weiss, *Introduction to Fourier Analysis on Euclidean Spaces*, Princeton University Press, Princeton, 1971.

[17] S. Strătilă, *Modular Theory in Operator Algebras*, Abacus Press, 1981.

[18] H. Upmeier, *Toeplitz C*-algebras on bounded symmetric domains*, Ann. Math. **119** (1984), 549–576.

[19] H. Upmeier, *Multivariable Toeplitz Operators and Index Theory*, Birkhäuser-Verlag, to appear.

FACHBEREICH MATHEMATIK, UNIVERSITÄT MARBURG, 35032 MARBURG, GERMANY
E-mail address: hagenb@mathematik.uni-marburg.de

FACHBEREICH MATHEMATIK, UNIVERSITÄT MARBURG, 35032 MARBURG, GERMANY
E-mail address: upmeier@mathematik.uni-marburg.de

Contemporary Mathematics
Volume **212**, 1998

On an Application of the Bochner–Martinelli Operator

A. M. Kytmanov and S. G. Myslivets

ABSTRACT. Some problems of holomorphic extension of functions from the boundary are considered using integral representations in several complex variables.

Let D be a bounded domain in $\mathbb{C}^n$ with smooth connected boundary ∂D. Consider the Bochner–Martinelli integral operator for a continuous function f given on ∂D:

$$M f(z) = \int_{\partial D} f(\zeta) \, U(\zeta, z), \qquad z \notin \partial D,$$

where the Bochner–Martinelli kernel is

$$U(\zeta, z) = \frac{(n-1)!}{(2\pi i)^n} \sum_{k=1}^{n} (-1)^{k-1} \frac{\overline{\zeta}_k - \overline{z}_k}{|\zeta - z|^{2n}} d\overline{\zeta}[k] \wedge d\zeta,$$

and $d\zeta = d\zeta_1 \wedge \cdots \wedge d\zeta_n$, $d\overline{\zeta}[k] = d\overline{\zeta}_1 \wedge \cdots \wedge d\overline{\zeta}_{k-1} \wedge d\overline{\zeta}_{k+1} \wedge \cdots \wedge d\overline{\zeta}_n$.

Suppose a complex line l has a form

$$l = \{\zeta : \zeta_k = z_k + b_k t, k = 1, \ldots, n, t \in \mathbb{C}\},$$
$$b = (b_1, \ldots, b_n) \in \mathbb{CP}^{n-1}.$$

Then

(1)
$$U(\zeta, z) = \frac{dt}{t} \wedge \lambda(b),$$

where $\lambda(b)$ is a differential $(n-1, n-1)$–form in $\mathbb{CP}^{n-1}$ which does not depend on t (see, for example, [**1**, **Ch. 4**]).

THEOREM 1. *If for a fixed nonnegative integer k, for all points $z \notin \partial D$, and for almost all complex lines l,*

(2)
$$\int_{\partial D \cap l} f(z + bt) t^k \, dt = 0,$$

1991 *Mathematics Subject Classification.* Primary 32A25, 32D15; Secondary 31B05.
Supported by RFFI, grant 93–11–00258

then the function f has a holomorphic continuation to D.

The proof of this theorem is based on two lemmas.

LEMMA 1. *If for a continuous function f given on ∂D the condition (2) holds for a fixed point z ∉ ∂D, then*

$$(3) \qquad \int_{\partial D} (\zeta - z)^\alpha f(\zeta) \, U(\zeta, z) = 0,$$

where $\alpha = (\alpha_1, \ldots, \alpha_n)$ is an arbitrary multiindex with the property $\|\alpha\| = \alpha_1 + \cdots + \alpha_n = k + 1$ and $(\zeta - z)^\alpha = (\zeta_1 - z_1)^{\alpha_1} \cdots (\zeta_n - z_n)^{\alpha_n}$.

This lemma is the direct corollary of (1) and Fubini's theorem.

LEMMA 2. *If for all points $z \notin \partial D$ the condition (3) holds, then the function f has a holomorphic continuation to D.*

PROOF. First of all we observe (by differentiation) that condition (3) is true for all multiindeces α with $0 < \|\alpha\| \leqslant k + 1$.

In particular we get the equality

$$(4) \qquad \int_{\partial D} (\zeta_k - z_k) f(\zeta) \, U(\zeta, z) = 0$$

for all $k = 1, \ldots, n$ and for all $z \notin \partial D$.

Applying to (4) the Laplace operator Δ we obtain

$$\frac{\partial}{\partial \bar{z}_k} M f(z) = 0 \quad \text{for all } z \notin \partial D, \ k = 1, \ldots, n.$$

Thus the Bochner–Martinelli integral $Mf(z)$ is holomorphic outside ∂D; therefore this integral vanishes outside $\overline{D}$, and by the jump theorem its boundary values coincide with f.   □

THEOREM 2. *Let condition (2) hold for all points $z \in V$ (V is some open set in D or outside $\overline{D}$) and for almost all complex lines l through z. Then the function f has a holomorphic extension to D.*

Theorem 2 is a generalization of Theorem 1. For its proof we need only a generalization of Lemma 2 to this case.

LEMMA 3. *If condition (3) holds for all $z \in V$, then the function f has a holomorphic extension in domain to the D.*

Our theorems are generalizations of the boundary Morera theorem from [2] and the theorem on functions with the property of one-dimensional holomorphic extension along complex lines [3]. These results were proved in [4] in detail.

Now, for fixed nonnegative integers $p_1, \ldots p_n$ we consider the class of algebraic curves l_p of the form

$$(5) \qquad l_p = \{\zeta \in \mathbb{C}^n : \zeta_j = z_j + b_j t^{p_j}, j = 1, \ldots, n, t \in \mathbb{C}\},$$

where $z, b \in \mathbb{C}^n$.

We say a function $f \in \mathcal{C}(\partial D)$ has *the property of one-dimensional holomorphic extension along algebraic curves l_p* if for each curve l_p (of the form (5)) there exists a function F_l such that: a) F_l is continuous on the set $\overline{D} \cap l_p$, (b) F_l is a holomorphic function (of t) in the interior of $\overline{D} \cap l_p$, (c) $F_l = f$ on $\partial D \cap l_p$.

THEOREM 3. *If the function $f \in C^\varepsilon(\partial D)$ ($\varepsilon > 0$) has the property of one-dimensional holomorphic extension along all algebraic curves l_p then f has a holomorphic extension to D, as a function of n complex variables.*

The proof of Theorem 3 uses the assertion that functions are representable by some Cauchy-Fantappie integral.

THEOREM 4. *If for the function $f \in C^\varepsilon(\partial D)$ ($\varepsilon > 0$) the Cauchy-Fantappie integral*

$$F(z) = \int_{\partial D} f(\zeta) \, U_\alpha(\zeta, z)$$

gives a continuous extension F of f into D, then the function F is holomorphic in D.

The Cauchy-Fantappie kernel $U_\alpha(\zeta, z)$ has the form

$$U_\alpha(\zeta, z) = \frac{(n-1)!}{(2\pi i)^n} \prod_{j=1}^{n} \alpha_j |\zeta_j - z_j|^{2\alpha_j - 2}$$

$$\times \sum_{k=1}^{n} (-1)^{k-1} \frac{\overline{\zeta}_k - \overline{z}_k}{\alpha_k |\zeta - z|^{2n}_{(\alpha)}} \, d\overline{\zeta}[k] \wedge d\zeta,$$

where

$$|\zeta - z|^2_{(\alpha)} = |\zeta_1 - z_1|^{2\alpha_1} + \cdots + |\zeta_n - z_n|^{2\alpha_n}.$$

The kernel $U(\zeta, z)$ is constructed with the help of functions $w_j = (\zeta_j - z_j)^{\alpha_j - 1} \cdot (\overline{\zeta}_j - \overline{z}_j)^{\alpha_j}$, $p_j \alpha_j = \text{const}$.

For $\alpha_j = 1$, $j = 1, \ldots, n$, we have that $U_\alpha(\zeta, z)$ is the Bochner–Martinelli kernel $U(\zeta, z)$, but in general the kernel U_α is not harmonic in ζ, z.

For the Bochner–Martinelli kernel Theorem 4 was proved earlier (see, for example, [1, Ch. 4]).

Theorem 3 is also a generalization of a theorem by Stout [3].

The proof of Theorem 3 is based on theorem 4 and the formula analogous to (1) for the Cauchy–Fantappie kernel $U_\alpha(\zeta, z)$.

The proof of Theorem 4 is more complicated. We need the jump theorem and the maximum modulus theorem for the Cauchy-Fantappie integral, the Cachy-Fantappie formula for smooth function (not only holomorphic) and so on. The proofs can be found in [5, 6] in detail.

At the end of this paper we formulate some problems.

PROBLEM 1. *Let $f \in C(\partial D)$ and suppose that for all points $z \notin \partial D$ and for almost all complex curves l_p through point z,*

$$\int_{\partial D \cap l_p} f(z_1 + b_1 t^{p_1}, \ldots, z_n + b_n t^{p_n}) \, dt = 0.$$

Then f has a holomorphic continuation into D.

This is an analogue of the boundary Morera theorem (see, [2] and Theorems 1, 2).

PROBLEM 2. *For which Cauchy-Fantappie kernels does Theorem 4 hold? Perhaps these are the kernels in the formula of multidimensional logarithmic residue for holomorphic mappings with one zero.*

About this problem see [1, **Ch. 5**].

References

1. A. M. Kytmanov, *The Bochner-Martinelli integral and its application*, Birkhäuser Verlag, Basel, Boston, Berlin, 1995.
2. J. Globevnik and E. L. Stout, *Boundary Morera theorems for holomorphic functions of several complex variables*, Duke Math. J. **64** (1991), no. 3, 571–615.
3. E. L. Stout, *The boundary values of holomorphic functions of several complex variebles*, Duke Math. J. **44** (1977), no. 1, 105–108.
4. A. M. Kytmanov and S. G. Myslivets, *On one boundary analogue of the Morera theorem*, Sibirsk. Mat. Zh. **36** (1995), no. 6, 1350–1353 (Russian).
5. A. M. Kytmanov and S. G. Myslivets, *On one criterion of the existence of holomorphic extension of functions in* $\mathbb{C}^2$, Multidimensional Complex Analysis, Krasnoyarsk State University, Krasnoyarsk, 1994, pp. 78–91 (Russian).
6. A. M. Kytmanov and S. G. Myslivets, *On functions which are representable by some Cauchy-Fantappie integral*, Complex Analysis and Differential Equations, Krasnoyarsk State University, Krasnoyarsk, 1996, pp. 96–112 (Russian).

AKADEMGORODOK 19-25, KRASNOYRASK, 660036, RUSSIA
E-mail address: simona@kgu.krasnoyarsk.su

Contemporary Mathematics
Volume **212**, 1998

Local Estimates for Fractional
Integral Operators and Potentials

N. K. Karapetyants

ABSTRACT. The smoothness problems for the Riesz potentials $I^\alpha \varphi$ in $L_p(B_1)$ and the fractional integrals $I_+^\alpha \varphi$ in $L_p(0,1)$ are investigated. The mean oscillation estimate and the local estimates for the Riesz potential on the unit ball B_1 are obtained. It is proved that the fractional integral operator establishes the isomorphism between Hölder spaces of variable order $H_0^{\mu(x)}[0,1]$ and $H_0^{\mu(x)+\alpha}[0,1]$. Here $0 < \alpha < 1$ and $\mu(x)$ satisfy an additional condition $|\mu(x+h) - \mu(x)| \le c|\ln h|^{-1}, 0 < h < 0,5$. Besides, the action of the Riesz potential and the singular integral operators in Hölder spaces of variable order is studied.

1. Introduction

We consider two problems connected with a smoothness of fractional integrals and Riesz potentials with a density in $L_p, 1 \le p < \infty$. One of these problems concern to he Riesz potential operator

$$I^\alpha \varphi = c \int_{B_1} |x - y|^{\alpha - n} \varphi(y) dy, \quad 0 < \alpha < n, \tag{1}$$

where B_1 is a unit ball in R^n with the center at the origin. It is well known that

$$I^\alpha : L_p(B_1) \to \begin{cases} L_q(B_1), & q = \dfrac{np}{n - \alpha p}, \ \alpha p < n, \\ H^{\alpha - \frac{n}{p}}(B_1), & \alpha p > n, \ \alpha \notin N, \\ BMO(B_1), & \alpha p = n. \end{cases}$$

(G. Hardy, J. Littlewood, S. Sobolev, du N. Plessis, J. Peetre). References and history may be found in the extensive monograph [1]. We give the mean oscillation for the Riesz potential in the form

$$m_{B_\varepsilon(t)}(I^\alpha \varphi) \le c\varepsilon^{\alpha - \frac{n}{p}} \|\varphi\|,$$

1991 *Mathematics Subject Classification.* Primary 26A33; Secondary 26A16.

The work was supported by RFFY (Russian Funds of Fundamental Investigations), grant No 94-01-00577-A.

where $B_\varepsilon(t)$ is a ball of radius ε, and centered at the point $t \in R^n$ and study local properties of $I^\alpha \varphi$ in terms of the local characteristic

$$\chi_{(r,c)}(f,\varepsilon) = \left(\frac{1}{\varepsilon^n} \int_{|x-c|<\varepsilon} |f(x)|^r dx \right)^{\frac{1}{r}}. \tag{2}$$

Another problem concerns fractional integrals in Hölder space of variable order. G. Hardy and J. Littlewood showed that the fractional integral operator

$$I_+^\alpha \varphi = \frac{1}{\Gamma(\alpha)} \int_0^x \frac{\varphi(t)dt}{(x-t)^{1-\alpha}}, \quad 0 < \alpha < 1, \tag{3}$$

establishes an isomorphism between spaces $H_0^\mu[0,1]$ and $H_0^{\mu+\alpha}[0,1], 0 < \alpha, \mu < 1, \mu + \alpha < 1$. Here $H_0^\mu[0,1]$ is a Hölder space of functions φ with the condition $\varphi(0) = 0$. It is known [2] that the singular integral operator does not preserve the class functions which have order μ everywhere and Hölder order $\mu_0 > \mu$ at a point $x_0 \in [0,1]$. The question is: does fractional integral preserve that local property? The answer is positive.

The main results of this report were obtained with A. I. Ginsburg. Local estimates for fractional integrals in the case $n = 1$ were earlier obtained in [3] (see also [1]).

2. Mean oscillation estimates for the Riesz potentials

First we give some necessary definitions.

DEFINITION 1. Let $BMO(B_1)$ denote a class of functions $f(x) \in L_1(B_1)$ such that the following norm is finite

$$\|f(x)\|_{BMO} = \sup_{t \in B_1, 0<\varepsilon<1} m_{B_\varepsilon(t)} f,$$

where $m_{B_\varepsilon(t)} f$ is the mean oscillation on $B_\varepsilon(t)$:

$$m_{B_\varepsilon(t)} f = \frac{1}{|B_\varepsilon(t)|} \int_{B_\varepsilon(t)} |f - f_{B_\varepsilon(t)}| dx$$

and $f_{B_\varepsilon(t)}$ denotes the mean of f over the ball $B_\varepsilon(t), |B_\varepsilon(t)|$ being its volume. Let further $VMO(B_1)$ be the subspace of functions from $BMO(B_1)$ defined as

$$VMO(B_1) = \left\{ f : f \in BMO(B_1), \lim_{\delta \to 0} \sup_{0<\varepsilon<\delta} m_{B_\varepsilon(t)} f = 0 \right\}$$

THEOREM 1. Let $\varphi \in L_p(B_1)$, $1 \le p < \infty, 0 < \alpha < \frac{n}{p} + 1$ and $f = I^\alpha \varphi$. Then

$$\sup_{t \in B_1, 0<\varepsilon<1} \varepsilon^{\frac{n}{p}-\alpha} m_{B_\varepsilon(t)} f \le c \|\varphi\|_p, \tag{4}$$

and

$$\lim_{\varepsilon \to 0} \varepsilon^{\frac{n}{p}-\alpha} m_{B_\varepsilon(t)} f = 0 \tag{5}$$

with the constant c depending on α and n only.

COROLLARY. *If $\alpha p = n$, then the operator $I^\alpha \varphi$ is bounded from $L_p(B_1)$ into $BMO(B_1)$ (and also into $VMO(B_1)$).*

We denote further

$$BMO^0(B_1) = \{f : f \in BMO(B_1), \quad A < \infty\},$$

where

$$A = \sup_{a \in \Sigma_{n-1}} \sup_{0 < \varepsilon < \frac{1}{2}} \left| \int_{B_\varepsilon(t) \cap B_1} f(x)dx \right|.$$

Let $\overline{f}$ be the continuation of $f(x)$ by zero outside B_1. In terms of $BMO^0(B_1)$ we may formulate the criterion for $\overline{f}(x) \in BMO(R^n)$ if $f(x) \in BMO(B_1)$. Namely, let $f \in BMO(B_1)$. Then $\overline{f} \in BMO(R^n)$ if and only if $f \in BMO^0(B_1)$

THEOREM 2. *Let $\varphi \in BMO^0(B_1)$, $\alpha p = n$ and*

$$\sup_{a \in \Sigma_{n-1}} (I^\alpha |\varphi|)(a) < \infty,$$

then $I^\alpha \varphi \in BMO^0(B_1)$.

3. Local estimates for the Riesz potentials

We need the following notations: $p_\alpha = np(n - \alpha p)^{-1}$, if $\alpha p < n$, $\rho_c = dist(c, d\Omega)$, Ω is a domain in R^n. The following theorem gives an estimate for the local characteristic $\chi_{r,c}(f, \varepsilon)$.

THEOREM 3. *Let $f = I^\alpha \varphi$ Riesz potential on the domain $\Omega, 0 < \alpha < n, \varphi \in L_p(\Omega)$, $1 \le p \le \infty$ if $mes\Omega < \infty$ and $1 \le p < \frac{n}{\alpha}$ if $mes\Omega = \infty$. Then*

$$\chi_{r,c}(f, \varepsilon) \le cw_{\alpha,p}(\varepsilon)\|\varphi\|_p,$$

where

$$w_{\alpha,p}(\varepsilon) = \begin{cases} \varepsilon^{\alpha - \frac{n}{p}} & \text{if } 1 < p < \frac{n}{\alpha}, 1 \le r \le p; \\ \varepsilon^{\alpha - n} & \text{if } p = 1, 1 \le r < p_\alpha; \\ (1 + |ln\varepsilon|)^{\frac{1}{p'}} & \text{if } p = \frac{n}{\alpha}, 1 \le r < \infty, mes\Omega < \infty; \\ 1 & \text{if } \frac{n}{\alpha} < p \le \infty, mes\Omega < \infty \end{cases}$$

for $c \in \Omega, \varepsilon < dist(c, d\Omega)$. For $c \notin \Omega, 0 < \alpha < \frac{n}{p}, 1 < p < \infty$, $\varepsilon < \rho_c = dist(c, d\Omega)$ the estimate is

$$\chi_{r,c}(f, \varepsilon) \le c\|\varphi\|_p \rho_c^{\alpha - \frac{n}{p}}, \varepsilon \to 0$$

It should be noted that this theorem may be used for a new description of the image $I^\alpha(L_p)$ in terms of behavior of norms $\|\varphi\|_r$ at infinity if $r \to \infty$. Also we note that under assumptions of this theorem, in case of $1 \le p < n/\alpha$, the following assertion holds:

$$\chi_{r,c}(f, \varepsilon) = o(w_{\alpha,p}(\varepsilon)), \varepsilon \to 0.$$

4. Fractional integrals in Hölder spaces of variable order

We introduce a class $H^{\mu(x)}[0,1]$ of Hölder functions of variable order $\mu(x)$. Let $\mu(x)$ be a nonnegative function defined on $[0,1]$ and let

$$\mu_- = \inf_{x\in[0,1]} \mu(x), \quad \mu_+ = \sup_{x\in[0,1]} \mu(x) \qquad (6)$$

We suppose that $\mu_- > 0$.

DEFINITION 2. By $H^{\mu(x)}[0,1]$ we denote the class of functions continuous on $[0,1]$, with the norm

$$\|f\|_\mu = \|f\|_{C[0,1]} + \sup_{x\neq y} \frac{|f(x) - f(y)|}{|x-y|^{\mu(x)}}$$

The subspace of functions $f(x) \in H^{\mu(x)}[0,1]$ for which

$$\lim_{\delta\to 0} \sup_{|x-y|<\delta} \frac{|f(x) - f(y)|}{|x-y|^{\mu(x)}} = 0,$$

is defined by $h^{\mu(x)}[0,1]$. It is interesting that the functions from $C^\infty[0,1]$ are dense in $h^{\mu(x)}[0,1]$ if $\mu_+ < 1$ and $\mu(x)$ satisfies the inequality

$$|\mu(x+h) - \mu(x)| \leq c|\ln h|^{-1}, \quad 0 < h < \frac{1}{2}. \qquad (7)$$

This condition plays an important role in our further considerations.

The fractional integral $I_+^\alpha(\varphi)$ is usually called a left-right integral operator. The expression

$$(D_{+,0}^\alpha)(x) = \frac{f(x)}{x^\alpha\Gamma(1-\alpha)} + \frac{\alpha}{\Gamma(1-\alpha)} \int_0^x \frac{f(x) - f(y)}{(x-y)^{1+\alpha}} dy \qquad (8)$$

is called Marchaud derivative which exists for sufficiently good functions (see [1] for details).

The next theorem (see [1], p. 131) gives a sufficient condition for the representation of a function $f(x)$ by the fractional integral with a density in $L_p[0,1]$.

THEOREM 4. If $f(x) \in L_p(0,1)$, $1 < p < \frac{1}{\alpha}$ and $t^{-\alpha-1}\omega(f,t) \in L_1(0,1)$ then $f = I^\alpha\varphi$, where $\varphi \in L_p(0,1)$.

The main result in this item is contained in the following theorem (see [4]).

THEOREM 5. Let $\mu(x)$ satisfy (7) and $0 < \alpha$, $\mu < 1$, $\alpha + \mu < 1$. Then the operator $I_+^\alpha\varphi$ establishes an isomorphism between the spaces $H_0^{\mu(x)}[0,1]$ and $H_0^{\mu(x)+\alpha}[0,1]$.

The proof of this theorem is based on Theorem 4 and two following theorems about mapping properties of fractional integration and differentiation operators.

THEOREM 6. If $\varphi \in H_0^{\mu(x)}[0,1]$, $0 < \mu_+$, $\alpha < 1$, $\alpha + \mu_+ < 1$, then $I^\alpha\varphi \in H_0^{\mu(x)}[0,1]$.

THEOREM 7. *If* $f(x) \in H^{\mu(x)+\alpha}$, $\alpha + \mu_+ < 1$ *and* $\mu(x)$ *satisfies* (7) *then* $D_+^\alpha f \in H_0^{\mu(x)}[0,1]$

It should be noted that Theorem 5 was proved [5] independently by S. Samko and B. Ross under the additional condition (7), but they considered also variable order $\alpha(x)$.

The assertion of Theorem 6 may be extended to the case of Volterra integral operator of convolution type

$$(K\varphi)(x) = \int_0^x k(x-t)\varphi(t)dt,$$

which is based on the estimate

$$|(\Delta_h K\varphi)(x)|$$

$$\leq c\left\{ \int_0^x t^{\mu(x)}|(\Delta_h k)(t)|dt + \int_0^h |k(t)|(h-t)^{\mu(x)}dt + x^{\mu(x)}|\int_x^{x+h} k(t)dt| \right\},$$

where $(\Delta_h f)(x) = f(x+h) - f(x)$.

In conclusion of this item we note that these results may be extended to the generalized Hölder spaces $\overline{H}^{\mu(x),k}[0,1]$ defined by a finite difference of the order k and under the that assumption $\mu_- \in (0,k]$. Here $\overline{H}^{\mu(x),k}[0,1]$ consists of functions f from $H^{\mu(x),k}[0,1]$ whose extension to R^1 by zero belongs to $H^{\mu(x),k}(-\infty,1)$. For $k > 1$ this class and the class $H^{\mu(x),k}[0,1] = f : f \in H^{\mu(x),k}[0,1], f(0) = 0$ are different. In contrast to the case $k = 1$, here we need the condition (7) both for the derivative $D_+^\alpha f$, and for the fractional integration operator $I_+^\alpha \varphi$ (cf. Theorem 6).

5. The Riesz potential in $H^{\mu(x),1}(B_1)$

By $H^{\mu(x),1}(B_1)$ we denote the class of functions continuous on B_1 and satisfying the condition

$$|(\Delta_h^1 f(x)| \leq c|h|^{\mu(x)},$$

where $x, x+h \in B_1$, $0 < \mu_-, \mu_+ < 1$

THEOREM 8. *Let* $\varphi(x) \in H^{\mu(x),1}(B_1)$, *and* $\varphi(x) = 0$ *for all* $x \in \Sigma_{n-1}$, $0 < \mu_-$, $\mu_+ + \alpha < 1$. *Then* $I^\alpha \varphi \in H^{\mu(x)+\alpha}(B_1)$.

The proof of this theorem is based on the simple inequality

$$\left| \int_{|y|<1} \frac{dy}{|x-y|^{n-\gamma}} - \int_{|y-h|<1} \frac{dy}{|x-y|^{n-\gamma}} \right| \leq c|h|^{\min(1,\gamma)}$$

6. The singular integral in $H^{\mu(x)}(\Gamma)$

Let

$$S\varphi = \frac{1}{2\pi i} \int_\Gamma \frac{\varphi(\tau)}{\tau - t} d\tau$$

be the singular integral operator on a smooth simple closed curve Γ.

Earlier we noted that the singular integral operator does not preserve the class of functions which have order μ everywhere and order $\mu_0 > \mu$ at a point $t_0 \in \Gamma$.

But it does preserve the space $H^{\lambda(x)}$ under the additional condition (7). Namely, the following theorem is valid.

THEOREM 9. *Let L be a smooth contour, $0 < \mu_-,\ \mu_+ < 1$ and $\mu(x)$ satisfy the condition (7). Then $S\varphi$ is bounded from $H^{\mu(x),1}(L)$ into $H^{\mu(x),1}(L)$.*

References

[1] S. G. Samko, A. A. Kilbas, and O. I. Marichev, *Integrals and Derivatives of Fractional Order, Theory and Applications*, Gordon and Breach Sci. Publ., 1993 (Russian edition by "Nauka i Tekhnika", Minsk, 1987).

[2] D. I. Mamedhanov and A. A. Nersesyan, *On a constructive characteristic for the class $H_\alpha^{\alpha+\beta}(x_0, [\pi, \pi])$*, Investigations on the theory of linear operators, Azerb. Gos. Univ., 1987, pp. 74–77 (Russian).

[3] N. Karapetyants and B. S. Rubin, *Riesz radial potentials on the disc and fractional integration operators*, DAN SSR **263** (1982), 1299–1302, (Russian); English translation Soviet Math. Dokl., **25** (1982), 522–525.

[4] N. K. Karapetyants and A. I. Ginsburg, *The fractional integrodifferentiation in the Hölder classes of variable order*, Dokl. RAN **333** (1994), 439–441 (Russian).

[5] B. Ross and S. Samko, *Fractional integration operator of variable order in the Hölder spaces $H^{\lambda(x)}$*, Intern. J. of Math. and Math. Sci. **18** (1995), 777–788.

DEPARTMENT OF MATHEMATICS, ROSTOV STATE UNIVERSITY, BOL'SHAYA SADOVAYA 105, ROSTOV-ON-DON, 344711, RUSSIA

E-mail address: nkarapet@ns.unird.ac.ru

Contemporary Mathematics
Volume **212**, 1998

Hankel Operators on Clifford Valued Bergman Space

Chun Li and Zhijian Wu

ABSTRACT. We study Hankel operators on Clifford valued Bergman space—the $L^2(\mathbb{R}_+^{n+1})$ closure of the null space of Dirac operator. Unlike the classical Bergman space on the disk of complex plane, this space contains no dense algebra and the multiplication is also not commutative. We characterize the harmonic symbol for which the associated Hankel operators are bounded. Some discussion on the general symbols and compact Hankel operators are also included.

1. Introduction

Hankel operators play an important role in modern analysis. The connection with moment problems, engineering design, system theory and approximation makes the study of Hankel operators more interesting. Recent research reveals that Hankel operators are closed related to paracommutators (see [7]) and $\overline{\partial}$ operators (see for example, [8], [13] and [21]). Therefore tools in harmonic analysis can be used effectively to study these operators. On the other hand, Clifford algebra is becoming popular in harmonic analysis and operator theory, because of its power and simplicity in dealing with functions of n real variables (see for example [6], [9], [10], [16]–[17], [19] and references therein). In this paper, we study Hankel operators on Clifford valued Bergman space. These are the generalized Hankel operators on monogenic function space. (For Hankel operators on analytic or holomorphic function spaces we refer the reader to [3], [1], [2], [4], [8], [11], [12], [14], [22]–[23] and references therein.) Our main object is to characterize the boundedness of Hankel operators in terms of the symbol functions. The technique used in this paper combines both Clifford and harmonic analysis.

Let $\mathbb{C}_{(n)}$ be the 2^n−dimensional algebra on $\mathbb{C}$ with the standard basis

$$\{\mathbf{e}_S = \mathbf{e}_{j_1}\mathbf{e}_{j_2}\cdots\mathbf{e}_{j_s}\}, \qquad 1 \le j_1 < j_2 < \cdots < j_s \le n\,;$$

where $S = \{j_1, \dots, j_s\}$ is any subset of $\{1, 2, \cdots, n\}$. The algebra in $\mathbb{C}_{(n)}$ follows the following rules

$$\mathbf{e}_0 = 1\,, \quad \mathbf{e}_j\mathbf{e}_k + \mathbf{e}_k\mathbf{e}_j = -2\delta_{jk}\,, \quad \overline{\mathbf{e}_j} = -\mathbf{e}_j\,, \quad \overline{\mathbf{e}_R\,\mathbf{e}_S} = \overline{\mathbf{e}_S}\,\overline{\mathbf{e}_R}\,.$$

1991 *Mathematics Subject Classification.* 32A37, 47B35, 47B47.

Key words and phrases. Clifford, Hankel, Bergman, monogenic, paracommutator, boundedness.

Here $1 \le j$, $k \le n$, δ_{jk} equals one if $j = k$ and zero otherwise; R and S are subsets of $\{1, 2, \ldots, n\}$. Suppose $\lambda = \sum_S \lambda_S \mathbf{e}_S$, $\mu = \sum_S \mu_S \mathbf{e}_S \in \mathbb{C}_{(n)}$, with $\{\lambda_S\}$ and $\{\mu_S\}$ are sets of complex numbers. The product and the dot product of λ and μ are defined, respectively, by

$$\lambda\mu = \sum_{S,R} \lambda_S \mu_R \mathbf{e}_S \mathbf{e}_R \qquad \text{and} \qquad \lambda \cdot \mu = \sum_S \lambda_S \overline{\mu_S}\,.$$

The magnitude of λ is $|\lambda| = (\lambda \cdot \lambda)^{1/2}$. We observe that $|\lambda\mu| \le C\,|\lambda|\,|\mu|$, where C is a positive constant depended only on n.

Suppose F is a function which takes value in $\mathbb{C}_{(n)}$. We can write

$$F = \sum_S f_S \mathbf{e}_S,$$

where $\{f_S\}$, the components of F, are complex valued functions. We say a Clifford valued function F is in L^r if $|F|$ is in L^r, or equivalently if its components are in L^r. For $F \in L^r(\mathbb{R}^{n+1}_+)$, the norm of F is defined by $\|F\|_r = \||F|\|_r$.

In this paper, we will always use upper case letters for Clifford valued functions (e.g. F, G etc.) and lower case letters for complex valued functions (e.g. f, g etc.).

Let $x = (x_1, \ldots, x_n)$ and $(x, y) \in \mathbb{R}^{n+1}_+$. The Dirac operator is defined by

$$\mathcal{D} = \sum_{j=0}^n \frac{\partial}{\partial x_j} \mathbf{e}_j, \qquad (x_0 = y).$$

Suppose $F = \sum_S f_S \mathbf{e}_S$. The left and right actions of $\mathcal{D}$ on F are respectively

$$\mathcal{D}F = \sum_0^n \sum_S \frac{\partial f_S}{\partial x_j} \mathbf{e}_j \mathbf{e}_S \qquad \text{and} \qquad \mathcal{D}_r F = F\mathcal{D} = \sum_0^n \sum_S \frac{\partial f_S}{\partial x_j} \mathbf{e}_S \mathbf{e}_j\,.$$

A Clifford valued function F on $\mathbb{R}^{n+1}_+$ is said to be left (or right) monogenic if $\mathcal{D}F = 0$ (or $\mathcal{D}_r F = 0$). Denote by LA and RA the spaces of all the left and right monogenic functions in $L^2(\mathbb{R}^{n+1}_+)$, respectively.

For F, G in $L^2(\mathbb{R}^{n+1}_+)$, define the pairing

$$\langle F, G \rangle = \int_{\mathbb{R}^{n+1}_+} F(x,y)\overline{G(x,y)}dxdy\,.$$

Note that $\langle \cdot, \cdot \rangle$ is an inner product for complex valued (but not for Clifford valued) $L^2(\mathbb{R}^{n+1}_+)$. However in Clifford valued $L^2(\mathbb{R}^{n+1}_+)$, we still have

$$\|F\|_2^2 \le |\langle F, F \rangle| \qquad \text{and} \qquad |\langle F, G \rangle| \le C\,\|F\|_2\,\|G\|_2\,.$$

For this reason, we will still use the word "pairing" and the corresponding concepts such as orthogonal, complement etc.

The orthogonal projection from $L^2(\mathbb{R}^{n+1}_+)$ onto RA will be denoted by $\mathcal{P}$. More precisely $\mathcal{P}$ satisfies

$$\langle \mathcal{P}(F), G \rangle = \langle F, G \rangle, \qquad \text{for all} \quad F \in L^2(\mathbb{R}^{n+1}_+) \quad \text{and} \quad G \in RA\,.$$

We note that an explicit integral formula for the projection $\mathcal{P}$ was found in [16]–[17], however we do not need it in this paper.

Let $\mathcal{M}_b$ denote the operator of multiplication by the function b. The big and small Hankel operators with symbol b are defined formally by

$$\mathcal{H}_b(F) = (\mathcal{I} - \mathcal{P})\mathcal{M}_b F \quad \text{and} \quad \mathbf{h}_b(F) = \overline{\mathcal{P}\mathcal{M}_b \overline{F}}.$$

The main results of this paper are

THEOREM 1. *Suppose* $b \in L^1(\mathbb{R}^{n+1}_+)$ *is complex valued and harmonic. Then* $\mathcal{H}_b$ *is bounded on RA if and only if b satisfies* $\sup_{(x,y)\in\mathbb{R}^{n+1}_+} \left| y\frac{\partial b}{\partial y}(x,y) \right| < \infty$.

THEOREM 2. *Suppose* $b \in L^1(\mathbb{R}^{n+1}_+)$ *is complex valued and harmonic. Then* $\mathbf{h}_b$ *is bounded on LA if and only if b satisfies* $\sup_{(x,y)\in\mathbb{R}^{n+1}_+} \left| y\frac{\partial b}{\partial y}(x,y) \right| < \infty$.

We note that Theorem 2 is different from the result for the boundedness of same type operators on harmonic Bergman space on $\mathbb{R}^{n+1}_+$.

Section 2 contains notation and preliminaries. Sections 3 and 4 prove the direct and inverse of our main results, respectively. Some remarks are gathered in Section 5. Throughout this paper, the letter "C" denotes a positive constant which may vary at each occurrence but is independent of the essential variables or quantities.

2. Preliminaries

The Fourier transform of a function F on $\mathbb{R}^n$ is defined by

$$\widehat{F}(\xi) = (2\pi)^{-n/2} \int_{\mathbb{R}^n} F(x)e^{-ix\cdot\xi}dx, \quad \xi \in \mathbb{R}^n.$$

For $F \in L^p(\mathbb{R}^n), p > 1$, we use $F(x,y)$ to mean the harmonic extension of F onto $\mathbb{R}^{n+1}_+$. More precisely

$$\widehat{F_y}(\xi) = \widehat{F}(\xi)e^{-y|\xi|} \quad (\text{ or } F(x,y) = p_y * F(x)),$$

where $e^{-y|\xi|} = \widehat{p_y}(\xi)$ and $p_y(x) = c_n y/\left(|x|^2 + y^2\right)^{(n+1)/2}$ is the Poisson kernel.

Suppose $F \in L^p(\mathbb{R}^n)$. It is well-known (see for example [6]) that the harmonic extension of F onto $\mathbb{R}^{n+1}_+$ is left, or right monogenic if and only if

$$\chi_-(\xi)\widehat{F}(\xi) = 0, \quad \text{or} \quad \widehat{F}(\xi)\chi_-(\xi) = 0,$$

respectively. Here $\chi_+(\xi)$ and $\chi_-(\xi)$ are the characteristic functions. The definition of $\chi_\pm(\xi)$ and some basic facts of them needed in this paper are gathered in the following

$$\chi_\pm(\xi) = \frac{|\xi| \pm \sum_{j=1}^n i\xi_j \mathbf{e}_j}{2|\xi|}, \quad \xi \neq 0;$$

$$\chi_+(\xi) + \chi_-(\xi) = 1, \quad \chi_+(\xi)\chi_-(\xi) = 0, \quad \overline{\chi_\pm(\xi)} = \chi_\pm(\xi), \quad \chi_\pm(\xi) \cdot \chi_\pm(\xi) = \frac{1}{2}.$$

Suppose $F \in L^2(\mathbb{R}^n)$ with $\widehat{F}(\xi)\chi_-(\xi) = 0$ (or $\chi_-(\xi)\widehat{F}(\xi) = 0$). Assume in further that $\widehat{F}(\xi)|\xi|^{-1/2} \in L^2(\mathbb{R}^n)$. Then the right (or left) monogenic function $F(x,y)$ satisfies

$$\|F\|_2^2 = \int_{\mathbb{R}^{n+1}_+} \left|\widehat{F_y}(\xi)\right|^2 d\xi dy = \int_{\mathbb{R}^{n+1}_+} \left|\widehat{F}(\xi)\right|^2 e^{-2y|\xi|}d\xi dy = \frac{1}{2}\int_{\mathbb{R}^n} \frac{\left|\widehat{F}(\xi)\right|^2}{|\xi|}d\xi.$$

Based on these reasons, it is not hard to conclude that the spaces RA and LA can be viewed as the completions of the sets

$$RA_0 = \{F(x,y) : \widehat{F_y}(\xi) = e^{-y|\xi|}\Theta(\xi), \ \Theta \in C_0^\infty(\mathbb{R}^n \setminus \{0\}) \text{ and } \Theta(\xi)\chi_-(\xi) = 0\},$$

$$LA_0 = \{F(x,y) : \widehat{F_y}(\xi) = e^{-y|\xi|}\Theta(\xi), \ \Theta \in C_0^\infty(\mathbb{R}^n \setminus \{0\}) \text{ and } \chi_-(\xi)\Theta(\xi) = 0\}$$

respectively, under the norm of

$$\|F\|_2^2 = \int_{\mathbb{R}_+^{n+1}} |F(x,y)|^2 \, dxdy = \frac{1}{2} \int_{\mathbb{R}^n} \frac{|\Theta(\xi)|^2}{|\xi|} d\xi.$$

We can regard the domain of the big (or small) Hankel operator as RA_0 (or LA_0). Therefore the big (or small) Hankel operator is bounded on RA (or LA) means that it can be extended linearly to a bounded operator on RA (or LA).

3. Direct results

Throughout this section, we assume that b is a complex valued harmonic function in $L^1(\mathbb{R}_+^{n+1})$ and denote by B the quantity $\sup_{(x,y)\in\mathbb{R}_+^{n+1}} \left| y\frac{\partial b}{\partial y}(x,y) \right|$. We remark that (see [20]) $|y \bigtriangledown b(x,y)| \leq CB$, if $b \in L^1(\mathbb{R}_+^{n+1})$ is harmonic.

PROOF OF THE DIRECT RESULT OF THEOREM 1. Suppose $\tau \in C_0^\infty(\mathbb{R}_+^{n+1})$ is a Clifford valued function and F is in RA_0. By the product rule of derivative, we have

$$\mathcal{D}_r(bF) = F(\mathcal{D}b) + b(\mathcal{D}_rF) = F\mathcal{D}b.$$

Therefore using integration by parts, we obtain

$$\langle \mathcal{H}_b(F), \overline{\mathcal{D}_r}\tau \rangle = -\langle \mathcal{D}_r(bF), \tau \rangle = -\langle F\mathcal{D}b, \tau \rangle.$$

This implies

$$\left| \langle \mathcal{H}_b(F), \overline{\mathcal{D}_r}\tau \rangle \right| \leq C \|F(\mathcal{D}b)y\|_2 \left\| \frac{\tau}{y} \right\|_2 \leq CB \|F\|_2 \left\| \frac{\tau}{y} \right\|_2.$$

We claim that the set $RA_0 + \{\overline{\mathcal{D}_r}\tau : \tau \in C_0^\infty(\mathbb{R}_+^{n+1})\}$ is dense in $L^2(\mathbb{R}_+^{n+1})$, therefore we only need to prove that the inequality

$$(3.1) \qquad\qquad \left\| \frac{\tau}{y} \right\|_2 \leq C \left\| \overline{\mathcal{D}_r}\tau \right\|_2$$

holds for all $\tau \in C_0^\infty(\mathbb{R}_+^{n+1})$. In fact, to see why the claim is true, we only need to show that for any $\Phi \in L^2(\mathbb{R}_+^{n+1})$, if the identity

$$(3.2) \qquad\qquad \langle \Phi, F + \overline{\mathcal{D}_r}\tau \rangle = 0$$

holds for all $F \in RA_0$ and $\tau \in C_0^\infty(\mathbb{R}_+^{n+1})$, then $\Phi = 0$. Without loss of generality, we may assume $\Phi \in C^\infty(\mathbb{R}_+^{n+1})$ (note that $C^\infty(\mathbb{R}_+^{n+1})$ is dense in $L^2(\mathbb{R}_+^{n+1})$). Fix $F = 0$ in (3.2), we obtain $\mathcal{D}_r\Phi = 0$. Therefore Φ is in RA. Fix $\tau = 0$ in (3.2), we obtain $\langle \Phi, F \rangle = 0$ for all $F \in RA_0$. this yields $\Phi = 0$.

For independent interests, we break the proof of (3.1) into next two lemmas. $\square$

LEMMA 3.1. *Suppose $\tau \in C_0^\infty(\mathbb{R}_+^{n+1})$. Then*

$$\int_{\mathbb{R}_+^{n+1}} \frac{|\tau(x,y)|^2}{y^2} \, dx \, dy \leq C \int_{\mathbb{R}_+^{n+1}} \left| \frac{\partial \tau}{\partial y}(x,y) \right|^2 \, dx \, dy \,.$$

REMARK. A similar result was proved in [8] on the unit disk of complex plane.

PROOF OF LEMMA 3.1. We start with the following easy estimate

$$\frac{\partial}{\partial y} |\tau(x,y)|^2 \leq C \, |\tau(x,y)| \left| \frac{\partial \tau}{\partial y}(x,y) \right| \,.$$

Integrating both sides of this inequality over the interval $[0,y]$, we obtain

$$|\tau(x,y)|^2 = \int_0^y \frac{\partial}{\partial s} |\tau(x,s)|^2 \, ds \leq C \int_0^y |\tau(x,s)| \left| \frac{\partial \tau}{\partial s}(x,s) \right| \, ds \,.$$

Integrating again the left and right sides of above inequality over $[0, \infty)$ with respect to the measure $y^{-2} dy$, then using Fubini's theorem, we get

$$\int_0^\infty \frac{|\tau(x,y)|^2}{y^2} \, dy \leq C \int_0^\infty \int_0^y |\tau(x,s)| \left| \frac{\partial \tau}{\partial s}(x,s) \right| \, ds \frac{dy}{y^2}$$

$$= C \int_0^\infty \left(\int_s^\infty \frac{dy}{y^2} \right) |\tau(x,s)| \left| \frac{\partial \tau}{\partial s}(x,s) \right| \, ds$$

$$= C \int_0^\infty |\tau(x,s)| \left| \frac{\partial \tau}{\partial s}(x,s) \right| \frac{ds}{s} \,.$$

Applying Schartz's inequality to the last integral above, we then derive

$$\int_0^\infty \frac{|\tau(x,y)|^2}{y^2} \, dy \leq C \int_0^\infty \left| \frac{\partial \tau}{\partial y}(x,y) \right|^2 \, dy \,.$$

The desired result is therefore followed.   □

LEMMA 3.2. *For any $\tau \in C_0^\infty(\mathbb{R}_+^{n+1})$, we have*

$$\left\| \frac{\partial \tau}{\partial y} \right\|_2 \leq C \left\| \overline{\mathcal{D}_r} \tau \right\|_2 \,.$$

PROOF. Denote the Fourier transform of $\tau(x,y)$ on $\mathbb{R}^{n+1}$ by $\widetilde{\tau}(\xi,t)$. We have clearly

$$\widetilde{\frac{\partial \tau}{\partial y}}(\xi,t) = it\widetilde{\tau}(\xi,t) \quad \text{and} \quad \widetilde{\frac{\partial \tau}{\partial x_j}}(\xi,t) = i\xi_j\widetilde{\tau}(\xi,t), \quad 1 \leq j \leq n \,.$$

For $\xi = (\xi_1, \cdots, \xi_n)$, denote $\xi = \sum_{j=1}^n \xi_j \mathbf{e}_j$. We have then

$$\widetilde{\overline{\mathcal{D}_r} \tau}(\xi,t) = i\widetilde{\tau}(\xi,t)(t - \xi) \,.$$

On the other hand, it is clear that $t^2 + |\xi|^2 = |t \pm \xi|^2 = (t \pm \xi)(t \mp \xi)$. Therefore we have

$$t \leq \frac{(t - \xi)(t + \xi)}{|t + \xi|} \,.$$

Using this estimate, we obtain

$$\left|\widetilde{\frac{\partial \tau}{\partial y}}(\xi,t)\right| = t\,|\widetilde{\tau}(\xi,t)|$$

$$\leq |\widetilde{\tau}(\xi,t)|\,\frac{(t-\xi)(t+\xi)}{|t+\xi|}$$

$$= \left|\widetilde{\tau}(\xi,t)\frac{(t-\xi)(t+\xi)}{|t+\xi|}\right|$$

$$\leq C\,|\widetilde{\tau}(\xi,t)(t-\xi)|\left|\frac{t+\xi}{|t+\xi|}\right|$$

$$= C\left|\widetilde{\mathcal{D}_r\tau}(\xi,t)\right|.$$

Therefore

$$\left\|\frac{\partial \tau}{\partial y}\right\|_2^2 = \int_{\mathbb{R}_+^{n+1}}\left|\widetilde{\frac{\partial \tau}{\partial y}}(\xi,t)\right|^2 d\xi dt \leq C\int_{\mathbb{R}_+^{n+1}}\left|\widetilde{\mathcal{D}_r\tau}(\xi,t)\right|^2 d\xi dt = C\left\|\widetilde{\mathcal{D}_r\tau}\right\|_2^2.$$

The proof is complete. $\square$

To prove the direct result of Theorem 2, we note first that the small Hankel operator $\mathbf{h}_b$ is bounded on LA if and only if for $G \in LA_0$ and $F \in RA_0$, the magnitude of the bilinear form

$$\overline{\langle \mathbf{h}_b(G),F\rangle} = \overline{\langle \mathcal{P}(b\overline{G}),F\rangle} = \overline{\langle b\overline{G},F\rangle} = \langle FG,b\rangle$$

is dominated by $C\,\|F\|_2\,\|G\|_2$. We then prove the following general result.

THEOREM 3.3. *Suppose $p > 1$, $1/p + 1/q = 1$ and $b \in L^1(\mathbb{R}_+^{n+1})$ is a complex valued harmonic function and satisfies*

$$B = \sup_{(x,y)\in\mathbb{R}_+^{n+1}}\left|y\frac{\partial b}{\partial y}(x,y)\right| < \infty.$$

Then the bilinear form $\langle FG,b\rangle$ is bounded on $\big(RA \cap L^p(\mathbb{R}_+^{n+1})\big)\times\big(LA \cap L^q(\mathbb{R}_+^{n+1})\big)$. More precisely the estimate

$$|\langle FG,b\rangle| \leq CB\,\|F\|_p\,\|G\|_q$$

holds for all $F \in RA_0$ and $G \in LA_0$.

We need the following standard result (see for example [20]).

LEMMA 3.4. *Suppose $b \in L^1(\mathbb{R}_+^{n+1})$ is harmonic, k is a nonnegative integer and $\delta > 0$. Then*
1. $\int_{\mathbb{R}_+^{n+1}}\left|y^k\frac{\partial^k}{\partial y^k}b(x,y)\right|dxdy \asymp \|b\|_1$;
2. $y^k\frac{\partial^k}{\partial y^k}b(x,y) \to 0$ *uniformly as $|x|\to\infty$ or $y\to\infty$ for $y > \delta$;*
3. *For $y_0 > 0$, $b(x,y+y_0) \in L^\infty(\mathbb{R}_+^{n+1})$;*
4. *the limit $\hat{b}(\xi) = \lim_{y\to 0+}\widehat{b_y}(\xi)$ exists and $\left|\hat{b}(\xi)\right| \leq C|\xi|\,\|b\|_1$.*

PROOF OF THEOREM 3.3. Suppose F and G are in RA_0 and LA_0, respectively. We only need to show that $\left|\left\langle FG, \bar{b}\right\rangle\right|$ (it is $\bar{b}$) is dominated by $CB\left\|F\right\|_p\left\|G\right\|_q$.

In the following, the use of Green's theorem and integration by parts need that the associate functions in (x, y) goes to zero when $y \to \infty$ or $|x| \to \infty$. We omit the details, because they are the consequences of Lemma 3.4 and some standard results about harmonic functions in $L^r(\mathbb{R}^{n+1}_+)$, $r > 1$ (see for example [15]). For $t > 0$ and a harmonic function f on $\mathbb{R}^{n+1}_+$, define $f_t(x, y) = f(x, t + y)$. It is clear that f_t is also a harmonic function on $\mathbb{R}^{n+1}_+$. We note that components of right or left monogenic functions are all harmonic. The following notations will be used frequently.

$$\nabla_x = \left(\frac{\partial}{\partial x_1}, \frac{\partial}{\partial x_2}, \ldots, \frac{\partial}{\partial x_n}\right) \qquad \text{and} \qquad \nabla = \left(\nabla_x, \frac{\partial}{\partial y}\right);$$

$$\triangle = \nabla \cdot \nabla = \nabla_x \cdot \nabla_x + \frac{\partial^2}{\partial y^2} = \sum_{j=1}^{n} \frac{\partial^2}{\partial x_j{}^2} + \frac{\partial^2}{\partial y^2};$$

$$\nabla \Phi \cdot \nabla \Psi = \frac{\partial \Phi}{\partial y}\frac{\partial \Psi}{\partial y} + \sum_{j=1}^{n} \frac{\partial \Phi}{\partial x_j}\frac{\partial \Psi}{\partial x_j} \qquad \text{and} \qquad |\nabla \Phi| = |\nabla \Phi \cdot \nabla \Phi|^{\frac{1}{2}}.$$

For fixed $t > 0$, by Green's theorem we have

$$\int_{\mathbb{R}^n} F(x, t)G(x, t)b(x, t)\, dx = \int_{\mathbb{R}^{n+1}_+} y\triangle\left(F_t(x, y)G_t(x, y)b_t(x, y)\right)\, dxdy.$$

Integrating both sides of the above formula over $[0, \infty)$ with respect to the measure dt, and then using Fubini's theorem, we obtain

$$\begin{aligned}
\left\langle FG, \bar{b}\right\rangle &= \int_{\mathbb{R}^{n+1}_+} F(x, t)G(x, t)b(x, t)\, dxdt \\
&= \int_0^\infty \int_{\mathbb{R}^{n+1}_+} y\triangle\left(F(x, t+y)G(x, t+y)b(x, t+y)\right)\, dxdydt \\
&= \int_0^\infty \int_t^\infty \int_{\mathbb{R}^n} (y-t)\triangle\left(F(x, y)G(x, y)b(x, y)\right)\, dxdydt \\
&= \frac{1}{2}\int_{\mathbb{R}^{n+1}_+} y^2\triangle\left(F(x, y)G(x, y)b(x, y)\right)\, dxdy.
\end{aligned}$$

Using the following identity, which combines the harmonicity of functions with the product rule of derivatives:

$$\triangle\left(FGb\right) = \triangle\left(FG\right)b + 2\nabla\left(FG\right)\cdot\nabla b = 2\left(\nabla F \cdot \nabla G\right)b + 2\nabla\left(FG\right)\cdot\left(\nabla b\right),$$

we obtain

$$(3.3) \qquad \left\langle FG, \bar{b}\right\rangle = \int_{\mathbb{R}^{n+1}_+} \nabla\left(FG\right)\cdot\nabla b\, y^2 dxdy + \int_{\mathbb{R}^{n+1}_+} \left(\nabla F \cdot \nabla G\right)b\, y^2 dxdy.$$

Since $F(x, y)$ and $G(x, y)$ are right and left monogenic on $\mathbb{R}^{n+1}_+$ respectively, we have the following identities

$$\frac{\partial F}{\partial y} = -\sum_{j=1}^{n} \frac{\partial F}{\partial x_j}\mathbf{e}_j \qquad \text{and} \qquad \frac{\partial G}{\partial y} = -\sum_{j=1}^{n} \mathbf{e}_j \frac{\partial G}{\partial x_j}.$$

Thus the second integral in (3.3) can be computed as

$$\int_{\mathbb{R}_+^{n+1}} (\nabla F \cdot \nabla G)\, b\, y^2 dxdy = \int_{\mathbb{R}_+^{n+1}} \frac{\partial F}{\partial y}\frac{\partial G}{\partial y} b\, y^2 dxdy$$

$$+ \int_{\mathbb{R}_+^{n+1}} (\nabla_x F \cdot \nabla_x G)\, b\, y^2 dxdy$$

$$= \int_{\mathbb{R}_+^{n+1}} \sum_{j,k=1}^{n} \frac{\partial F}{\partial x_j}\mathbf{e}_j\mathbf{e}_k\frac{\partial G}{\partial x_k} b\, y^2 dxdy$$

$$+ \int_{\mathbb{R}_+^{n+1}} \sum_{j=1}^{n} \frac{\partial F}{\partial x_j}\frac{\partial G}{\partial x_j} b\, y^2 dxdy\,.$$

Integration by parts in x allows us to transform the derivatives from F onto G and then to continue the above computation to

$$= -\int_{\mathbb{R}_+^{n+1}} \sum_{j,k=1}^{n} F\mathbf{e}_j\mathbf{e}_k\frac{\partial^2 G}{\partial x_k \partial x_j} b\, y^2 dxdy + \int_{\mathbb{R}_+^{n+1}} \sum_{j=1}^{n} F\mathbf{e}_j\frac{\partial G}{\partial y}\frac{\partial b}{\partial x_j} y^2 dxdy$$

$$- \int_{\mathbb{R}_+^{n+1}} \sum_{j=1}^{n} F\frac{\partial^2 G}{\partial x_j^2} b\, y^2 dxdy - \int_{\mathbb{R}_+^{n+1}} F\sum_{j=1}^{n} \frac{\partial G}{\partial x_j}\frac{\partial b}{\partial x_j} y^2 dxdy\,.$$

Note that

$$\sum_{j,k=1}^{n} \mathbf{e}_j\mathbf{e}_k\frac{\partial^2 G}{\partial x_k \partial x_j} = -\sum_{j=1}^{n} \frac{\partial^2 G}{\partial x_j^2} \qquad \text{and} \qquad \sum_{j=1}^{n} \mathbf{e}_j\frac{\partial b}{\partial x_j} = \mathcal{D}b - \frac{\partial b}{\partial y}\,.$$

We then refine the result of the above computation as

$$\int_{\mathbb{R}_+^{n+1}} (\nabla F \cdot \nabla G)\, b\, y^2 dxdy = \int_{\mathbb{R}_+^{n+1}} F\, (\mathcal{D}b)\frac{\partial G}{\partial y} y^2 dxdy - \int_{\mathbb{R}_+^{n+1}} F \nabla G \cdot \nabla b\, y^2 dxdy\,.$$

Do the similar computation as above by transforming the derivatives from G to F, we can obtain

$$\int_{\mathbb{R}_+^{n+1}} (\nabla F \cdot \nabla G)\, b\, y^2 dxdy = \int_{\mathbb{R}_+^{n+1}} \frac{\partial F}{\partial y}\, (\mathcal{D}b)\, G\, y^2 dxdy - \int_{\mathbb{R}_+^{n+1}} \nabla F \cdot (G \nabla b)\, y^2 dxdy\,.$$

By these two formulas, together with (3.3), we conclude

$$\langle FG, \bar{b} \rangle = \frac{1}{2}\int_{\mathbb{R}_+^{n+1}} \left(\nabla (FG) \cdot \nabla b + F\,(\mathcal{D}b)\frac{\partial G}{\partial y} + \frac{\partial F}{\partial y}\,(\mathcal{D}b)\,G \right) y^2 dxdy\,.$$

By Hölder inequality, we get

$$|\langle FG, b \rangle| \leq C \int_{\mathbb{R}_+^{n+1}} (|\nabla F|\,|G| + |F|\,|\nabla G|)\,|\nabla b|\, ydxdy$$

$$\leq C \sup_{(x,y)\in\mathbb{R}_+^{n+1}} |y \nabla b(x,y)|\,\|F\|_p\,\|G\|_q\,.$$

Here we employ a standard result for harmonic functions $F \in L^r(\mathbb{R}^{n+1}_+)$ which says if $r > 1$ then $\|(\nabla F)y\|_r \asymp \|F\|_r$. Another standard result for harmonic function on $\mathbb{R}^{n+1}_+$ says (see [15]):

$$\sup_{(x,y)\in\mathbb{R}^{n+1}_+} \left| y\frac{\partial b}{\partial x_j}(x,y) \right| \leq C \sup_{(x,y)\in\mathbb{R}^{n+1}_+} \left| y\frac{\partial b}{\partial y}(x,y) \right|, \qquad j = 1, 2, \ldots, n.$$

This implies $\sup_{(x,y)\in\mathbb{R}^{n+1}_+} |y\nabla b(x,y)| \leq CB$. The proof is complete. $\qquad\qquad\square$

4. Inverse results

To prove the inverse results, we use the results and techniques established in [7] on the study of the paracommutators. For a complex valued function b defined on $\mathbb{R}^n$, the paracommutator with symbol b and kernel $A(\xi,\eta)$ is the operator defined by the following bilinear form on $C_0^\infty(\mathbb{R}^n) \times C_0^\infty(\mathbb{R}^n)$

$$(4.1) \qquad \langle T_b(A)f, g \rangle_{L^2(\mathbb{R}^n)} = \int_{\mathbb{R}^n}\int_{\mathbb{R}^n} \hat{f}(\eta)\hat{b}(\xi - \eta)A(\xi,\eta)\hat{g}(\xi)\, d\eta d\xi.$$

Assume that the kernel $A(\xi,\eta)$ meets certain regularities (see detail later), Janson and Peetre proved in [7] that the boundedness of the paracommutator $T_b(A)$ depends on the smoothness of the symbol b. We need some notations and definitions for these detail.

For $U, V \subseteq \mathbb{R}^n$, the set of Schur multipliers on $U \times V$ is denoted by $M(U \times V)$. This set contains all the functions $\phi \in L^\infty(U \times V)$ that admits the representation

$$(4.2) \qquad\qquad \phi(\xi,\eta) = \int_X \alpha(\xi,x)\beta(\eta,x)\, d\mu(x)$$

for some $\sigma-$finite measure space (X,μ) and measurable functions α on $U \times X$ and β on $V \times X$ with

$$\int_X \|\alpha(\cdot,x)\|_{L^\infty(U)}\, \|\beta(\cdot,x)\|_{L^\infty(V)}\, d\mu(x) < \infty.$$

The norm of the Schur multiplier ϕ, denoted by $\|\phi\|_{M(U\times V)}$, is defined by the minimum of the left hand side of the above inequality over all representations (4.2).

For $k = 0, \pm 1, \pm 2, \cdots$, consider the dyadic decomposition of $\mathbb{R}^n\backslash\{0\}$:

$$\Delta_k = \{\xi \in \mathbb{R}^n : 2^k \leq |\xi| < 2^{k+1}\}, \qquad \tilde{\Delta}_k = \Delta_{k-1} \cup \Delta_k \cup \Delta_{k+1}.$$

Let $\hat{\psi}$ be a test function with support in some annulus $\{\xi : r < |\xi| < R\}$ such that $\inf\{|\hat{\psi}(\xi)| : \xi \in \Delta_0\} > 0$, and define ψ_k by

$$\widehat{\psi_k}(\xi) = \hat{\psi}(2^{-k}\xi), \qquad k = 0, \pm 1, \pm 2, \cdots.$$

The Besov space B_∞ is defined by

$$b \in B_\infty \iff \sup_k\{\|\psi_k * b\|_{L^\infty}\} < \infty.$$

The quantity in the right hand side above is the B_∞ norm of b. It is standard that different choices of ψ give the same space and equivalent norms.

The properties of the kernel $A(\xi,\eta)$ are classified as following:

A0. Homogeneity, if $A(r\xi, r\eta) = A(\xi,\eta)$ for all $r \neq 0$ and $\xi, \eta \in \mathbb{R}^n$.

A1. Boundedness, if $\|A\|_{M(\Delta_j \times \Delta_k)} < C$ for all $j, k \in \mathbb{Z}$.

A3. Zero on the diagonal, if there exist $\gamma, \delta > 0$ such that $\|A\|_{M(B \times B)} \leq C \left(r / |\xi_0| \right)^\gamma$, where $B = Ball(\xi_0, r) = \{\xi : |\xi - \xi_0| < r\}$ and $0 < r < \delta |\xi_0|$.

A4. Nondegeneracy, if there exists no $\xi \neq 0$ such that $A(\xi + \eta, \eta) = 0$ for a.e. η.

The following results are proved in [7].

THEOREM A. (Janson and Peetre). *Suppose that A satisfies A0, A1, A3 and A4. Then the boundedness of the paracommutator (4.1) on $L^2(\mathbb{R}^n) \times L^2(\mathbb{R}^n)$ implies*

$$\|b\|_{B_\infty} \leq C \|T_b(A)\|_{(L^2(\mathbb{R}^n), L^2(\mathbb{R}^n))} \, .$$

LEMMA B. *If $\varphi \in C^{n+1}(\tilde{\Delta}_j \times \tilde{\Delta}_k)$, then*

$$\|\varphi\|_{M_{\Delta_j \times \Delta_k}} \leq C \sup_{|\alpha| + |\beta| \leq n+1} \sup_{\xi \in \Delta_j, \, \eta \in \Delta_k} |\xi|^{|\alpha|} |\eta|^{|\beta|} \left| D_\xi^\alpha D_\eta^\beta \varphi(\xi, \eta) \right| \, .$$

LEMMA C. *If $\varphi \in C_0^\infty(0, \infty)$, then $\varphi(|\eta|/|\xi|) \in M(\mathbb{R}^n \times \mathbb{R}^n)$.*

We now transform the boundedness of Hankel operators on RA or LA to the boundedness of certain paracommutators on $L^2(\mathbb{R}^n) \times L^2(\mathbb{R}^n)$.

Let $g(x) \mapsto R(x, y)$ and $f(x) \mapsto F(x, y)$ be two maps defined respectively by

$$\widehat{R_y}(\xi) = (1 - 2y|\xi|)|\xi|^{1/2} e^{-y|\xi|} \overline{\hat{g}(\xi)}; \qquad g \in C_0^\infty(\mathbb{R}^n) \, ,$$

$$\widehat{F_y}(\xi) = |\xi|^{1/2} e^{-y|\xi|} \hat{f}(\xi) \chi_+(\xi) \, , \qquad f \in C_0^\infty(\mathbb{R}^n) \, .$$

It is easy to see that these maps can be extended to bounded linear maps from $L^2(\mathbb{R}^n)$ into $L^2(\mathbb{R}_+^{n+1})$. Indeed, $C_0^\infty(\mathbb{R}^n)$ is dense in $L^2(\mathbb{R}^n)$ and straightforward computation yields the following identities

$$\|F\|_2^2 = \frac{1}{2} \int_0^\infty \left| \hat{f}(\xi) \right|^2 d\xi \qquad \text{and} \qquad \|R\|_2^2 = \frac{1}{4} \int_0^\infty |\hat{g}(\xi)|^2 \, d\xi \, .$$

Moreover, we have also that F lies in $LA_0 \cap RA_0$ and R lies in the complementary of $RA \cup LA$ in the sense of $\langle R, G \rangle = 0$ for any $G \in RA$ or LA. In fact, one can see this by assuming $\widehat{G_y}(\xi) = |\xi|^{1/2} e^{-y|\xi|} \widehat{G}(\xi)$, then do the computation

$$\langle G, R \rangle = \int_{\mathbb{R}_+^{n+1}} |\xi|^{1/2} e^{-y|\xi|} \widehat{G}(\xi) (1 - 2y|\xi|) |\xi|^{1/2} e^{-y|\xi|} \hat{g}(\xi) d\xi dy$$

$$= \int_{\mathbb{R}^n} \hat{G}(\xi) \hat{g}(\xi) \int_0^\infty (1 - 2y|\xi|) e^{-2y|\xi|} |\xi| dy d\xi$$

$$= 0 \, .$$

Now assume $b \in L^1(\mathbb{R}_+^{n+1})$ is harmonic. By Lemma 3.4, $\hat{b}(\xi)$ exists. Therefore $\widehat{b_y}(\xi) = e^{-y|\xi|} \hat{b}(\xi)$. For the big Hankel operator $\mathcal{H}_b$, we consider the pairing

$$\langle \mathcal{H}_b(F), R \rangle = \langle bF, R \rangle = \int_0^\infty \langle b_y F_y, R_y \rangle_{L^2(\mathbb{R}^n)} dy \, .$$

By Plancherel's formula, we have

$$\langle bF, R\rangle = \int_0^\infty \int_{\mathbb{R}^n} \int_{\mathbb{R}^n} \widehat{b_y}(\xi - \eta)\widehat{F_y}(\eta)\overline{\widehat{R_y}(\xi)}d\eta d\xi dy$$

$$= \int_0^\infty \int_{\mathbb{R}^n} \int_{\mathbb{R}^n} e^{-y|\xi - \eta|}\hat{b}(\xi - \eta)$$

$$\times |\eta|^{1/2}e^{-y|\eta|}\hat{f}(\eta)\chi_+(\eta)(1 - 2y|\xi|)|\xi|^{1/2}e^{-y|\xi|}\hat{g}(\xi)\, d\eta d\xi dy.$$

Using the integral formula

$$\int_0^\infty e^{-y|\xi - \eta|}e^{-y|\eta|}(1 - 2y|\xi|)e^{-y|\xi|}dy = \frac{|\eta| - |\xi| + |\xi - \eta|}{(|\xi| + |\eta| + |\xi - \eta|)^2},$$

we can continue the above computation as

$$= \int_{\mathbb{R}^n} \int_{\mathbb{R}^n} \frac{(|\eta| - |\xi| + |\xi - \eta|)|\eta|^{1/2}|\xi|^{1/2}}{(|\xi| + |\eta| + |\xi - \eta|)^2}\hat{b}(\xi - \eta)\chi_+(\eta)\hat{f}(\eta)\hat{g}(\xi)\, d\xi d\eta.$$

Note that the boundedness of the big Hankel operator implies that the scalar part of $\langle \mathcal{H}_b(F), R\rangle$ is dominated by $C\|F\|_2\|R\|_2 = C\|f\|_{L^2(\mathbb{R}^n)}\|g\|_{L^2(\mathbb{R}^n)}$. Thus the boundedness of the big Hankel operator implies that the following bilinear form of f and g, for $f, g \in C_0^\infty(\mathbb{R}^n)$,

$$(4.3) \qquad \int_{\mathbb{R}^n} \int_{\mathbb{R}^n} \frac{(|\eta| - |\xi| + |\xi - \eta|)|\eta|^{1/2}|\xi|^{1/2}}{(|\xi| + |\eta| + |\xi - \eta|)^2}\hat{b}(\xi - \eta)\hat{f}(\eta)\hat{g}(-\xi)\, d\xi d\eta,$$

can be extended linearly to a bounded operator on $L^2(\mathbb{R}^n) \times L^2(\mathbb{R}^n)$.

For the small Hankel operator $\mathbf{h}_b$, we know the boundedness implies that the pairing $\langle FG, \bar{b}\rangle$ is bounded by $C\|F\|_2\|G_2\|$ for $F \in RA_0$ and $G \in LA_0$. By Plancherel's formula, together with the fact that $\overline{\widehat{G_y}(\xi)} = \widehat{G_y}(-\xi)$, we have

$$\langle FG, \bar{b}\rangle = \langle bF, \overline{G}\rangle = \int_0^\infty \int_{\mathbb{R}^n} \int_{\mathbb{R}^n} \widehat{b_y}(\xi - \eta)\widehat{F_y}(\eta)\widehat{G_y}(-\xi)\, d\eta d\xi\, dy.$$

Suppose further that

$$\widehat{F_y}(\xi) = |\xi|^{1/2}e^{-y|\xi|}\hat{f}(\xi)\chi_+(\xi)$$

and

$$\widehat{G_y}(\xi) = |\xi|^{1/2}e^{-y|\xi|}\hat{g}(-\xi)\chi_+(\xi), \qquad f, g \in C_0^\infty(\mathbb{R}^n).$$

Then $\|F\|_2 = (1/\sqrt{2})\|f\|_{L^2(\mathbb{R}^n)}$ and $\|G\|_2 = \frac{1}{\sqrt{2}}\|g\|_{L^2(\mathbb{R}^n)}$. Using the fact that

$$\int_0^\infty e^{-y|\eta|}e^{-y|-\xi|}e^{-y|\xi - \eta|}dy = (|\xi| + |\eta| + |\xi - \eta|)^{-1},$$

we have therefore

$$\langle FG, \bar{b}\rangle = \int_{\mathbb{R}^n} \int_{\mathbb{R}^n} \frac{\chi_+(\eta)\chi_-(\xi)}{|\xi| + |\eta| + |\xi - \eta|}|\eta|^{1/2}|\xi|^{1/2}\hat{b}(\xi - \eta)\hat{f}(\eta)\hat{g}(\xi)\, d\eta d\xi.$$

A straightforward computation yields

$$\chi_+(\eta)\chi_-(\xi) = \frac{1}{4}\left(1 - \frac{\xi \cdot \eta}{|\xi||\eta|}\right) - \frac{i}{4}\sum_{j=1}^n \left(\frac{\xi_j}{|\xi|} - \frac{\eta_j}{|\eta|}\right)\mathbf{e}_j + \frac{1}{4}\sum_{j<k}\left(\frac{\eta_j\xi_k - \eta_k\xi_j}{|\xi||\eta|}\mathbf{e}_{jk}\right).$$

Hence the boundedness of the small Hankel operator implies that the bilinear form

$$(4.4) \qquad \int_{\mathbb{R}^n} \int_{\mathbb{R}^n} \frac{\hat{b}(\xi - \eta)|\xi|^{1/2}|\eta|^{1/2}}{|\xi| + |\eta| + |\xi - \eta|} \left(1 - \frac{\xi \cdot \eta}{|\xi||\eta|}\right) \hat{f}(\eta)\hat{g}(\xi) \, d\eta d\xi$$

is bounded on $L^2(\mathbb{R}^n) \times L^2(\mathbb{R}^n)$.

It is easy to see that bilinear forms (4.3) and (4.4) are the paracommutators $T_b(A_j)$, $j = 1, 2$ respectively with

$$A_1(\xi, \eta) = \frac{(|\eta| - |\xi| + |\xi - \eta|)|\eta|^{1/2}|\xi|^{1/2}}{(|\xi| + |\eta| + |\xi - \eta|)^2},$$

$$A_2(\xi, \eta) = \frac{|\xi|^{1/2}|\eta|^{1/2}}{|\xi| + |\eta| + |\xi - \eta|} \left(1 - \frac{\xi \cdot \eta}{|\xi||\eta|}\right).$$

We can now proof the inverse results of our main theorems. Together with Lemma C, we know the boundedness of bilinear forms (4.3) and (4.4) implies the boundedness of the paracommutators T_b with kernels $\varphi(|\eta|/|\xi|)A_j(\xi, \eta)$, $j = 1, 2$. Here $\varphi \in C_0^\infty(0, \infty)$, $\text{supp}\{\varphi\} = [4, 8]$. It is easy to see that these kernels satisfy A0, A3, A4 and A1 for $|j - k|$ small. To verify $\varphi(|\eta|/|\xi|)A_1(\xi, \eta)$ satisfies A1 for $|j - k|$ large, by the theory of Schur multiplier, we only need to check $A_1(\xi, \eta)$ itself satisfies A1 for $|j - k|$ large. This is a consequence of Lemma B. Similarly, for $\varphi(|\eta|/|\xi|)A_2(\xi, \eta)$, we only need to verify $A(\xi, \eta) = \frac{|\xi|^{1/2}|\eta|^{1/2}}{|\xi| + |\eta| + |\xi - \eta|}$ satisfies A1 for j and k apart, and $\phi(\xi, \eta) = 1 - \frac{\xi \cdot \eta}{|\xi||\eta|}$ is in $M(\mathbb{R}^n \times \mathbb{R}^n)$. The first one is again a consequence of Lemma B. For the second one, we have indeed

$$\left\| 1 - \frac{\xi \cdot \eta}{|\xi||\eta|} \right\|_{M(\mathbb{R}^n \times \mathbb{R}^n)} \leq 1 + \left\| \sum_{j=1}^n \frac{\xi_j}{|\xi|} \frac{\eta_j}{|\eta|} \right\|_{M(\mathbb{R}^n \times \mathbb{R}^n)} \leq 1 + n.$$

Thus, by Theorem A, we obtain that b is in B_∞. It is standard that if $b \in B_\infty$ and the harmonic function $b(x, y)$ in $L^1(\mathbb{R}_+^{n+1})$, then $b(x, y)$ satisfies $y\left|\frac{\partial b}{\partial y}(x, y)\right| < C$.

5. Remarks

Ideas and techniques developed in previous sections are enough to obtain the following results about the compactness of Hankel operators.

THEOREM 3. *Suppose $b \in L^1(\mathbb{R}_+^{n+1})$ is complex valued and harmonic. Then $\mathcal{H}_b$ is compactness on RA if and only if b satisfies $\lim_{y \to 0} \left|y\frac{\partial b}{\partial y}(x, y)\right| = 0$.*

THEOREM 4. *Suppose $b \in L^1(\mathbb{R}_+^{n+1})$ is complex valued and harmonic. Then $\mathbf{h}_b$ is compact on LA if and only if b satisfies $\lim_{y \to 0} \left|y\frac{\partial b}{\partial y}(x, y)\right| = 0$.*

REMARK. For Theorem 4, the similar result on harmonic Bergman space is that $b \equiv 0$ (see [20]).

We note that it is also possible to characterize the Hankel operators in Schatten $p-$class. However more technical detail are needed.

For the general symbol $b \in L^1(\mathbb{R}_+^{n+1})$, we still do not have a characterization for big and small Hankel operators. It is not hard to see that if b satisfies certain smoothness condition, then the corresponding big Hankel operator is bounded. However it is hard to prove the inverse. Luecking's idea for big Hankel operator (see [8], also [21]) is hard to push through because of the fact that the product of two monogenic functions is not monogenic in general. We believe more understanding on the product of monogenic functions and the reproducing formula are needed.

References

[1] J. Arazy, S. Fisher, S. Janson, and J. Peetre, *Membership of Hankel operators on the ball in unitary ideals*, J. London Math. Soc. (2) **43** (1991), no. 3, 485-508.

[2] J. Arazy, S. Fisher, and J. Peetre, *Hankel operators on weighted Bergman spaces*, Amer. J. Math. **110** (1988), 989-1054.

[3] S. Axler, *The Bergman space, the Bloch space and commutators of multiplication operators*, Duke Math. J. **53** (1986), 315-332.

[4] C.A. Berger, L.A. Coburn, and K. Zhu, *BMO on the Bergman spaces of the classical domains*, Bull. Amer. Math. Soc. **17** (1987), 133-136.

[5] R. Coifman and R. Rochberg, *Representation theorems for holomorphic and harmonic functions*, Astérisque **77** (1980), 11-65.

[6] J. Gilbert, and M. Murray, *Clifford Algebras and Dirac operators in Harmonic Analysis*, C.U.P., Cambridge, 1991.

[7] S. Janson, and J. Peetre, *Paracommutators-boundedness and Schatten-von Neumann properties*, Trans. Amer. Math. Soc. **305** (1988), 467-504.

[8] D. H. Luecking, *Characterizations of certain classes of Hankel operators on the Bergman spaces of the unit disk*, J. Funct. Anal. **110** (1992), 247-271.

[9] C. Li, A. Mcintosh, and S. Semmes, *Convolution singular integrals on Lipschitz surfaces*, J. of A.M.S. **5** (1992), 455–481.

[10] A. McIntosh, *Clifford algebras and the higher dimensional Cauchy integral*, Approximation Theory and Function Spaces, Banach Center Publications, Warsaw, Poland **22** (1989), 253-267.

[11] V.V. Peller, *Vectorial Hankel operators, commutators and related operators of the Schatten-Von Neumann class γ_p*, Int. Eq. Op. Theory **5** (1982), 244-272.

[12] S.C. Power, *Hankel operators on Hilbert space*, Pitman Publishing Inc, Boston, 1982.

[13] L. Peng, R. Rochberg, and Z. Wu, *Orthogonal polynomials and middle Hankel operators on Bergman spaces*, Studia Mathematica **102 (1)** (1992), 57-75.

[14] R. Rochberg, *Decomposition theorems for Bergman spaces and their applications*, Operator and Function Theory (S. C. Power, ed.), Reidel, Dordrecht, 1985, pp. 225-278.

[15] E. M. Stein, *Singular Integrals and Differentiability Properties of Functions*, Princeton Univ. Press, Princeton Univ., 1970.

[16] M. V. Shapiro, and N. L. Vasilevski, *On the Bergmann kernel function in the Clifford analysis*, Clifford Algebras and their applications in Mathematical Physics (1993), 183-192.

[17] _____, *On the Bergmann kernel function in hyperholomorphic analysis*, Acta Appl. Math. (1995) (to appear).

[18] A. Torchinsky, *Real–variable methods in harmonic analysis*, Academic press, Inc., New York, 1986.

[19] Z. Wu, *Clifford algebra, Hardy space and compensated compactness*, Clifford Algebra in Analysis and Related Topics (John Ryan, ed.), Studies in Adv. Math. Series, CRC Press, Boca Raton, 1996, pp. 217–238.

[20] _____, *Commutators oand related operators on harmonic Bergman space of $\mathbb{R}^{n+1}_+$*, J. Funct. Anal. (1996) (to appear).

[21] J. Wang, and Z. Wu, *Minimum solution of $\overline{\partial}^{k+1}$ and middle Hankel operators*, J. Funct. Anal. **118** (1993), No. 1, 167-187.

[22] K. Zhu, *Schatten class Hankel operators on the Bergman space of the unit ball*, Amer. J. Math. **113** (1991), 147-168.

[23] _____, *Operator Theory in Function Spaces*, Marcel Dekker, Inc., New York and Basel, 1990.

SCHOOL OF MATH., PHYS., MSCQUARIE UNIVERSITY, NSW 2109, AUSTRALIA
E-mail address: chun@macadam.mpce.mq.edu.au

DEPARTMENT OF MATHEMATICS, UNIVERSITY OF ALABAMA, TUSCALOOSA, AL 35487, U.S.A.
E-mail address: zwu@mathdept.as.ua.edu

Contemporary Mathematics
Volume **212**, 1998

Weitzenböck Type Formulas and Joint Seminormality

Mircea Martin and Norberto Salinas

ABSTRACT. This paper proposes an approach to joint seminormality based on analogies to the theory of Dirac and Laplace operators on Clifford vector bundles.

Introduction

This note is centered on two Weitzenböck type identities in multi-dimensional operator theory and some of their applications. The pattern we are going to pursue is explained in Section 1, where we very briefly survey the classical Weitzenböck formula and Bochner's method.

Our specific goal is to employ Bochner type techniques in the study of jointly seminormal tuples of commuting Hilbert space operators. The relevant definitions are reviewed in Section 2 and the main results are stated in Section 3. Section 4 is mostly concerned with some applications.

Before proceeding with a detailed presentation, we want to mention that the theory of seminormal operators has a fairly old history that could be tracked down, for instance, by summing up [21], [6], [25], and [20]. Its extension to systems of operators, despite some significant contributions (cf. [17], [15], [26], as well as [13] and the references therein), seems to be a difficult undertaking. We propose in the present article an approach based on some analogies with the much better understood theory of Dirac operators on Clifford vector bundles. In this regard it is not at all surprising that the Clifford algebras are encountered quite soon in some of our makeups, either implicitly, or explicitly. A few remarks in the final part of Section 4 will illustrate the point.

1. The classical Weitzenböck formula and Bochner's method

This section provides an informal very short review focused on the two related topics announced in its title. For a historical perspective on the genesis of Bochner's method we refer the reader to [3], [4], and [17]. Elaborate accounts on the current status of the large industry developed around more general types of Dirac and Laplace operators can be found in the excellent monographs [2], [16], and [18].

1991 *Mathematics Subject Classification.* Primary 47B20, 47A13; Secondary 47A63.
Supported in part by NSF Grant DMS 9301187.

To begin with, suppose M is a compact and connected riemannian manifold and let $\Lambda^*(TM^*) = \bigoplus_{p \geq 0} \Lambda^p(TM^*)$ denote the smooth vector bundle whose fiber at each point $x \in M$ equals the exterior algebra $\Lambda^*(TM^*_x) = \bigoplus_{p \geq 0} \Lambda^p(TM^*_x)$ of the cotangent space at x. The bundle $\Lambda^*(TM^*)$ comes in with two natural first order differential operators, namely, the exterior differential d, and the covariant derivative ∇ corresponding to the Levi-Civita connection. We let d* and ∇^* denote the formal adjoints of d and ∇, respectively. Three other basic differential operators acting on the same bundle are the *Dirac operator* $D = \mathrm{d} + \mathrm{d}^*$, its square $\Delta = \mathrm{d}\mathrm{d}^* + \mathrm{d}^*\mathrm{d}$, referred to as the *Hodge laplacian*, and the so-called *connection laplacian* $\Delta_\mathrm{c} = \nabla^*\nabla$.

The classical Weitzenböck formula, usually written down as

$$(1.1) \qquad\qquad \Delta = \Delta_\mathrm{c} + \mathfrak{R},$$

consists of an explicit and simple description of the remainder $\mathfrak{R}$. The point is that the operator $\mathfrak{R}$ turns out to be of order zero and, more importantly, it can be entirely expressed in terms of the curvature tensor of M.

These nice features of $\mathfrak{R}$ provide the basis of a method discovered by Bochner that yields estimates of the Betti numbers of M. Specifically, since $\mathfrak{R}$ is of order zero, it makes sense to consider the linear operators

$$\mathfrak{R}^{(p)}_x : \Lambda^p(TM^*_x) \to \Lambda^p(TM^*_x),$$

for any $p \geq 0$ and every $x \in M$. Further, let us assume that for a fixed $p \geq 1$ we have

$$(1.2) \qquad\qquad \mathfrak{R}^{(p)}_x \geq 0, \quad x \in M.$$

Then, from Weitzenböck formula (1.1) one may conclude that any p-form in the kernel of Δ also lies in the kernel of Δ_c. In other words, the space of all harmonic p-forms on M is contained in the space of all forms that are parallel with respect to the Levi-Civita connection. Employing Hodge theorem we easily get that the p-th Betti number of M does not exceed $m!/p!(m-p)!$, where m stands for the dimension of M. Moreover, if there exists at least one point $x \in M$ such that $\mathfrak{R}^{(p)}_x$ is a positive definite operator, then the p-th Betti number of M vanishes.

The positivity condition (1.2) is often reached under appropriate positivity assumptions on the curvature tensor. For instance, the operator $\mathfrak{R}^{(1)}_x$ is just a disguise of the Ricci curvature operator of M. Therefore, in case $p = 1$ condition (1.2) merely means that M is a riemannian manifold of non-negative Ricci curvature.

2. Joint seminormality

For later convenience, we next recollect the four already in use definitions of jointly seminormal tuples of operators. We begin with the case of a single operator. Suppose $\mathcal{H}$ is an infinite-dimensional complex Hilbert space and let $\mathcal{L}(\mathcal{H})$ be the algebra of all linear bounded operators on $\mathcal{H}$. An operator $T \in \mathcal{L}(\mathcal{H})$ is called *seminormal* whenever the commutator $[T, T^*] = TT^* - T^*T$ of T and its adjoint T^* is semidefinite. More precisely, T is called *hyponormal*, or *cohyponormal*, provided that $[T, T^*] \leq 0$, or $[T, T^*] \geq 0$, respectively. Clearly, a given operator is hyponormal if and only if its adjoint is cohyponormal.

Assume next that $T = (T_1, T_2, \ldots, T_n)$ is a commuting n-tuple of operators on $\mathcal{H}$. The collection of all commutators $[T_i, T_j^*]$, $1 \le i, j \le n$ may be employed in at least two different ways to manufacture a multi-dimensional substitute of the commutator used in the single operator case. Specifically, we let $\mathfrak{C}_{\mathrm{L}}(T)$ and $\mathfrak{C}_{\mathrm{R}}(T)$ denote the operators on $\mathcal{H} \otimes \mathbb{C}^n$ defined by

$$(2.1) \qquad \mathfrak{C}_{\mathrm{L}}(T)\xi \otimes e_j = \sum_{i=1}^{n} [T_i, T_j^*]\xi \otimes e_i, \quad 1 \le j \le n, \ \xi \in \mathcal{H},$$

and, respectively,

$$(2.2) \qquad \mathfrak{C}_{\mathrm{R}}(T)\xi \otimes e_j = \sum_{i=1}^{n} [T_j, T_i^*]\xi \otimes e_i, \quad 1 \le j \le n, \ \xi \in \mathcal{H},$$

where $\{e_1, e_2, \ldots, e_n\}$ is the standard basis for $\mathbb{C}^n$.

The terminology we are going to adopt throughout this article is summarized in the next definition.

DEFINITION 1. A commuting n-tuple $T = (T_1, T_2, \ldots, T_n)$ of operators on $\mathcal{H}$ is said to be *left seminormal* whenever the operator $\mathfrak{C}_{\mathrm{L}}(T)$ is semidefinite. In case $\mathfrak{C}_{\mathrm{L}}(T) \le 0$ the tuple T is called *left hyponormal*, and when $\mathfrak{C}_{\mathrm{L}}(T) \ge 0$ the tuple T is called *left cohyponormal*.

The concepts of *right seminormality, right hyponormality,* and *right cohyponormality* are introduced by imposing similar conditions on $\mathfrak{C}_{\mathrm{R}}(T)$.

Before proceeding with a couple of remarks we should mention that the notions of joint hyponormality and joint cohyponormality first introduced and studied in [1] and [26], respectively, correspond in our terminology to left hyponormality and left cohyponormality. In the same regard, what we call right hyponormality has been previously considered in [26] under the name of t-hyponormality.

The first simple remark we want to make refers to the relationship between a tuple $T = (T_1, T_2, \ldots, T_n)$ and its adjoint $T^* = (T_1^*, T_2^*, \ldots, T_n^*)$. Explicitly, we easily check that T is left hyponormal, or left cohyponormal, if and only if T^* is right cohyponormal, or right hyponormal, respectively.

The second observation is concerned with an alternative description of the operators $\mathfrak{C}_{\mathrm{L}}(T)$ and $\mathfrak{C}_{\mathrm{R}}(T)$. By identifying the space $\mathcal{H} \otimes \mathbb{C}^n$ with the orthogonal direct sum of n copies of $\mathcal{H}$, we can represent $\mathfrak{C}_{\mathrm{L}}(T)$ and $\mathfrak{C}_{\mathrm{R}}(T)$ by the $n \times n$ matrices

$$(2.3) \qquad\qquad \mathfrak{M}_{\mathrm{L}}(T) = ([T_i, T_j^*])_{i,j=1}^{n},$$

and, respectively,

$$(2.4) \qquad\qquad \mathfrak{M}_{\mathrm{R}}(T) = ([T_j, T_i^*])_{i,j=1}^{n}.$$

We may think of $\mathfrak{M}_{\mathrm{R}}(T)$ as the transpose of $\mathfrak{M}_{\mathrm{L}}(T)$, and in this way the terminology proposed in [26] gains a clear motivation. The reason we prefer our terminology will be explained below.

3. Two Weitzenböck type formulas

All the subsequent computations are carried out at the level of the Koszul complex associated with a given tuple $T = (T_1, T_2, \ldots, T_n)$ of operators on $\mathcal{H}$. Our goal is two-fold: we first want to detect natural counterparts of the classical Weitzenbröck formula (1.1), and then to investigate some consequences of such formulas under positivity assumptions similar to (1.2). We start with a brief review of some basic constructions in several variable spectral theory.

Given $n \geq 2$, we let $\Lambda_n^* = \bigoplus_{p=0}^{n} \Lambda_n$ denote the exterior algebra of $\mathbb{C}^n$. The inner product on $\mathbb{C}^n$ extends naturally to an inner product on Λ_n^* such that Λ_n^p is orthogonal to Λ_n^q when $p \neq q$. For each $v \in \mathbb{C}^n$ we define the *creation operator* $\varepsilon(v)$ in $\mathcal{L}(\Lambda_n^*)$ as the linear extension of

$$(3.1) \qquad \varepsilon(v)(v_1 \wedge \cdots \wedge v_p) = v \wedge v_1 \wedge \cdots \wedge v_p, \quad v_1, \ldots, v_p \in \mathbb{C}^n.$$

Its adjoint is called the *annihilation operator* associated with v and its given by

$$(3.2) \quad \varepsilon(v)^*(v_1 \wedge \cdots \wedge v_p) = \sum_{k=1}^{p} (-1)^{k-1} \langle v, v_k \rangle v_1 \wedge \cdots \wedge v_{k-1} \wedge v_{k+1} \wedge \cdots \wedge v_p.$$

Furthermore, for any $v, w \in \mathbb{C}^n$ we have

$$(3.3) \qquad \varepsilon(v)\varepsilon(w) + \varepsilon(w)\varepsilon(v) = 0,$$

and

$$(3.4) \qquad \varepsilon(v)\varepsilon(w)^* + \varepsilon(w)^*\varepsilon(v) = \langle v, w \rangle I,$$

where I stands for the identity operator on Λ_n^*. In particular, if $\varepsilon_i = \varepsilon(e_i)$ and $\varepsilon_i^* = \varepsilon(e_i)^*$ are the operators coresponding to the i-th standard basis vector of $\mathbb{C}^n$, then

$$(3.5) \qquad \varepsilon_i \varepsilon_j + \varepsilon_j \varepsilon_i = 0,$$

and

$$(3.6) \qquad \varepsilon_i \varepsilon_j^* + \varepsilon_j^* \varepsilon_i = \delta_{ij} I,$$

for all $1 \leq i, j \leq n$.

Suppose next that $T = (T_1, T_2, \ldots, T_n)$ is a commuting n-tuple of operators on a fixed infinite-dimensional Hilbert space $\mathcal{H}$ and $\omega = (\omega_1, \omega_2, \ldots, \omega_n)$ is a point in $\mathbb{C}^n$. We set $\omega - T = (\omega_1 - T_1, \omega_2 - T_2, \ldots, \omega_n - T_n)$ and let $d_{\omega - T} : \mathcal{H} \otimes \Lambda_n^* \to \mathcal{H} \otimes \Lambda_n^*$ denote the linear operator defined by

$$(3.7) \qquad d_{\omega - T} = \sum_{k=1}^{n} (\omega_k - T_k) \otimes \varepsilon_k.$$

It readily follows that $d_{\omega - T}^2 = 0$, hence

$$(3.8) \qquad \mathcal{K}^*(\omega, T) : \cdots \to \mathcal{H} \otimes \Lambda_n^{p-1} \xrightarrow{d_{\omega - T}} \mathcal{H} \otimes \Lambda_n^p \xrightarrow{d_{\omega - T}} \mathcal{H} \otimes \Lambda_n^{p+1} \to \cdots$$

is a cochain complex called the *Koszul complex of T at ω*. We associate to $d_{\omega - T}$ the operators $D(\omega, T)$ and $\Delta(\omega, T)$ in $\mathcal{L}(\mathcal{H} \otimes \Lambda_n^*)$ by setting

$$(3.9) \qquad D(\omega, T) = d_{\omega - T} + d_{\omega - T}^*,$$

and

$$(3.10) \qquad \Delta(\omega, T) = D(\omega, T)^2 = d_{\omega - T} d_{\omega - T}^* + d_{\omega - T}^* d_{\omega - T}.$$

In a fine analogy to Section 1, they will be referred to as the *Dirac operator* and the *full laplacian* of T at ω, respectively. Further, we let $\Delta_{\mathrm{L}}(\omega, T)$ and $\Delta_{\mathrm{R}}(\omega, T)$ denote the operators on $\mathcal{H}$ defined as

$$(3.11) \qquad \Delta_{\mathrm{L}}(\omega, T) = \sum_{k=1}^{n} (\omega_k - T_k)^*(\omega_k - T_k),$$

and

$$(3.12) \qquad \Delta_{\mathrm{R}}(\omega, T) = \sum_{k=1}^{n} (\omega_k - T_k)(\omega_k - T_k)^*.$$

They are called the *left laplacian*, and the *right laplacian of T at ω*, respectively. We notice that the full laplacian $\Delta(\omega, T)$ is homogeneous of degree zero with respect to the grading of $\mathcal{H} \otimes \Lambda_n^*$ so it makes sense to consider its restrictions

$$\Delta(\omega, T)^{(p)} = \Delta(\omega, T) | \mathcal{H} \otimes \Lambda_n^p, \quad 0 \le p \le n.$$

Under the canonical isomorphisms $\mathcal{H} \otimes \Lambda_n^0 \cong \mathcal{H}$ and $\mathcal{H} \otimes \Lambda_n^n \cong \mathcal{H}$ we have that $\Delta(\omega, T)^{(0)} \cong \Delta_{\mathrm{L}}(\omega, T)$ and $\Delta(\omega, T)^{(n)} \cong \Delta_{\mathrm{R}}(\omega, T)$.

The three laplacians associated to T and ω are related by two formulas analogous to formula (1.1). We list them below.

THEOREM. *If $T = (T_1, T_2, \ldots, T_n)$ and $\omega = (\omega_1, \omega_2, \ldots, \omega_n)$, then*

$$(3.13) \qquad \Delta(\omega, T) = \Delta_{\mathrm{L}}(\omega, T) \otimes I + \mathfrak{R}_{\mathrm{L}}(T),$$

and

$$(3.14) \qquad \Delta(\omega, T) = \Delta_{\mathrm{R}}(\omega, T) \otimes I + \mathfrak{R}_{\mathrm{R}}(T),$$

where the remainders $\mathfrak{R}_{\mathrm{L}}(T)$ and $\mathfrak{R}_{\mathrm{R}}(T)$ are explicitly given by

$$(3.15) \qquad \mathfrak{R}_{\mathrm{L}}(T) = \sum_{i,j=1}^{n} [T_i, T_j^*] \otimes \varepsilon_i \varepsilon_j^*,$$

and

$$(3.16) \qquad \mathfrak{R}_{\mathrm{R}}(T) = -\sum_{i,j=1}^{n} [T_j, T_i^*] \otimes \varepsilon_i^* \varepsilon_j.$$

PROOF. Indeed, from the previous definitions and based on (3.6) we successively get

$$\mathfrak{R}_{\mathrm{L}}(T) = \Delta(\omega, T) - \Delta_{\mathrm{L}}(\omega, T) \otimes I$$

$$= \sum_{i,j} \{ (\omega_i - T_i)(\omega_j - T_j)^* \otimes \varepsilon_i \varepsilon_j^* + (\omega_j - T_j)^*(\omega_i - T_i) \otimes \varepsilon_j^* \varepsilon_i \}$$

$$\qquad - \sum_{k} (\omega_k - T_k)^*(\omega_k - T_k) \otimes I$$

$$= \sum_{i,j} \{ [(\omega_i - T_i), (\omega_j - T_j)^*] \otimes \varepsilon_i \varepsilon_j^* + (\omega_j - T_j)^*(\omega_i - T_i) \otimes \delta_{ij} I \}$$

$$\qquad - \sum_{k} (\omega_k - T_k)^*(\omega_k - T_k) \otimes I$$

$$= \sum_{i,j} [T_i, T_j^*] \otimes \varepsilon_i \varepsilon_j^*.$$

Equation (3.16) follows similarly.

The interested reader may find two other identities resembling (3.14) and (3.15) in [14, Theorem 4].

For a later use, just as in Section 1, we set

(3.17) $\mathfrak{R}_{\mathrm{L}}(T)^{(p)} = \mathfrak{R}_{\mathrm{L}}(T)|\mathcal{H} \otimes \Lambda_n^p \in \mathcal{L}(\mathcal{H} \otimes \Lambda_n^p), \quad 0 \le p \le n,$

and

(3.18) $\mathfrak{R}_{\mathrm{R}}(T)^{(p)} = \mathfrak{R}_{\mathrm{R}}(T)|\mathcal{H} \otimes \Lambda_n^p \in \mathcal{L}(\mathcal{H} \otimes \Lambda_n^p), \quad 0 \le p \le n.$

The next two results make clear the significance of some semidefiniteness assumptions analogous to (1.2).

PROPOSITION 1. *The following four conditions are equivalent:*

(i) $\mathfrak{R}_{\mathrm{L}}(T) \ge 0$ *(resp. ≤ 0) as an operator in $\mathcal{L}(\mathcal{H} \otimes \Lambda_n^*)$;*
(ii) $\mathfrak{R}_{\mathrm{L}}(T)^{(p)} \ge 0$ *(resp. ≤ 0) as an operator in $\mathcal{L}(\mathcal{H} \otimes \Lambda_n^p)$ for any $0 \le p \le n$;*
(iii) $\mathfrak{R}_{\mathrm{L}}(T)^{(1)} \ge 0$ *(resp. ≤ 0) as an operator in $\mathcal{L}(\mathcal{H} \otimes \Lambda_n^1)$;*
(iv) T *is left cohyponormal (resp. left hyponormal).*

PROOF. The equivalence between (i) and (ii) and the fact that (ii) implies (iii) are obvious. Observe next that by identifying Λ_n^1 with $\mathbb{C}^n$ the operator $\mathfrak{R}_{\mathrm{L}}(T)^{(1)}$ equals the operator $\mathfrak{C}_{\mathrm{L}}(T)$ defined in Section 2. Therefore, condition (iii) could be reformulated as $\mathfrak{C}_{\mathrm{L}}(T) \ge 0$ (resp. ≤ 0), so (iii) and (iv) are indeed equivalent.

To conclude the proof it suffices to show that (iv) implies (i). To this end we first choose a unitary operator $U : \mathcal{H} \otimes \mathbb{C}^n \to \mathcal{H}$. Such a choice is possible because $\mathcal{H}$ is infinite-dimensional. Let $R = (R_1, R_2, \ldots, R_n)$ be the n-tuple of operators on $\mathcal{H}$ whose entries are uniquely determined by

(3.19) $R_j^* \xi = U\{\varepsilon \mathfrak{C}_{\mathrm{L}}(T)\}^{1/2}(\xi \otimes e_j), \ 1 \le j \le n, \ \xi \in \mathcal{H},$

where ε equals 1 or -1 according as $\mathfrak{C}_{\mathrm{L}}(T)$ is positive or negative semidefinite. From (3.19) and (2.1) we get

$$\begin{aligned}
\langle R_i R_j^* \xi, \xi \rangle_{\mathcal{H}} &= \langle R_j^* \xi, R_i^* \xi \rangle_{\mathcal{H}} \\
&= \langle U\{\varepsilon \mathfrak{C}_{\mathrm{L}}(T)\}^{1/2}(\xi \otimes e_j), U\{\varepsilon \mathfrak{C}_{\mathrm{L}}(T)\}^{1/2}(\xi \otimes e_i) \rangle_{\mathcal{H}} \\
&= \varepsilon \langle \mathfrak{C}_{\mathrm{L}}(T)(\xi \otimes e_j), \xi \otimes e_i \rangle_{\mathcal{H} \otimes \mathbb{C}^n} \\
&= \varepsilon \left\langle \sum_{k=1}^n [T_k, T_j^*] \xi \otimes e_k, \xi \otimes e_i \right\rangle_{\mathcal{H} \otimes \mathbb{C}^n} \\
&= \varepsilon \langle [T_i, T_j^*] \xi, \xi \rangle_{\mathcal{H}},
\end{aligned}$$

for any $1 \le i, j \le n$ and all $\xi \in \mathcal{H}$. Therefore

(3.20) $[T_i, T_j^*] = \varepsilon R_i R_j^*, \quad 1 \le i, j \le n.$

We may now associate to R an operator d_R in $\mathcal{L}(\mathcal{H} \otimes \Lambda_n^*)$ by setting

(3.21) $d_R = \sum_{k=1}^n R_k \otimes \varepsilon_k.$

It remains to observe that from (3.20) we have

$$(3.22) \qquad\qquad \mathfrak{R}_{\mathrm{L}}(T) = \varepsilon d_R d_R^*,$$

an equality that clearly implies condition (i).

PROPOSITION 2. *The following four conditions are equivalent:*

(i) $\mathfrak{R}_{\mathrm{R}}(T) \geq 0$ *(resp. ≤ 0) as an operator in $\mathcal{L}(\mathcal{H} \otimes \Lambda_n^*)$;*

(ii) $\mathfrak{R}_{\mathrm{R}}(T)^{(p)} \geq 0$ *(resp. ≤ 0) as an operator in $\mathcal{L}(\mathcal{H} \otimes \Lambda_n^p)$ for any $0 \leq p \leq n$;*

(iii) $\mathfrak{R}_{\mathrm{R}}(T)^{(n-1)} \geq 0$ *(resp. ≤ 0) as an operator in $\mathcal{L}(\mathcal{H} \otimes \Lambda_n^{n-1})$;*

(iv) T *is right hyponormal (resp. right cohyponormal).*

PROOF. The proof goes along the same lines as the proof of Proposition 1, with a minor adjustment. More precisely, we indentify $\mathbb{C}^n$ with Λ_n^{n-1} by mapping the i-th basis vector $e_i \in \mathbb{C}^n$ into $(-1)^{i-1} e_1 \wedge \cdots \wedge e_{i-1} \wedge e_{i+1} \wedge \cdots \wedge e_n \in \Lambda_n^{n-1}$. Under this identification the operator $\mathfrak{R}_{\mathrm{R}}(T)^{(n-1)}$ on $\mathcal{H} \otimes \Lambda_n^{n-1}$ corresponds to the operator $-\mathfrak{C}_R(T)$ on $\mathcal{H} \otimes \mathbb{C}^n$, so conditions (iii) and (iv) are equivalent. The rest of the proof is left to our reader.

Propositions 1 and 2 provide a motivation for the terminology introduced by Definition 1. The equivalence between conditions (iii) and (iv) in these propositions shows that we could check the left or right seminormality of T merely by looking at the left or right side of the Koszul complex. On the other hand let us observe that $\mathfrak{R}_{\mathrm{L}}^{(1)}(T)$ is the first non-trivial component of $\mathfrak{R}_{\mathrm{L}}(T)$ because clearly $\mathfrak{R}_{\mathrm{L}}^{(0)} = 0$. Likewise, $\mathfrak{R}_{\mathrm{R}}^{(n-1)}(T)$ is the last non-trivial component of $\mathfrak{R}_{\mathrm{R}}(T)$.

4. Applications and concluding remarks

In this section we are going to point out some consequences of the Weitzenböck type formulas (3.13) and (3.14). It goes without saying that all these consequences occur in pairs.

4.1. We begin with a very simple observation about the kernel of the laplacian $\Delta(\omega, T)$ corresponding to a left cohyponormal tuple $T = (T_1, T_2, \ldots, T_n)$, namely, we notice that

$$(4.1) \qquad \mathrm{Ker}\, \Delta(\omega, T) = \{\mathrm{Ker}\, \Delta_{\mathrm{L}}(\omega, T) \otimes \Lambda_n^*\} \cap \mathrm{Ker}\, \mathfrak{R}_{\mathrm{L}}(T),$$

for any $\omega \in \mathbb{C}^n$. In particular, from (4.1) we deduce that $\mathrm{Ker}\, \Delta(\omega, T)$ is finite dimensional if and only if $\mathrm{Ker}\, \Delta_{\mathrm{L}}(\omega, T)$ is finite dimensional. Such a situation occurs, for instance, in case T is a tuple in the Cowen-Douglas class $\mathcal{B}_k(\Omega)$ and $\omega \in \Omega$, where $k \geq 1$ is an integer and $\Omega \subset \mathbb{C}^n$ is an open set (see [10], [11] for more details). Following [12, Section 6] we may use (4.1) to characterize the Taylor essential spectrum of a left cohyponormal tuple. A similar characterization is also at hand for right hyponormal tuples. The precise statement is given below.

PROPOSITION 3. *Suppose T is left cohyponormal, or right hyponormal. Then $\omega \in \mathbb{C}^n$ is a point in the Taylor essential spectrum of T, if and only if $Ker\, \Delta_{\mathrm{L}}(\omega, T)$, or $Ker\, \Delta_{\mathrm{R}}(\omega, T)$, is infinite-dimensional, respectively.*

4.2. We next recall the definitions of some other spectral sets associated to a tuple $T = (T_1, T_2, \ldots, T_n)$ of Hilbert space operators.

DEFINITION 2. (cf. [22]). The *Taylor spectrum* $\sigma(T)$ of T is the set of all points $\omega \in \mathbb{C}^n$ such that the Koszul complex $\mathcal{K}^*(\omega, T)$ has a non-trivial cohomology.

DEFINITION 3. (cf. [5]). The *left spectrum* $\sigma_{\mathrm{L}}(T)$ (resp. *right spectrum* $\sigma_{\mathrm{R}}(T)$) of T consists of all points $\omega \in \mathbb{C}^n$ for which the exactness of the Koszul complex $\mathcal{K}^*(\omega, T)$ fails at $\mathcal{H} \otimes \Lambda_n^0$ (resp. $\mathcal{H} \otimes \Lambda_n^n$).

Each of the three spectra defined above has an alternative description in terms of laplacians associated to T, namely, a point $\omega \in \mathbb{C}^n$ lies in $\sigma(T)$, or $\sigma_{\mathrm{L}}(T)$, or $\sigma_{\mathrm{R}}(T)$, if and only if $\Delta(\omega, T)$, or $\Delta_{\mathrm{L}}(\omega, T)$, or $\Delta_{\mathrm{R}}(\omega, T)$ is not invertible, respectively. For details we may refer the reader to [12] or [23], [24].

Obviously, $\sigma_{\mathrm{L}}(T) \subseteq \sigma(T)$ and $\sigma_{\mathrm{R}}(T) \subseteq \sigma(T)$. Formulas (3.13) and (3.14) suggest a Bochner type reasoning that might be adopted to find sufficient conditions under which the left or right spectrum of T equals its Taylor spectrum. All we need is to employ the simple fact that the sum of two positive semidefinite Hilbert space operators is invertible whenever at least one of that operators is invertible. In view of Propositions 1 and 2 and based on the previous comments we get the following result.

PROPOSITION 4. *If T is left cohyponormal, or right hyponormal, then $\sigma_{\mathrm{L}}(T) = \sigma(T)$, or $\sigma_{\mathrm{R}}(T) = \sigma(T)$, respectively.*

For a different proof of Proposition 4 we refer to [26, Section 4] where this result was established for the first time.

4.3. As another application of formulas (3.13) and (3.14) we next indicate the multi-dimensional analoques of some techniques developed in the single operator case in [7], [8], [9], and [19]. We focus our attention exclusively on the class of left cohyponormal tuples.

If $T = (T_1, T_2, \ldots, T_n)$ is a left cohyponormal tuple and $\omega = (\omega_1, \omega_2, \ldots, \omega_n)$ is a given point in $\mathbb{C}^n$, then clearly

$$(4.2) \qquad \Delta(\omega, T) \geq \mathfrak{R}_{\mathrm{L}}(T),$$

and

$$(4.3) \qquad \Delta(\omega, T) \geq (\omega_i - T_i)^*(\omega_i - T_i) \otimes I, \quad 1 \leq i \leq n.$$

Let $R = (R_1, R_2, \ldots, R_n)$ be the n-tuple defined by formulas (3.19). Using (3.10) and (3.22) we may write (4.1) as

$$D(\omega, T)^2 \geq d_R d_R^*.$$

Consequently, there exists a unique contraction $C(\omega) \in \mathcal{L}(\mathcal{H} \otimes \Lambda_n^*)$ such that

$$(4.4) \qquad C(\omega)^* D(\omega, T) = d_R^*, \quad C(\omega)^* | \mathrm{Ker} D(\omega, T) = 0.$$

A similar reasoning yields a collection of contractions $K_i(\omega) \in \mathcal{L}(\mathcal{H} \otimes \Lambda_n^*)$, $1 \leq i \leq n$, satisfying

$$(4.5) \qquad K_i(\omega) D(\omega, T) = (\omega_i - T_i) \otimes I, \quad K_i(\omega) | \mathrm{Ker} D(\omega, T) = 0.$$

All these contractions are connected by the identity

$$(4.6) \qquad C(\omega) C(\omega)^* + \sum_{i=1}^n K_i(\omega)^* K_i(\omega) = P(\omega),$$

where $P(\omega)$ stands for the orthogonal projection onto the closure of the range of $D(\omega, T)$.

The proof starts by observing that (4.6) clearly holds on $\operatorname{Ker} \Delta(\omega, T)$. On the other hand, from (4.4) and (4.5) we get that formula (3.13) is equivalent to

$$\Delta(\omega, T) \left\{ C(\omega)C(\omega)^* + \sum_{i=1}^{n} K_i(\omega)^* K_i(\omega) \right\} \Delta(\omega, T) = \Delta(\omega, T)P(\omega)\Delta(\omega, T).$$

The last identity implies that (4.6) holds on the closure of the range of $\Delta(\omega, T)$, hence on the orthogonal complement of $\operatorname{Ker} \Delta(\omega, T)$ in $\mathcal{H} \otimes \Lambda_n^*$. The proof of (4.6) is complete.

4.4. The factorizations indicated in (4.4) and (4.5) above are not the only ones we could take advantage of. There are at least two other possibilities. Instead of $D(\omega, T)$ we can use either the square root of $\Delta(\omega, T)$, or the skew-hermitian operator $\hat{D}(\omega, T)$ defined by

$$(4.7) \qquad\qquad \hat{D}(\omega, T) = d_{\omega - T} - d_{\omega - T}^*,$$

where $d_{\omega - T}$ is the differential of the Koszul complex of T at ω. The later choice works because

$$(4.8) \qquad\qquad \Delta(\omega, T) = \hat{D}(\omega, T)^* \hat{D}(\omega, T) = \hat{D}(\omega, T)\hat{D}(\omega, T)^*.$$

Of course we get different contractions in each of the two cases, but still those contractions satisfy (4.6).

There is another reason that singles out $\hat{D}(\omega, T)$ as a better choice than $D(\omega, T)$. A significant part of the theory of non-normal operators, including the seminormal ones, has been initially developed as a theory of pairs of self-adjoint operators. Hence, it is natural to ask whether or not an analogous approach is available in a multi-dimensional setting. We conjecture that for tuples of self-adjoint operators a reasonable definition of seminormality may be possible by relying on Clifford algebras. To make a point, we let

$$(4.9) \qquad \omega_i - T_i = (\lambda_i - X_i) + \sqrt{-1}(\lambda_{i+1} - X_{i+1}), \quad 1 \le i \le n,$$

be the Cartesian form of the i-th entry of $\omega - T$. Further, set $X = (X_1, X_2, \ldots, X_{2n})$ and $\lambda = (\lambda_1, \lambda_2, \ldots, \lambda_{2n})$. An easy computation shows that

$$(4.10) \qquad\qquad \hat{D}(\omega, T) = \sum_{k=1}^{2n} (\lambda_k - X_k) \otimes \gamma_k,$$

where $\gamma_k \in \mathcal{L}(\Lambda_n^*)$, $1 \le k \le 2n$, are given by

$$(4.11) \qquad\qquad \gamma_i = \varepsilon_i - \varepsilon_i^*, \quad 1 \le i \le n,$$

and

$$(4.12) \qquad\qquad \gamma_{i+n} = \sqrt{-1}(\varepsilon_i + \varepsilon_i^*), \quad 1 \le i \le n.$$

From (3.5) and (3.6) we get

$$(4.13) \qquad\qquad \gamma_k \gamma_l + \gamma_l \gamma_k = -2\delta_{kl} I, \quad 1 \le k, l \le 2n.$$

Consequently, the algebra generated in $\mathcal{L}(\Lambda_n^*)$ by $\{\gamma_k : 1 \le k \le 2n\}$ is a realization of the complex Clifford algebra $\mathfrak{A}_m(\mathbb{C})$ in $m = 2n$ generators. As a matter of fact, in this way we get the entire algebra $\mathcal{L}(\Lambda_n^*)$, since $\mathcal{L}(\Lambda_n^*)$ has the same complex dimension as $\mathfrak{A}_m(\mathbb{C})$.

As a closing remark we want to say that although all the previous definitions and results could be now reformulated in terms of X and the Clifford algebra $\mathfrak{A}_m(\mathbb{C})$, however the tuple X defined above is rather peculiar. It has an even number of entries and these entries inherit some special properties from the strong assumption on T to be a commuting tuple. Presumably the best way of dealing with more general tuples of self-adjoint operators might originate from some different Weitzenböck type formulas. We are planning to pursue such an approach elsewhere.

References

[1]   A. Athavale, *On joint hyponormality of operators*, Proc. Amer. Math. Soc. **103** (1988), 417–423.

[2]   N. Berline, E. Getzler, and M. Vergne, *Heat Kernels and Dirac Operators*, Grundlehren der mathematischen Wissenschaften, vol. 298, Springer-Verlag, Berlin, Heidelberg, New York, 1992.

[3]   S. Bochner, *Curvature and Betti numbers. I, II*, Ann. of Math. **49; 50** (1948; 1949), 379–390; 79–93.

[4]   S. Bochner and K. Yano, *Curvature and Betti Numbers,* Ann. of Math. Studies, **32**, Princeton University Press, Princeton, 1953.

[5]   J. Bunce, *The joint spectrum of commuting non-normal operators*, Proc. Amer. Math. Soc. **29** (1971), 499–505.

[6]   K. F. Clancey, *Seminormal Operators*, Lecture Notes in Math. **742**, Springer-Verlag, Berlin, Heidelberg, New York, 1979.

[7]   K. F. Clancey, *Toeplitz models for operators with one-dimensional self-commutator*, Operator Theory: Advances and Applications, **11**, Birkhäuser Verlag, Basel, 1983, pp. 81–107.

[8]   K. F. Clancey, *The Cauchy transform of the principal function associated with a non-normal operator*, Indiana Univ. Math. J. **34** (1985), 21–32.

[9]   K. F. Clancey, and B. L., *Local spectra of seminormal operators*, Trans. Amer. Math. Soc. **280** (1983), 415–428.

[10] M. J. Cowen and R. G. Douglas, *Complex geometry and operator theory*, Acta Math. **141** (1978), 187–261.

[11] M. J. Cowen and R. G. Douglas, *Operators possessing an open set of eigenvalues*, Colloquia Math., **35**, North Holland, 1980, pp. 323–341.

[12] R. E. Curto, *Applications of several complex variables to multiparameter spectral theory*, Pitman Research Notes in Mathematics Series, **278**, Longman, 1988, pp. 25–90.

[13] R. E. Curto, *Joint hyponormality: a bridge between hyponormality and subnormality*, Symposia Pure Math., **51**, 1990, pp. 69–91.

[14] R. E. Curto and R. Jian, *A matricial identity involving the self-commutator of a commuting n-tuple*, Proc. Amer. Math. Soc. **121** (1994), 461–464.

[15] R. E. Curto, P. S. Muhly, and J. Xia, *Hyponormal pairs of commuting operators*, Operators Theory: Advances and Applications, **35**, Birkhäuser Verlag, Basel, 1988, pp. 1–22.

[16] J. E. Gilbert and M. A. M. Murray, *Clifford Algebras and Dirac Operators in Harmonic Analysis*, Cambridge Studies in Advanced Mathematics, **26**, Cambridge University Press, 1991.

[17] R. Goldberg, *Curvature and Homology*, Academic Press, New York, London, 1962.

[18] H. B. Lawson and M.-L. Michelsohn, *Spin Geometry*, Princeton Mathematical Series, **38**, Princeton University Press, Princeton, 1989.

[19] M. Martin and M. Putinar, *A unitary invariant for hyponormal operators*, J. Funct. Anal. **72** (1987), 297–232.

[20] M. Martin and M. Putinar, *Lectures on Hyponormal Operators,* Operator Theory: Advances and Applications, **39**, Birkhäuser Verlag, Basel, 1989.

[21] C. R. Putnam, *Commutation Properties of Hilbert Space Operators and Related Topics*, Springer-Verlag, Berlin, Heidelberg, New York, 1967.

[22] J. L. Taylor, *A joint spectrum for several commuting operators*, J. Funct. Anal. **6**, 172–191.

[23] F.-H. Vasilescu, *A characterization of the joint spectrum in Hilbert spaces*, Rev. Roumaine Math. Pures Appl. **22** (1977), 1003–1009.

[24] F.-H. Vasilescu, *A multidimensional spectral theory in C^*-algebras*, Banach Center Publ., **8**, 1982, pp. 471–491.

[25] D. Xia, *Spectral Theory of Hyponormal Operators*, Operator Theory: Advances and Applications, **10**, Birkhäusen Verlag, Basel, 1983.

[26] D. Xia, *On some classes of hyponormal tuples of commuting operators*, Operator Theory: Advances and Applications, **48**, Birkhäusen Verlag, Basel, 1990, pp. 423–448.

DEPARTMENT OF MATHEMATICS, BAKER UNIVERSITY, BALDWIN CITY, KANSAS 66006
E-mail address: mmartin@harvey.bakeru.edu

DEPARTMENT OF MATHEMATICS, KANSAS UNIVERSITY, LAWRENCE, KANSAS 66045
E-mail address: norberto@kuhub.cc.ukans.edu

Contemporary Mathematics
Volume **212**, 1998

C^*-Algebras of Pseudodifferential Operators and Limit Operators

V. S. Rabinovich

ABSTRACT. We consider the C^*-algebra **A** of operators acting on $L_2(\mathbf{R}^n, \mathbf{C}^N)$ generated by pseudodifferential operators in the L. Hörmander class $OPS^0_{0,0}$. The operator symbol of $A \in \mathbf{A}$ is introduced as a family of the limit operators of A. The conditions for $A \in \mathbf{A}$ to be a Fredoholm operator or to be a local invertible operator, are formulated in terms of operator symbols. Examples are given when these conditions have an explicit form.

In this paper we touch few aspects of the limit operators method and its applications to the investigation of the Fredholm property of operators in the C^*-algebra **A** generated by pseudodifferential operators in the L. Hörmander class $OPS^0_{0,0}(\mathbf{R}^{2n})$. We also investigate the problem of local invertibility of pseudodifferential operators. It should be noted that the limit operators for the investigation of the Fredholm property of pseudodifferential operators were introduced in [1, 2].

We emphasize that operators in the class $OPS^0_{0,0}$ are bounded in $L_2(\mathbf{R}^n)$, but their commutators are not compact, hence the standard technique (see for instance [3–5]) to investigate the Fredholm property can not be applied to them. The main feature which makes the limit operators method different from others is in the fact that this method imposes no conditions other than the boundedness of pseudodifferential operators on $L_2(\mathbf{R}^n)$.

We introduce a C^*-algebra $\hat{\mathbf{A}}$ of operator symbols which are families of the limit operators. It is established that $\mathbf{A} / \mathcal{K}(L_2(\mathbf{R}^n, \mathbf{C}^N))$ is isomorphic to $\hat{\mathbf{A}}$, where $\mathcal{K}(L_2(\mathbf{R}^n, \mathbf{C}^N))$ is the ideal of compact operators on $L_2(\mathbf{R}^n, \mathbf{C}^N)$.

In Section 1 we consider some subalgebras of **A**, for which the symbol algebra $\hat{\mathbf{A}}$ has quite effective description. The results of this section have many points in common with the monograph [6] and the papers [7–9].

In Section 2 we consider the problem of local invertibility of operators $A \in \mathbf{A}$ at infinitely distant points of $\mathbf{R}^n$ compactified by the "sphere" of points at infinity. The local operator symbol at an infinitely distant point is defined to be a set of limit operators corresponding to this point. The condition of local invertibility of $A \in \mathbf{A}$ is given in terms of the local operator symbol.

1991 *Mathematics Subject Classification.* Primary 35S05, 47G30.

We also consider a similar problem of local invertibility at the origin for operators in the C^*-algebra generated by Mellin pseudodifferential operators on $\mathbf{R}_+$.

In Section 3 we give an application of this result to the investigation of the C^*-algebra generated by singular integral operators on composed contours with whirl points, and with coefficients which can have discontinuties of the second kind.

It should be noted that there are many works devoted to algebras generated by singular integral operators with piece-wise continuous coefficients on piece-wise Lyapunov contours (see for instance [10–12]). The Banach algebra generated by singular integral operators with piece-wise continuous coefficients on closed contours with logarithmic whirl points was investigated in the paper [13]. The paper [14] is devoted to the C^*-algebra generated by singular integral operators on composed contours with whirl points, and with coefficients which have slowly oscillating discontinuities.

By using the Mellin pseudodifferential operators technique and the limit operators method, we give effective conditions for the Fredholm property of singular integral operators in wider classes than in the previous works.

1. Fredholmness of pseudodifferential operators and limit operators

1. Let $\mathcal{H}_1, \mathcal{H}_2$ denote complex Hilbert spaces, and $\mathcal{L}(\mathcal{H}_1, \mathcal{H}_2)$ be the space of bounded linear operators mapping $\mathcal{H}_1$ into $\mathcal{H}_2$. Let also $\mathcal{K}(\mathcal{H}_1, \mathcal{H}_2)$ be the space of compact operators acting from $\mathcal{H}_1$ into $\mathcal{H}_2$. We shall write $\mathcal{L}(\mathcal{H})$, $\mathcal{K}(\mathcal{H})$ when $\mathcal{H}_1 = \mathcal{H}_2 = \mathcal{H}$.

We say that $a(x, \xi) \in S^0_{0,0}(\mathbf{R}^{2n}) = S^0(\mathbf{R}^{2n})$ if $a(x, \xi) \in C^\infty(\mathbf{R}^{2n})$ and

$$| \partial_x^\beta \partial_\xi^\alpha a(x, \xi) | \le C_{\alpha\beta}$$

for all multi-indices α, β.

We say that $a(x, \xi) \in S^0(\mathbf{R}^{2n})$ is slowly varying when $(x, \xi) \to \infty$ (see [5]) if

$$(1) \qquad \lim_{(x,\xi) \to \infty} \partial_x^\beta \partial_\xi^\alpha a(x, \xi) = 0, \quad \forall (\alpha, \beta) \ne 0.$$

The class of matrix-functions $a(x, \xi) = (a_{i,j}(x, \xi))_{i,j=1}^N$, where $a_{i,j}(x, \xi)$ satisfy (1), is denoted by $L^0(\mathbf{R}^{2n}, \mathbf{C}^{N \times N})$. We say that a matrix-function $a(x, \xi) \in \mathcal{I}^0(\mathbf{R}^{2n}, \mathbf{C}^{N \times N})$ if the elements $a_{ij}(x, \xi)$ satisfy the condition (1) for all multi-indices α, β.

Given a matrix-function $a(x, \xi) \in S^0(\mathbf{R}^{2n}, \mathbf{C}^{N \times N})$, as usual, we associate the pseudodifferential operator with the symbol $a(x, \xi)$

$$Au = a(x, D)u = (2\pi)^{-n} \int a(x, \xi)\widehat{u}(\xi)e^{i(x,\xi)}d\xi, \quad u \in C_0^\infty(\mathbf{R}^n, \mathbf{C}^N),$$

where $\widehat{u}(\xi)$ is the Fourier transform of u. Here the integral is taken over the entire space. This is always implied when the limits of integration are not indicated.

The classes of operators with symbols in $S^0(\mathbf{R}^{2n}, \mathbf{C}^{N \times N})$, $L^0(\mathbf{R}^{2n}, \mathbf{C}^{N \times N})$, $\mathcal{I}^0(\mathbf{R}^{2n}, \mathbf{C}^{N \times N})$ are denoted by $OPS^0(N)$, $OPL^0(N)$, $OP\mathcal{I}^0(N)$, respectively.

PROPOSITION 1.1 [3, 4, 5].

(a) *An operator* $A \in OPS^0(N)$ *is bounded in* $L_2(\mathbf{R}^n, \mathbf{C}^N)$ *and*

$$\|A\|_{L(L_2(\mathbf{R}^n, \mathbf{C}^N))} \le C \sum_{|\alpha| \le [n/2]+1, |\beta| \le [n/2]+1} \sup_{x,\xi} \left\|\partial_x^\beta \partial_\xi^\alpha a(x, \xi)\right\|_{L(\mathbf{C}^N)}.$$

(b) *Let* $A, B \in OPS^0(N)$. *Then* $AB \in OPS^0(N)$ *and the symbol* $\sigma_{AB}(x,\xi)$ *of* AB *is given by the formula*

$$(2) \qquad \sigma_{AB}(x,\xi) = (2\pi)^{-n} \int \int a(x,\xi+\eta)b(x+y,\xi)e^{-i(y,\eta)}\,dy\,d\eta.$$

(c) *Let* $A \in OPS^0(N)$. *Then* $A^* \in OPS^0(N)$ *and*

$$(3) \qquad \sigma_{A^*}(x,\xi) = (2\pi)^{-n} \int \int a^*(x+y,\xi+\eta)e^{-i(y,\eta)}\,dy\,d\eta.$$

(The integrals in the formulas (2)–(3) are understood as oscillatory integrals.)
   (d) *Let* $A \in OPS^0(N)$ *and be invertible on* $L_2(\mathbf{R}^n, \mathbf{C}^N)$. *Then* $A^{-1} \in OPS^0(N)$.
   (e) *Let* $A, B \in OPL^0(N)$. *Then* $AB \in OPL^0(N)$ *and*

$$\sigma_{AB}(x,\xi) = a(x,\xi)b(x,\xi) + t(x,\xi)$$

where $t(x,\xi) \in \mathcal{I}^0(\mathbf{R}^{2n}, \mathbf{C}^{N\times N})$.
   (f) *Let* $A \in OPL^0(N)$. *Then* $A^* \in OPL^0(N)$ *and*

$$\sigma_{A^*}(x,\xi) = a^*(x,\xi) + t(x,\xi),$$

where $t(x,\xi) \in \mathcal{I}^0(\mathbf{R}^{2n}, \mathbf{C}^{N\times N})$.
   (g) $OPI^0(N) \subset \mathcal{K}(L_2(\mathbf{R}^n, \mathbf{C}^N))$.

2. Let $a(x,\xi) \in S^0(\mathbf{R}^{2n}, \mathbf{C}^{N\times N})$ and $(\mathbf{Z}^{2n} \ni) (p_k, q_k) \to \infty$. Let us consider the functional sequence $a(x+p_k, \xi+q_k)$. This sequence is bounded in the space $C^\infty(\mathbf{R}^{2n})$, therefore, it has a subsequence $a(x+p_{k_j}, \xi+q_{k_j})$ convergent to the limit function $\tilde{a}(x,\xi)$ in the topology of $C^\infty(\mathbf{R}^{2n}, \mathcal{L}(\mathbf{C}^N))$, i.e., uniformly on compact subsets of $\mathbf{R}^{2n}$ together with all derivatives. It is easy to see that $\tilde{a}(x,\xi)$ $\in S^0(\mathbf{R}^{2n}, \mathbf{C}^{N\times N})$.

DEFINITION 1.1. The operator $\tilde{A} = \tilde{a}(x,D)$ is called the limit operator for A $\in OPS^0(N)$, defined by the sequence (p_{k_j}, q_{k_j}). The set of all limit operators of A is called the operator symbol of A and denoted by $\Sigma(A)$.

Let T_p be the translation operator $T_p u(x) = u(x-p)$, $p \in \mathbf{R}^n$, e_q be the operator of multiplication by the exponential $\exp i(x,q)$, $q \in \mathbf{R}^n$ and $U_h = T_p e_q$, $h = (p,q)$. It is evident that these operators are unitary in $L_2(\mathbf{R}^n, \mathbf{C}^N)$.

PROPOSITION 1.2.
   (a) *Let* $A, B \in OPS^0(N)$, $\tilde{A}$ *and* $\tilde{B}$ *be the limit operators defined by a sequence* (p_k, q_k). *Then* $\tilde{A} + \tilde{B}$, $\tilde{A}\tilde{B}$ *are the limit operators of* $A+B$, AB *defined by the same sequence* (p_k, q_k).
   (b) *Let* $\tilde{A}$ *be the limit operator defined by a sequence* $h_k = (p_k, q_k)$. *Then for any function* $\Psi(x,\xi) \in C_0^\infty(\mathbf{R}^{2n})$:

$$\lim_{k\to\infty} \left\| (U_{h_k}^{-1} A U_{h_k} - \tilde{A})\Psi(x,D) \right\| = \lim_{k\to\infty} \left\| \Psi(x,D)(U_{h_k}^{-1} A U_{h_k} - \tilde{A}) \right\| = 0.$$

   (c) s-$\lim_{k\to\infty} U_{h_k}^{-1} A U_{h_k} = \tilde{A}$,
   (d)

$$\left\| \tilde{A} \right\| \leq \|A\|$$

for any limit operator $\widetilde{A} \in \sigma(A)$.

This proposition follows from Proposition 1.1 (a), (b), (c). (See [1, 2] for the details of the proof).

THEOREM 1.1 [1, 2]. *Let $A \in OPS^0(N)$. Then the following statements are equivalent:*

(a) *A is a Fredholm operator in $L_2(\mathbf{R}^n, \mathbf{C}^N)$;*

(b) *All the limit operators $\widetilde{A} \in \Sigma(A)$ are invertible on $L_2(\mathbf{R}^n, \mathbf{C}^N)$;*

(c) *All the limit operators $\widetilde{A} \in \Sigma(A)$ are uniformly invertible on $L_2(\mathbf{R}^n, \mathbf{C}^N)$, that is there exists a constant $C > 0$ such that $\left\| \widetilde{A}^{-1} \right\| \leq C$ for each $\widetilde{A} \in \Sigma(A)$;*

(d) *There exists an operator $R \in OPS^0(N)$ such that $RA = I + T_1$, $AR = I + T_2$, where the operators $T_1, T_2 \in OP\mathcal{T}^0(N)$.*

PROPOSITION 1.3. *Let $A \in OPS^0(N)$. Then*

$$(4) \qquad \||A\|| = \inf_{T \in \mathcal{K}(L_2(\mathbf{R}^n, \mathbf{C}^N))} \|A - T\| = \sup_{\widetilde{A} \in \Sigma(A)} \left\| \widetilde{A} \right\|.$$

This proposition follows from Theorem 1.1.

Let $\mathbf{A}, \mathbf{B}, \mathbf{I}$ be the C^*-subalgebras in $\mathcal{L}(L_2(\mathbf{R}^n, \mathbf{C}^N))$ generated by operators in the classes $OPS^0(N), OPL^0(N), OP\mathcal{T}^0(N)$, respectively. It is evident that $\mathbf{I}$ is a two-sided ideal in $\mathbf{A}$ and $\mathbf{B}$.

PROPOSITION 1.4.

(a) $\mathbf{I} = \mathcal{K}(L_2(\mathbf{R}^n, \mathbf{C}^N))$,

(b) *The C^*-algebras $\mathbf{A}$ and $\mathbf{B}$ are irreducible.*

Let us define the limit operators for an operator $A \in \mathbf{A}$. There exists a sequence of operators $A_m \in OPS^0(N)$ such that $\|A - A_m\| \to 0$ when $m \to \infty$. Let a sequence $h_j = (p_j, q_j) \to \infty$, then there exists a subsequence $h_{j_k} = (p_{j_k}, q_{j_k})$ defining the limit operators $\widetilde{A}_m$ for each m. From Proposition 1.2 (d) there follows that $\widetilde{A}_m$ is a fundamental sequence in $\mathcal{L}(L_2(\mathbf{R}^n, \mathbf{C}^N))$, therefore, it has the limit $\widetilde{A}$. This operator is called the limit operator defined by the sequence h_{j_k}. The set of all limit operators of A is called the operator symbol and denoted by $\Sigma(A)$. Thus the operator symbol of A is the collection $\{\widetilde{A}\}$ of its limit operators. Let $\hat{\mathbf{A}}$ be the set of operator symbols $\Sigma(A)$. Algebraic operations on $\hat{\mathbf{A}}$ are defined by the formulas

$$\Sigma(A) + \Sigma(B) = \{\widetilde{A} + \widetilde{B}\}, \quad \Sigma(A)\Sigma(B) = \{\widetilde{A}\widetilde{B}\},$$

where $\{\widetilde{A} + \widetilde{B}\}$, $\{\widetilde{A}\widetilde{B}\}$ are the sets of operators: $\widetilde{A} + \widetilde{B}$, $\widetilde{A}\widetilde{B}$ with $\widetilde{A}$ and $\widetilde{B}$ being the limit operators of A, B defined by the same sequences.

Let us introduce the operation of conjugation in $\mathbf{A}$ by means of the formula $\Sigma(A)^* = \Sigma(A^*)$. The norm in $\mathbf{A}$ is defined by the formula

$$\|\Sigma(A)\|_{\hat{A}} = \sup_{\widetilde{A} \in \Sigma(A)} \left\| \widetilde{A} \right\|.$$

THEOREM 1.2.

(a) $\hat{\mathbf{A}}$ *is a* $\mathbf{C}^*$*-algebra.*

(b) *The mapping $\Sigma : \mathbf{A} \to \hat{\mathbf{A}}$ is the epimorphism. The kernel of Σ is the ideal $\mathcal{K}(L_2(\mathbf{R}^n, \mathbf{C}^N))$.*

(c) $A \in \mathbf{A}$ is a Fredholm operator on $L_2(\mathbf{R}^n, \mathbf{C}^N)$ if and only if its operator symbol $\Sigma(A)$ is invertible in $\hat{\mathbf{A}}$, i.e., each limit operator $\widetilde{A} \in \Sigma(A)$ is invertible in $L_2(\mathbf{R}^n, \mathbf{C}^N)$.

(d) A regularizator R of the Fredholm operator $A(\in \mathbf{A})$ belongs to $\mathbf{A}$.

It should be noted that the condition (c) is equivalent to that all the limit operators $\widetilde{A} \in \Sigma(A)$ are uniformly invertible.

3. We will give applications of Theorem 1.2 to C^*-algebras generated by some subclasses in $OPS^0(N)$.

Let us introduce the notation. Let $\mathcal{A}$ be a C^*-algebra. Then $C_b(\mathbf{R}^m, \mathcal{A})$ is the C^*-algebra of continuous bounded mappings $a : \mathbf{R}^m \to \mathcal{A}$ with the norm

$$\|a\|_{C_b(\mathbf{R}^m, \mathcal{A})} = \sup_{\mathbf{R}^m} \|a(x)\|_{\mathcal{A}},$$

$C_0(\mathbf{R}^m, \mathcal{A})$ is the $*$-subalgebra in $C_b(\mathbf{R}^m, \mathcal{A})$ of vector-functions vanishing at the infinity, $\Omega(\mathbf{R}^m, \mathcal{A})$ is the $*$-subalgebra of vector-functions $a \in C_b(\mathbf{R}^m, \mathcal{A})$ such that

$$\lim_{x \to \infty} \sup_{|h| \leq 1} \|a(x + h) - a(x)\|_{\mathcal{A}} = 0.$$

We use the notation $\Omega^N(\mathbf{R}^m) = \Omega(\mathbf{R}^m, \mathcal{L}(C^N))$.

The following theorem gives a description of the C^*-algebra $\hat{\mathbf{B}}$.

THEOREM 1.3. The C^*-algebra $\hat{\mathbf{B}}$ is isomorphic to the C^*-algebra

$$\hat{\Omega}^N(\mathbf{R}^{2n}) = \Omega^N(\mathbf{R}^{2n})/C_0(\mathbf{R}^{2n}, \mathcal{L}(\mathbf{C}^N)).$$

The proof is based on the following property: a limit operator $\tilde{A}$ of $A \in \mathbf{B}$ is either a translation invariant operator $\tilde{a}(D)$ with a symbol $\tilde{a}(\xi) \in \Omega^N(\mathbf{R}^n)$ or an operator of multiplication by a function $\tilde{a}(x) \in \Omega^N(\mathbf{R}^n)$.

Therefore, the operator symbol of $A \in \mathbf{B}$ can be identified with the factor-class $\hat{a}(x, \xi) \in \hat{\Omega}^N(\mathbf{R}^{2n})$, and the necessary and sufficient condition for $A \in \mathbf{B}$ to be a Fredholm operator is

$$\lim_{R \to \infty} \inf_{|(x,\xi)| > R} |\det a(x, \xi)| > 0,$$

where $a(x, \xi)$ is a representative of the factor-class $\hat{a}(x, \xi)$.

By $\mathcal{D}(N)$ we denote the C^*-algebra of operators acting on $\mathcal{L}(L_2(\mathbf{R}^n, \mathbf{C}^N))$ generated by operators of the form

$$(5) \qquad A = \sum_{1 \leq i,j \leq M} a_i B_{ij} T_{h_j},$$

where $a_i \in CAP(\mathbf{R}^n)$, which is the space of uniform almost-periodic functions on $\mathbf{R}^n$, $B_{ij} = b_{ij}(x, D) \in OPL^0(N)$, T_{h_j} are shift operators.

Let $\mathcal{U}_1(N)$ be the C^*-subalgebra of $D(N)$ generated by operators of the form

$$A = \sum_{1 \leq i,j \leq M} a_i b_{ij} T_{h_j},$$

where $a_i \in CAP(\mathbf{R}^n)$, b_{ij} are multiplication operators by matrix-functions in $\Omega^N(\mathbf{R}^{2n})$, $\mathcal{U}_2(N)$ is the C^*-subalgebra of $D(N)$ generated by operators of the form

$$A = \sum_{1 \leq i,j \leq M} a_i b_{ij}(D) T_{h_j},$$

where $a_i \in CAP(\mathbf{R}^n)$ and $b_{ij}(D)$ are pseudodifferential operators with symbols $b_{ij}(\xi) \in \Omega^N(\mathbf{R}^n)$. We set $\mathcal{U}_3(N) = \mathcal{U}_1(N) \cap \mathcal{U}_2(N)$. It is easy to see that $\mathcal{U}_3(N)$ is generated by operators of multiplication by functions in $CAP(\mathbf{R}^n) \otimes L(\mathbf{C}^N)$ and shift operators.

We define mappings: $\pi_j : \mathcal{D}(N) \to \hat{\Omega}(R^n, U_j(N))$, $j = 1, 2$ for operators of the form (5) by the formulas:

$$\pi_1(A) = \sum_{1 \le i,j \le M} a_i(x)\hat{b}_{ij}^{(1)}(x,\xi)T_{h_j} \in \hat{\Omega}(\mathbf{R}_\xi^n, \mathcal{U}_1(N)),$$

where $\hat{\Omega}(\mathbf{R}_\xi^n, \mathcal{U}_1(N)) = \Omega(\mathbf{R}_\xi^n, \mathcal{U}_1(N))/C_0(\mathbf{R}_\xi^n, \mathcal{U}_1(N))$,

$$\pi_2(A) = \sum_{1 \le i,j \le M} a_i(x)\hat{b}_{ij}^{(2)}(x,D)T_{h_j} \in \hat{\Omega}(\mathbf{R}_x^n, \mathcal{U}_2(N))$$

where $\hat{\Omega}(\mathbf{R}_x^n, \mathcal{U}_2(N)) = \Omega(\mathbf{R}_x^n, \mathcal{U}_2(N))/C_0(\mathbf{R}_x^n, \mathcal{U}_2(N))$, and $\hat{b}_{ij}^{(1)}(x,\xi)$ is the factor-class in $C_b(\mathbf{R}_x^n) \otimes \hat{\Omega}^N(\mathbf{R}_\xi^n)$ corresponding to $b_{ij}(x,\xi)$. The factor-class $\hat{b}_{ij}^{(2)}(x,\xi)$ is defined similarly.

It should be noted that the mapping π_j, $(j = 1,2)$ can be extended to a morphism of C^*-algebras.

THEOREM 1.4. *The C^*-algebra $\hat{\mathcal{D}}(N)$ is isomorphic to the subalgebra*

$$\hat{\Omega}(\mathbf{R}_\xi^n, \mathcal{U}_1(N)) \oplus \hat{\Omega}(\mathbf{R}_x^n, \mathcal{U}_2(N))$$

consisting of all pairs (a_1, a_2) such that

$$(I_{\mathbf{R}_\xi^n} \otimes \pi_2)(a_1) = (I_{\mathbf{R}_x^n} \otimes \pi_1)(a_2).$$

The proof of this theorem is based on the analysis of structure of the limit operators of $A \in \mathcal{D}(N)$. From Theorem 1.4 there follows the criterion for Fredholmness of an operator $A \in \mathcal{D}(N)$.

THEOREM 1.5. *$A \in \mathcal{D}(N)$ is a Fredholm operator on $L_2(\mathbf{R}^n, C^N)$ if and only if the mappings $\pi_j(A)$, $j = 1, 2$ are invertible.*

4. We can reduce the investigation of the Fredholm property of non zero order pseudodifferential operators in the Sobolev spaces to the investigation of zero order operators acting on $L_2(\mathbf{R}^n, \mathbf{C}^N)$, and reformulate Theorem 1.5 for this case.

We shall give two examples.

EXAMPLE 1. Let us consider the partial differential-difference operator on $\mathbf{R}^n$

$$(6) \quad Au(x) = \sum_{|\alpha| \le m} \sum_{j=1}^k a_{\alpha j}(x)D^\alpha u(x - h_{\alpha j}), \quad h_{\alpha j} \in \mathbf{R}^n, \quad a_{\alpha j}(x) \in \Omega^N(\mathbf{R}^n),$$

where $u(x) \in H^m(\mathbf{R}^n, \mathbf{C}^N)$.

It follows from Theorem 1.5 the operator $A : H^m(\mathbf{R}^n, \mathbf{C}^N) \to L_2(\mathbf{R}^n, \mathbf{C}^N)$ is a Fredholm operator if and only if:

(a)

$$\lim_{R \to \infty} \inf_{|x| \ge R, \xi \in \mathbf{R}^n} \left| \det \sum_{|\alpha| \le m} \sum_{j=1}^k a_{\alpha j}(x)\xi^\alpha e^{i(\xi, h_{\alpha,j})} \right| > 0,$$

(b) For arbitrary $\xi \in \mathbf{S}^{n-1}$ the difference operator

$$A_\xi = \sum_{|\alpha| \leq m} \sum_{j=1}^{k} a_{\alpha j}(x) \xi^\alpha T_{h_{\alpha j}} : L_2(\mathbf{R}^n, \mathbf{C}^N) \to L_2(\mathbf{R}^n, \mathbf{C}^N)$$

is invertible.

EXAMPLE 2. Let us consider a partial differential operator A on $\mathbf{R}^n$ with oscillatory coefficients

$$Au(x) = \sum_{|\alpha| \leq m} \sum_{j=1}^{k} a_{\alpha j}(x) e^{i(x, h_{\alpha j})} D^\alpha u(x), \quad h_{\alpha j} \in \mathbf{R}^n,$$

$$a_{\alpha j}(x) \in \Omega^N(\mathbf{R}^n), \qquad u(x) \in H^m(\mathbf{R}^n, \mathbf{C}^N).$$

It follows from Theorem 1.5 that the operator $A : H^m(\mathbf{R}^n, \mathbf{C}^N) \to L_2(\mathbf{R}^n, \mathbf{C}^N)$ is a Fredholm operator if and only if:

(a)

$$\inf_{\substack{\xi \in \mathbf{S}^{n-1} \\ x \in \mathbf{R}^n}} \left| \det \sum_{|\alpha| \leq m} \sum_{j=1}^{k} a_{\alpha j}(x) \xi^\alpha e^{i(x, h_{\alpha, j})} \right| > 0.$$

(b) There exists $R > 0$ such that the almost-periodic operator

$$A_y = \sum_{|\alpha| \leq m} \sum_{j=1}^{k} a_{\alpha j}(y) e^{i(x, h_{\alpha j})} D_x^\alpha$$

is invertible from $H^m(\mathbf{R}^n, \mathbf{C}^N)$ into $L_2(\mathbf{R}^n, \mathbf{C}^N)$ for any $y : |y| \geq R$ and

$$\sup_{|y| \geq R} \left\| A_y^{-1} \right\| < \infty.$$

It should be noted that there are not effective conditions of invertibility of the difference and almost-periodic operators on $\mathbf{R}^n$, but there are effective necessary conditions (see for instance [15, 16]) and there are sufficient conditions only for some classes of such operators.

2. Local invertibility of pseudodifferential operators and limit operators

2.1. Pseudodifferential operators on $\mathbf{R}^n$. Let $\widetilde{\mathbf{R}}^n$ be the compactification of $\mathbf{R}^n$ which is homeomorphic to the unit ball $B = \{x \in \mathbf{R}^n : |x| \leq 1\}$, and let $\mathcal{M}$ $(\subset \widetilde{\mathbf{R}}^n)$ be the set of infinitely distant points. Neighborhoods of an infinitely distant point $\eta \in \mathcal{M}$ are the sets $\widetilde{V}_{\Gamma,R} = \Gamma \cap B_R' \cup \mathcal{M}_\Gamma$, where Γ is an open cone with centre at the origin containing the ray corresponding to the point η, $B_R' = \{x \in \mathbf{R}^n : |x| > R\}$, $\mathcal{M}_\Gamma$ is a subset of infinitely distant points, corresponding to the rays lying in Γ. Let Ω be the trace of the cone Γ on the unit sphere $\mathbf{S}^{n-1}$, and $\omega_\eta (\in \mathbf{S}^{n-1})$ be the point corresponding to the point $\eta \in \mathcal{M}$. Let $V_{\Gamma,R} = \Gamma \cap B_R'$ and $\chi_{\Gamma,R}$ be the characteristic function of the set $V_{\Gamma,R}$.

DEFINITION 2.1. We say that an operator $A \in \mathcal{L}(L_2(\mathbf{R}^n, C^N))$ is locally invertible at the point $\eta \in \mathcal{M}$, if there exist a neighborhood $\widetilde{V}_{\Gamma,R} \ni \eta$ and operators $B'_\eta, B''_\eta \in \mathcal{L}(L_2(\mathbf{R}^n, C^N))$ such that

$$ B'_\eta A \chi_{\Gamma,R} = \chi_{V_{\Gamma,R}}, \qquad \chi_{\Gamma,R} A B''_\eta = \chi_{V_{\Gamma,R}}. $$

DEFINITION 2.2. Let $\eta \in \mathcal{M}$, $A \in \mathbf{A}$. We denote by $\Sigma_\eta(A)$ the set of all the limit operators of A corresponding to all sequences $h_m = (p_m, q_m)$ such that $p_m \to \eta$. The set $\Sigma_\eta(A)$ is called the local operator symbol at the infinitely distant point η.

Let $\mathcal{I}_\eta$ be the two-sided ideal in $\mathbf{A}$ generated by operators with matrix-symbols $a(x, \xi)$ satisfying the condition: $\lim_{x \to \infty} a(x, \xi) = 0$ uniformly with respect to $\xi \in \mathbf{R}^n$.

THEOREM 2.1. (a) Let $A \in \mathbf{A}$. Then

(7)
$$ \|\|A\|\|_\eta = \inf_{T \in \mathcal{I}_\eta} \|A - T\| = \sup_{\tilde{A} \in \Sigma_\eta(A)} \left\| \tilde{A} \right\|. $$

(b) An operator $A \in \mathbf{A}$ is locally invertible at the point $\eta \in \mathcal{M}$ if and only if A is invertible in the factor-algebra $\mathbf{A}/\mathcal{I}_\eta$.

(c) An operator $A \in \mathbf{A}$ is locally invertible at the point $\eta \in \mathcal{M}$ if and only if an arbitrary limit operator $\tilde{A} \in \Sigma_\eta(A)$ is invertible on $L_2(\mathbf{R}^n, \mathbf{C}^N)$.

Let $A = a(x, D) \in OPL^0(N)$. Then the equality (7) takes the form

(8)
$$ \|\|A\|\|_\eta = \lim_{\substack{R \to \infty \\ \Omega \to \omega}} \sup_{\substack{x \in \Gamma_{\Omega,R} \\ \xi \in \mathbf{R}^n}} \|a(x, \xi)\|_{\mathcal{L}(\mathbf{C}^N)}. $$

It is easy to see that the right-hand side in (8) is the norm in the factor-algebra $\Omega^N(\mathbf{R}^{2n})/\Omega_0^N(\mathbf{R}^{2n})$, where $\Omega_0^N(\mathbf{R}^{2n}) = \{a(x, \xi) \in \Omega^N(\mathbf{R}^{2n}) : \lim_{x \to \infty} a(x, \xi) = 0$ uniformly with respect to $\xi \in \mathbf{R}^n\}$.

Passage to the limit in the equality (8) corresponds a local matrix symbol $\hat{a}(x, \xi) \in \Omega^N(\mathbf{R}^{2n})/\Omega_0^N(\mathbf{R}^{2n})$ for an operator $A \in \mathbf{B}$.

THEOREM 2.2. $A \in \mathbf{B}$ is a locally invertible operator at the point $\eta \in \mathcal{M}$ if and only if its symbol $\hat{a}(x, \xi)$ is invertible in the factor-algebra $\Omega^N(\mathbf{R}^{2n})/\Omega_0^N(\mathbf{R}^{2n})$. This condition is equivalent to the following one:

$$ \lim_{\substack{R \to \infty \\ \Omega \to \omega}} \sup_{\substack{\eta x \in \Gamma_{\Omega,R} \\ \xi \in \mathbf{R}^n}} |\det a(x, \xi)| > 0, $$

where $a(x, \xi)$ is a representative of the class $\hat{a}(x, \xi)$.

As example of application of the Theorem 2.1 we give the necessary and sufficient conditions of local invertibility of the differential-difference operator (6).

THEOREM 2.3. The differential-difference operator

$$ A : H^m(\mathbf{R}^n, \mathbf{C}^N) \to L_2(\mathbf{R}^n, \mathbf{C}^N) $$

is a locally invertible operator at the infinitely distant point $\eta \in \mathcal{M}$ if and only if:

(a)

$$\lim_{R\to\infty} \sup_{\substack{x\in\Gamma_{\Omega,R} \\ \xi\in\mathbf{R}^n}} \mid \det \sum_{|\alpha|\le m} \sum_{j=1}^k a_{\alpha j}(x)\omega_\eta^\alpha e^{i(\xi,h_{\alpha,j})} \mid > 0,$$

where ω_η $(\in \mathbf{S}^{n-1})$ is the point corresponding to $\eta \in \mathcal{M}$.

(c) The difference operator

$$A_{\omega_\eta} = \sum_{|\alpha|\le m} \sum_{j=1}^k a_{\alpha j}(x)\omega_\eta^\alpha T_{h_{\alpha j}} : L_2(\mathbf{R}^n, C^N) \to L_2(\mathbf{R}^n, C^N),$$

is invertible.

2.2. Mellin pseudodifferential operators. The Mellin transform of a vector-function $u(t) \in C_0^\infty(\mathbf{R}_+, \mathbf{C}^N)$ is defined by the formula

$$Mu(\lambda) = (2\pi)^{-1/2} \int_{\mathbf{R}_+} t^{-i\lambda-1} u(t)dt, \quad \lambda \in \mathbf{C},$$

and the inverse Mellin transform is given by the integral

$$u(t) = (2\pi)^{-1/2} \int_{R_\mu} (Mu)(\lambda) t^{i\lambda} d\lambda,$$

where $\mathbf{R}_\mu = \{\lambda \in \mathbf{C} : Im\lambda = \mu\}$ is a straight line in the complex plain $\mathbf{C}$. It should be noted that M is a unitary isomorphism from $L_{2,\delta}(\mathbf{R}_+, \mathbf{C}^N)$ into $L_2(\mathbf{R}_{\delta+1/2}, \mathbf{C}^N)$, where $L_{2,\delta}(\mathbf{R}_+, \mathbf{C}^N)$ and $L_2(\mathbf{R}_{\delta+1/2}, \mathbf{C}^N)$ are the Hilbert spaces with norms:

$$\|u\|_{L_{2,\delta}(\mathbf{R}_+, \mathbf{C}^N)} = \left(\int_{\mathbf{R}_+} t^{2\delta} \|u(t)\|_{\mathbf{C}^N}^2 dt\right)^{1/2},$$

$$\|u\|_{L_2(\mathbf{R}_\mu, \mathbf{C}^N)} = \left(\int_{\mathbf{R}_\mu} \|u(\lambda)\|_{\mathbf{C}^N}^2 d\lambda\right)^{1/2}.$$

A matrix-function $a(t,\lambda)$ belongs to the class $E(\mathbf{R}_+ \times \mathbf{R}_\mu, \mathbf{C}^{N\times N})$ if

$$\mid a \mid_{k,l} = \sup_{(t,\lambda)\in\mathbf{R}_+\times\mathbf{R}_\mu} \sum_{0\le\beta\le l,\, 0\le\alpha\le k} \left\|(t\partial_t)^\beta \partial_\lambda^\alpha a(t,\lambda)\right\|_{L(\mathbf{C}^N)} < \infty$$

for all $k, l \in \mathbf{Z}_+$.

We say that a matrix-function $a(t,\lambda) \in E(\mathbf{R}_+ \times \mathbf{R}_\mu, \mathbf{C}^{N\times N})$ is slowly varying at the origin, if

$$\lim_{t\to+0}(t\partial_t)a(t,\lambda) = 0,$$

uniformly with respect to $\lambda \in \mathbf{R}_\mu$. This class is denoted by $\mathcal{E}(\mathbf{R}_+ \times \mathbf{R}_\mu, \mathbf{C}^{N\times N})$.

The class of matrix-functions $a(t,\lambda) \in E(\mathbf{R}_+ \times \mathbf{R}_\mu, \mathbf{C}^{N\times N})$ such that

$$\lim_{t\to+0} a(t,\lambda) = 0,$$

uniformly with respect to $\lambda \in \mathbf{R}_\mu$ is denoted by $\mathcal{I}(\mathbf{R}_+ \times \mathbf{R}_\mu, \mathbf{C}^{N\times N})$.

The operator

$$\mathcal{A}u(t) = a(t, \mathcal{D}_t)u = (2\pi)^{-1} \int_{\mathbf{R}_\mu} a(t,\lambda)(Mu)(\lambda) t^{i\lambda} d\lambda,$$

where $u \in C_0^\infty(\mathbf{R}_+, \mathbf{C}^N)$, $a(t, \lambda) \in E(\mathbf{R}_+ \times \mathbf{R}_\mu, \mathbf{C}^{N \times N})$ is called the Mellin pseudodifferential operator with symbol $a(t, \lambda)$.

We denote by $OPE_\mu(N)$ the class of Mellin pseudodifferential operators with symbols in $E(\mathbf{R}_+ \times \mathbf{R}_\mu, \mathbf{C}^{N \times N})$. The notations $OP\mathcal{E}_\mu(N)$, $OP\mathcal{I}_\mu(N)$ have the similar sense.

It is easy to reformulate Proposition 1.1 and Theorem 1.1 and other results of Section 1 for the Mellin pseudodifferential operators. We just restrict ourselves by the criterion for local invertibility at the origin.

DEFINITION 2.3. We say that a linear operator $A : L_{2,\delta}(\mathbf{R}_+, \mathbf{C}^N) \to L_{2,\delta}(\mathbf{R}_+, \mathbf{C}^N)$ is locally invertible at the origin, if there exist both a characteristic function χ_r of some segment $[0, r]$ $(r > 0)$ and operators $B', B" \in L(L_{2,\delta}(\mathbf{R}_+, \mathbf{C}^N))$ such that

$$B' A \chi_r = \chi_r, \qquad \chi_r A B" = \chi_r.$$

Let $a(t, \lambda) \in E(\mathbf{R}_+ \times \mathbf{R}_\mu, \mathbf{C}^{N \times N})$, and $(\mathbf{R}_+ \ni) \; p_k \to +0$. One can see that the sequence $a(p_k t, \lambda)$ has a subsequence $a(p_{k_m} t, \lambda)$ tending to a matrix-function $\tilde{a}(t, \lambda) \in E(\mathbf{R}_+ \times \mathbf{R}_\mu, \mathbf{C}^{N \times N})$ in the topology of $C^\infty(\mathbf{R}_+, \mathcal{L}(\mathbf{C}^N))$.

DEFINITION 2.4. The operator $\tilde{a}(t, \mathcal{D}_t) = \tilde{A}$ is called the limit operator defined by the sequence $p_{k_m} \to +0$. The local operator symbol at the origin is defined to be the set of all the limit operators defined by the sequences tending to zero. It is denoted by $\Sigma_0(A)$.

Let us denote by $\mathbf{E}(\delta, N)$ the C^*−algebra of operators acting on $L_{2,\delta}(\mathbf{R}_+, \mathbf{C}^N)$ generated by operators in $OPE_{\delta+1/2}(N)$, and let $\mathcal{I}_0(\delta, N)$ be the two-sided ideal in $\mathbf{E}(\delta, N)$ generated by operators in $OP\mathcal{I}_{\delta+1/2}(N)$.

PROPOSITION 2.1. (a) $A \in \mathbf{E}(\delta, N)$ is a local invertible operator at the origin if and only if the factor-class $\hat{A}$ corresponding to A is invertible in the factor-algebra $\hat{\mathbf{E}}(\delta, N) = \mathbf{E}(\delta, N)/\mathcal{I}_0(\delta, N)$.

(b) Let $A \in OPE_{\delta+1/2}(N)$. Then

$$(9) \qquad \||A\||_0 = \left\| \hat{A} \right\|_{\hat{\mathbf{E}}(\delta,N)} = \sup_{\Sigma_0(A)} \left\| \tilde{A} \right\|.$$

Applying the equality (9) we can define the limit operators at the origin for an operator in the C^*-algebra $\mathbf{E}(\delta, N)$. Therefore, we can correspond the local operator symbol $\Sigma_0(A)$ to an operator $A \in \mathbf{E}(\delta, N)$, and the equality (9) holds for operators $A \in \mathbf{E}(\delta, N)$.

THEOREM 2.4. Let $A \in \mathbf{E}(\delta, N)$. Then the following statements are equivalent:

(a) The operator $A : L_{2,\delta}(\mathbf{R}_+, \mathbf{C}^N) \to L_{2,\delta}(\mathbf{R}_+, \mathbf{C}^N)$ is a locally invertible operator at the origin;

(b) Any limit operator $\tilde{A} \in \Sigma_0(A)$ is invertible in $L_{2,\delta}(\mathbf{R}_+, \mathbf{C}^N)$;

(c) All the limit operators $\tilde{A} \in \Sigma_0(A)$ are uniformly invertible;

(d) There exist left and right locally inverse operators at the origin belonging to $\mathbf{E}(\delta, N)$.

We denote by $\mathbf{F}(\delta, N)$ the C^*-algebra of operators acting in $L_{2,\delta}(\mathbf{R}_+, \mathbf{C}^N)$ generated by operators in the class $OP\mathcal{E}_{\delta+1/2}(N)$. The equivalent conditions (b) and (c) of the previous theorem for operators from C^*-algebra $\mathbf{F}(\delta, N)$ take the effective form.

Let us introduce the following notations. We denote by $\Lambda(\delta, N)$ the subalgebra of the algebra $C_b(\mathbf{R}_+ \times \mathbf{R}_{\delta+1/2}, \mathcal{L}(\mathbf{C}^N))$ consisting of the matrix-functions $a(t, \lambda)$ such that

$$(10) \qquad \lim_{t \to +0} \|a(pt, \lambda) - a(t, \lambda)\|_{\mathcal{L}(\mathbf{C}^N)} = 0,$$

uniformly with respect to $(p, \lambda) \in K \times \mathbf{R}_{\delta+1/2}$ where K $(\subset \mathbf{R}_+)$ is a compact. Denote by $C_{01}(\mathbf{R}_+ \times \mathbf{R}_{\delta+1/2}, \mathcal{L}(\mathbf{C}^N)) = C_{01}(\delta, N)$ the two-sided ideal of $C_b(\mathbf{R}_+ \times \mathbf{R}_{\delta+1/2}, \mathcal{L}(\mathbf{C}^N))$ consisting of the matrix-functions $a(t, \lambda)$ such that

$$\lim_{t \to +0} \|a(t, \lambda)\|_{\mathcal{L}(\mathbf{C}^N)} = 0$$

uniformly with respect to $\lambda \in \mathbf{R}_{\delta+1/2}$. Let $\hat{\Lambda}(\delta, N) = \Lambda(\delta, N)/C_{01}(\delta, N)$.

For an operator $A = a(t, \mathcal{D}_t) \in OP\mathcal{E}_{\delta+1/2}(N)$ the equality (9) has the form

$$(11) \qquad \|\|A\|\|_0 = \|\hat{a}(t, \lambda)\|_{\hat{\Lambda}(\delta, N)},$$

where $\hat{a}(t, \lambda)$ is the factor-class corresponding to the matrix-symbol $a(t, \lambda)$.

Passage to the limit in (11) defines the local operator symbol $\sigma_0(A) \in \hat{\Lambda}(\delta, N)$ at the origin for an arbitrary operator $A \in \mathbf{F}(\delta, N)$.

THEOREM 2.5. $A \in \mathbf{F}(\delta, N)$ *is a locally invertible operator at the origin if and only if its symbol* $\sigma(A)(t, \lambda)$ *is invertible in the factor-algebra* $\hat{\Lambda}(\delta, N)$. *This condition is equivalent to the following one:*

$$\lim_{r \to 0} \inf_{(0,r) \times \mathbf{R}_{\delta+1/2}} |\det a(t, \lambda)| > 0,$$

where $a(t, \lambda)$ *is a representative of the factor-class* $\sigma(A)(t, \lambda) \in \hat{\Lambda}(\delta, N)$.

3. The C^*-algebra of singular integral operators on composed contours with whirl points

DEFINITION 3.1. We will say that a simple nonclosed rectifiable oriented arc $\gamma \subset \mathbf{C}$ with endpoints x_1 and x_2 belongs to the class $\mathcal{L}_0$ if the following conditions are satisfied:

(i) $\gamma \setminus \{x_1, x_2\}$ is a locally Lyapunov arc.

(ii) Let

$$x - x_j = \varphi_j(t) = t \exp i\omega_j(t), \quad t \in [0, t_j], \quad j = 1, 2$$

be the parametrization of γ in a neighborhood of the endpoint x_j. We suppose that

$$\omega_j(t) = \nu_j \ln t + \theta_j(t),$$

where $\nu_j \in \mathbf{R}$, $\theta_j(t)$ is a real-valued function belonging to $C^\infty((0, t_j])$ and

$$(12) \qquad \lim_{t \to +0} (t\partial/\partial t)^k \theta_j(t) = 0, \quad \forall k \in \mathbf{N}.$$

For example, the functions: $|\ln t|^\alpha$, $\sin |\ln t|^\alpha$, $\alpha \in (0, 1)$ is satisfied the conditions (12).

The union of finitely many arcs $\gamma_j \in \mathcal{L}_0$ each pair of which have at most endpoints in common is called a composed curve. Let $F \subset \Gamma$ be the set of the endpoints. The points of the set F are called nodes. Let $z \in F$, $N = N(z)$ be number of arcs which make up the node z, and the parametrization of the arc γ_j in a neigborhood of node z is:

(13) $x - z = \phi_j(t) = t\omega_j(t), \quad t \in [0, t_j], \quad \omega_j(t) = \nu lnt + \theta_j(t), \quad j = 1, \ldots, N,$

where $\theta_j(t)$ are real-valued functions satisfying the condition (12). We suppose that there exists $t_0 > 0$ such that $\theta_j(t) \in (\alpha_j, \beta_j)$ when $t \in (0, t_0)$, and

(14) $0 \le \alpha_1 < \beta_1 < \alpha_2 < \beta_2 < \cdots < \alpha_N < \beta_N < 2\pi.$

DEFINITION 3.2. We denote by $PSV(\Gamma, F)$ the C^*-algebra obtained by the closing in the sup-norm on Γ the set of bounded continuously differentiable functions on $\Gamma \backslash F$ such that

(15) $\lim_{\Gamma \ni x \to x_j} |x - x_j| \, a'(x) = 0, \quad \forall x_j \in F.$

Functions from the class $PSV(\Gamma, F)$ can have discontinuities of the second kind at the nodes x_j, and (15) is the condition of slow-oscillation of $a(x)$ in these points.
Let

$$w(x) = \prod_{m=1}^{l} |x - x_m|^{\beta_m}, \quad x_m \in F, \quad \beta_m \in (-1/2, 1/2)$$

be the power weight. We denote by $L_2(\Gamma, w)$ the Hilbert space with norm

$$\|u\|_{L_2(\Gamma, w)} = \left(\int_{\Gamma} (w(x) |u(x)|)^2 |dx| \right)^{1/2}.$$

Let

$$S_{\Gamma} u(x) = \frac{1}{\pi i} \int_{\Gamma} \frac{u(y) dy}{y - x}, \quad x \in \Gamma$$

be the singular integral operator on the Γ. The operator S_{Γ} is bounded in $L_2(\Gamma, w)$ because Γ is the union of finitely many of the Carleson arcs.

Let $a(x) \in PSV(\Gamma, F)$. With a point $x \in \Gamma \backslash F$ we associate the local representative a^x (see for instance [12]), which is an operator of multiplication by the constant $a(x)$ acting in $L_2(\mathbf{R})$, and with a point $x \in F$ we associate the operator a^x, which is the multiplication operator by the matrix-function $diag\{a_1(t), a_2(t), \ldots, a_N(t)\}$, where $a_k(t) = a \mid_{\gamma_k} (x + \phi_k(t))$. Here $z = x + \phi_k(t), \, k = 1, \ldots, N$ are the parametrizations of curves γ_k, which make up the node $x(\in F)$, t runs a small neighborhood of the origin. The operator a^x acts in the space $L_{2,\beta_x}(\mathbf{R}_+, \mathbf{C}^N)$.

Let us consider local representatives of the operator $S_{\Gamma} : L_2(\Gamma, w) \to L_2(\Gamma, w)$. It is well-known that the local representative S_{Γ}^x at a point $x \in \Gamma \backslash F$ is the singular integral operator $S_{\mathbf{R}} : L_2(\mathbf{R}) \to L_2(\mathbf{R})$ with symbol $sgn\xi, \, \xi \in \mathbf{R}$.

Let $x \in F$. Then applying a change of variables and the calculation of the Mellin pseudodifferential operators, one can show that the local representative of S_{Γ} at the point x is the operator $S_{\Gamma}^x \in \mathbf{F}(\beta_x, N)$. It is the operator with matrix symbol:

$$\Omega^x(t, \lambda) = \left(\Omega_{jk}^x(t, \lambda) \right)_{j,k=1}^{N}, \quad N = N(x),$$

where $\Omega_{jk}^x(t, \lambda)$ are given by the formulas:

(16) $\Omega_{jk}^x(t, \lambda) = \pm \dfrac{\exp\left[\lambda(\theta_k(t) - \theta_j(t) - \pi) \right]}{\sinh \pi \lambda}, \quad k > j, \quad \lambda \in L_{\nu, \beta_x + 1/2},$

$$(17) \qquad \Omega_{jk}^x(t,\lambda) = \pm \frac{\exp\left[\lambda(\theta_k(t) - \theta_j(t) + \pi)\right]}{\sinh \pi\lambda}, \quad k < j, \quad \lambda \in L_{\nu, \beta_x + 1/2},$$

$$(18) \qquad \Omega_{jj}^x(t,\lambda) = \pm \coth \pi\lambda, \quad \lambda \in L_{\nu, \beta_x + 1/2}$$

where

$$L_{\nu, \beta_x + 1/2} = \{w = u + iv \in \mathbf{C} : v = -\nu u + \beta_x + 1/2\}$$

is the straight line in the complex plain $\mathbf{C}$ with the slope $-\nu$ passing through the point $(0, \beta_x + 1/2)$. Here the sign $+$ is taken if the curve γ_k goes out from the node $x \in F$, and the sign $-$ is taken if the γ_k comes in the node.

The formulas (16)–(18) are similar to the formulas for the local symbol of singular integral operators in a node made up by piece-wise Lyapunov curves (see for instance [11]), but there are the differences:

1) the parameter λ runs along the sloping straight line $L_{\nu, \beta_x + 1/2}$,

2) the symbol $\Omega^x(t, \lambda)$ depends on the variables t, λ.

DEFINITION 3.3. We denote by $\mathbf{M}(\Gamma, w)$ the C^*-algebra of operators acting in $L_2(\Gamma, w)$ generated by operators

$$(19) \qquad A = \sum_k \prod_j A^{kj}$$

where A^{kj} is either the singular integral operator S_Γ or an operator of multiplication by a function $a(x) \in PSV(\Gamma, F)$, where k, j run a finite set of integers.

The C^*- algebra $\mathbf{M}(\Gamma, w)$ is nonreducible, and it contains the two-sided ideal $\mathcal{K}(L_2(\Gamma, w))$ of compact operators.

Let

$$\sigma^x(A) = \sum_k \prod_j \sigma^x(A^{kj}),$$

where $\sigma^x(A^{kj})$ is the local symbol. Here $\sigma^x(A) \in \mathbf{F}(\beta_x, N)$, if $x \in F$ and $\sigma^x(A) = (\sigma_+^x(A), \sigma_-^x(A)) \in \mathbf{C} \oplus \mathbf{C}$, if $x \in \Gamma \backslash F$.

THEOREM 3.1. *An operator A (19) is the Fredholm operator in $L_2(\Gamma, w)$ if and only if:*

(a) *$\sigma_\pm^x(A) \neq 0$ for any point $x \in \Gamma \backslash F$,*

(b)

$$(20) \qquad \lim_{r \to 0} \inf_{t \in (0, r), \, \lambda \in L_\nu, \, \beta_x + 1/2} |\det \sigma^x(A)(t, \lambda)| > 0$$

for any point $x \in F$.

It should be noted that the condition (b) is equivalent to the invertibility of the factor-class $\hat{\sigma}^x(A)$ in $\hat{\Lambda}(\beta_x, N) = \Lambda(\beta_x, N)/C_{01}(\beta_x, N)$.

Let us introduce the notations:

$$\|\hat{\sigma}^x(A)\| = \max(|\, \sigma_+^x(A) \,|, |\, \sigma_-^x(A) \,|), \quad x \in \Gamma \backslash F,$$

$$\|\hat{\sigma}^x(A)\| = \|\hat{\sigma}^x(A)\|_{\hat{\Lambda}(\beta_x, N)}, \quad x \in F.$$

PROPOSITION 3.1. *Let A be an operator of the form* (19) *and*

$$\||A|\| = \inf_{T \in \mathcal{K}(L_2(\Gamma, w))} \|A - T\|$$

be the essential norm of A. Then

(21)
$$\sup_{x \in \Gamma} \|\hat{\sigma}^x(A)\| \leq \||A|\| \leq C \sup_{x \in \Gamma} \|\hat{\sigma}^x(A)\|.$$

Passage to the limit in the left-hand side of the inequality (21) permits to define the symbol

$$\sigma(A) = \{\hat{\sigma}^x(A)\}$$

for an arbitrary operator $A \in \mathbf{M}(\Gamma, w)$.

We denote by $\mathbf{N}(\Gamma, w)$ the C^*-algebra of symbols of operators in $\mathbf{M}(\Gamma, w)$. We introduce pointwise operations on $\mathbf{N}(\Gamma, w)$, and the operation of conjugation

$$* : \{\hat{\sigma}^x(A)\} \to \{\hat{\sigma}^x(A)\}_{x \in \Gamma}^*,$$

and the norm

$$\|\sigma(A)\| = \sup_{x \in \Gamma} \|\hat{\sigma}^x(A)\|.$$

THEOREM 3.2. (a) *The mapping $\Sigma : \mathbf{M}(\Gamma, w) \to \mathbf{N}(\Gamma, w)$ associating to $A \in \mathbf{M}(\Gamma, w)$ its symbol $\sigma(A) \in \mathbf{N}(\Gamma, w)$ is the epimorphism of the C^*-algebras. Its kernel is the ideal $\mathcal{K}(L_2(\Gamma, w))$.*

(b) $A \in \mathbf{M}(\Gamma, w)$ *is a Fredholm operator if and only if the symbol $\sigma(A)$ is invertible in the algebra $\mathbf{N}(\Gamma, w)$. This condition is equivalent to the following ones:*

(1) $\sigma_{\pm}^x(A) \neq 0 \; \forall \; x \in \Gamma \backslash F$,
(2) *the condition (20) holds for any point $x \in F$.*

(c) *Let $A \in \mathbf{M}(\Gamma, w)$ be the Fredholm operator. Then its regularizator $R \in \mathbf{M}(\Gamma, w)$.*

References

[1] B. V. Lange, V. S. Rabinovich, *Pseudodifferential Operators on $\mathbf{R}^n$ and Limit Operators*, Math. USSR, Sbornik **57** (1987), 183–194.

[2] V. S. Rabinovich, *Fredholm Property of Pseudodifferential Operators on $\mathbf{R}^n$ in the Scale of the Spaces $L_{2,p}$*, Translation from Sibirskii Mathematicheskii Zhurnal **29** (1988), 635–646.

[3] L. Hörmander, *The Analysis of Linear Partial Differential Operators. III: Pseudo-Differential Operators*, Springer-Verlag, Berlin, Heidelberg, New York, Tokyo, 1985.

[4] M. E. Taylor, *Pseudodifferential Operators*, Princeton University Press, Princeton, New Jersey, 1981.

[5] V. S. Rabinovich, *The Noether Property for the Pseudodifferential Operators with Symbols in the Class $S_{\rho,\delta}^m$* $(0 \leq \delta = \rho < 1)$, Matematicheskie Zametki **27** (1980), 457–467.

[6] H. O. Cordes, *Spectral Theory of Linear Differential Operators and Comparison Algebras.* London Mathematical Society, Lecture Notes Series 76, Cambridge University Press, 1987.

[7] H. O. Cordes and S. T. Melo, *An Algebra of Singular Integral Operators with Kernels of Bounded Oscillation, and Application to Periodic Differential Operators*, Journal of Differential Equations **75** (1988), 216–238.

[8] S. T. Melo, *A Comparison Algebra on a Cylinder with Semi-Periodic Multiplications*, Pacific Journal of Mathematics **146** (1990), 281–304.

[9] V. S. Rabinovich, *On the Algebra Generated by Pseudodifferential Operators on $\mathbf{R}^n$, Operators of Multiplications by Almost-Periodic Functions, and Shift Operators*, Soviet Math. Dokl **25** (1982), 498–502.

[10] A. Böttcher and B. Silberman, *Analysis of Toeplitz Operators*, Springer-Verlag, Berlin, 1990.

[11] S. Roch and B. Silberman, *Algebras of Convolution Operators and their Image in the Calkin Algebras*, Report Math.90-05, Karl-Weierstrass -Inst. Math., Akad. Wiss. DDR, Berlin (1990).

[12] I. B. Simonenko and Chin Ngok Min, *Local Principal in the Theory of One Dimensional Singular Integral Equations with Piece-Wise Continuous Coefficients*, Rostov University, Rostov-on-Don (1986 (In Russian)).

[13] A. Böttcher and Yu. I. Karlovich, *Toeplitz and Singular Integral Operators on Carleson Curves with Logarithmic Whirl Points*, Techniche Universitat Chemnitz-Zwickau, 1994.

[14] V. S. Rabinovich, *Singular Integral Operators on a Complex Contours with Oscillating Tangent, and Pseudodifferential Mellin Operators*, Soviet Math. Dokl **44** (1992), 791–796.

[15] M. A. Shubin, *Almost-Periodic Functions and Partial Differential Equations*, Uspekhi Mat. Nauk **33** (1978), 3–47.

[16] L. A. Coburn, R. D. Moyer, and I. M. Singer, *C^*-algebras of Almost-Periodic Pseudodifferential Operators*, Acta Math. **139:3-4** (1973), 279–307.

DEPARTMENT OF MATHEMATICS AND MECHANICS OF ROSTOV STATE UNIV., UL. ZORGE 5, 344104, ROSTOV-ON-DON, RUSSIA

E-mail address: rabinov@ns.unird.ac.ru

Contemporary Mathematics
Volume **212**, 1998

Bargmann Projection, Three-Valued Functions and Corresponding Toeplitz Operators

Enrique Ramírez de Arellano and Nikolai Vasilevski

ABSTRACT. We describe the C^*-algebra generated by the Bargmann projection and the multiplication operators by three–valued functions. The key point of this investigation is the description of one important specific case of the C^*-algebra generated by three orthogonal projections.

The algebra generated by Toeplitz operators with three–valued symbols acting on the Fock space, and spectra of Toeplitz operators are considered as well.

1. Introduction

1.1. The investigation of the C^*-algebra generated by an orthogonal projection and piecewise continuous functions (in the case when this projection commutes with multiplication operators by continuous functions up to a compact operator) is quite standard (see, for example, [8, 13]) and splits basically into the two following steps: the application of the local principle for C^*-algebras [5, 12] and the description of model (or local) C^*-algebras generated by an orthogonal projection and the multiplication operator by the characteristic function of an appropriate measurable set. The last step is normally based on the well known description of the C^*-algebra generated by two orthogonal projections [7, 9, 14].

However, if one would like to change piecewise continuous functions for functions having three cluster points on some set, the second step fails. There is no general description of the C^*-algebra generated by three orthogonal projections, even in the needed specific case: the product of two of them is equal to zero. In Section 2 we study one important special case of the C^*-algebra generated by three orthogonal projections. This allows us to carry out the desired step. In Section 3 we describe the C^*-algebra generated by the Bargmann projection and two others: the multiplication operators by characteristic functions of two different measurable sets. In Section 4 we show how the structure of the algebra depends on these measurable sets. The corresponding Toeplitz operator algebra and the spectrum of Toeplitz operators are considered in Section 5.

1991 *Mathematics Subject Classification.* Primary 47B35; Secondary 47D25.

Keywords: *Orthogonal projections, Toeplitz operators, Fock space, operator algebras.*

This work was partially supported by CONACYT Project 4069-E9404, México.

1.2. We use the following standard notation: $z = x + iy = (z_1, \ldots, z_n) \in \mathbb{C}^n$; $\overline{z} = (\overline{z}_1, \ldots, \overline{z})$ with the usual notion of the complex conjugation; for $z, w \in \mathbb{C}^n$, $z \cdot w = z_1 w_1 + \ldots + z_n w_n$; $|z|^2 = |z_1|^2 + \ldots + |z_n|^2 (= z \cdot \overline{z})$; $(x, y) = (x_1, \ldots, x_n, y_1, \ldots, y_n) \in \mathbb{R}^{2n} = \mathbb{R}^n \oplus \mathbb{R}^n$. Denote by $d\mu_n(z)$ the following Gaussian measure over $\mathbb{C}^n$:

$$d\mu_n(z) = \pi^{-n} e^{-z \cdot \overline{z}} dv(z),$$

where $dv(z) = dx\, dy$ is the usual Euclidean volume measure on $\mathbb{C}^n = \mathbb{R}^{2n}$.

Denote by $L_2(\mathbb{C}^n, d\mu_n)$ the Hilbert space of μ square-integrable functions on $\mathbb{C}^n$ with the following inner product:

$$\langle f, \varphi \rangle = \int_{\mathbb{C}^n} f(z) \overline{\varphi(z)} \, d\mu_n(z).$$

The closed subspace $F^2(\mathbb{C}^n)$ of $L_2(\mathbb{C}^n, d\mu_n)$ consisting of all entire functions is usually called the Fock [2, 6] or the Segal–Bargmann space [1, 11].

The orthogonal Bargmann projection

$$P_n : L_2(\mathbb{C}^n, d\mu_n) \to F_2(\mathbb{C}^n)$$

is given [3] by the formula

(1.1) $$(P_n \varphi)(z) = \int_{\mathbb{C}^n} \varphi(\zeta) e^{\overline{\zeta} \cdot z} d\mu_n(\zeta).$$

2. Special case of the algebra generated by three orthogonal projections

2.1. Let H be a Hilbert space, $l_0 \in H$ with $\|l_0\| = 1$. Denote by L_0 the one-dimensional subspace in H generated by l_0, and let

$$P\varphi = \langle \varphi, l_0 \rangle l_0 : H \to L_0$$

be the one-dimensional orthogonal projection of H onto L_0.

Consider two (other) orthogonal projections Q_1 and Q_2 on H with the property

(2.1) $$Q_1 \cdot Q_2 = Q_2 \cdot Q_1 = 0$$

We are going to describe the C^*-algebra $\mathcal{R} = \mathcal{R}(\mathbb{C}; P, Q_1, Q_2)$ with identity $I : H \to H$, generated by the three orthogonal projections: the one-dimensional projection P and projections Q_1, Q_2 with the property (2.1).

2.2. Introduce $Q_3 = I - Q_1 - Q_2$ and denote by M_1, M_2, M_3 the images of Q_1, Q_2, Q_3 respectively. Then,

$$H = M_1 \oplus M_2 \oplus M_3.$$

In what follows we assume that L_0 *is not orthogonal* to each M_k, otherwise the problem reduces directly to the case of two projections.

Denote by m_k the unit vector in the direction of the orthogonal projection of l_0 onto M_k, and by M_k' the one-dimensional space, generated by m_k. Then

$$l_0 = \alpha_1 \cdot m_1 + \alpha_2 \cdot m_2 + \alpha_3 \cdot m_3,$$

where $|\alpha_1|^2 + |\alpha_2|^2 + |\alpha_3|^2 = 1$, and $\alpha_1 \alpha_2 \alpha_3 \neq 0$.

With respect to the orthogonal decomposition

$$H = (M_1 \ominus M_1') \oplus (M_2 \ominus M_2') \oplus (M_3 \ominus M_3') \oplus (M_1' \oplus M_2' \oplus M_3'),$$

the projections at hand can be represented as

$$Q_1 = I + 0 + 0 + \begin{pmatrix} 1 & 0 & 0 \\ 0 & 0 & 0 \\ 0 & 0 & 0 \end{pmatrix},$$

$$Q_2 = 0 + I + 0 + \begin{pmatrix} 0 & 0 & 0 \\ 0 & 1 & 0 \\ 0 & 0 & 0 \end{pmatrix},$$

$$Q_3 = 0 + 0 + I + \begin{pmatrix} 0 & 0 & 0 \\ 0 & 0 & 0 \\ 0 & 0 & 1 \end{pmatrix},$$

$$P = 0 + 0 + 0 + \begin{pmatrix} \alpha_1\overline{\alpha_1} & \alpha_1\overline{\alpha_2} & \alpha_1\overline{\alpha_3} \\ \alpha_2\overline{\alpha_1} & \alpha_2\overline{\alpha_2} & \alpha_2\overline{\alpha_3} \\ \alpha_3\overline{\alpha_1} & \alpha_3\overline{\alpha_2} & \alpha_3\overline{\alpha_3} \end{pmatrix}.$$

2.3. THEOREM. *The algebra* $\mathcal{R} = \mathcal{R}(\mathbb{C}; P, Q_1, Q_2)$ *is isomorphic to* $\mathbb{C}^3 \oplus \mathrm{Mat}_3(\mathbb{C})$.

The isomorphism

$$\nu : \mathcal{R} \longrightarrow \mathbb{C}^3 \oplus \mathrm{Mat}_3(\mathbb{C})$$

is generated by the following mapping of the projections

$$\nu : P \longmapsto \left((0,0,0), \begin{pmatrix} \gamma_1 & \sqrt{\gamma_1\gamma_2} & \sqrt{\gamma_1\gamma_3} \\ \sqrt{\gamma_1\gamma_2} & \gamma_2 & \sqrt{\gamma_2\gamma_3} \\ \sqrt{\gamma_1\gamma_3} & \sqrt{\gamma_2\gamma_3} & \gamma_3 \end{pmatrix} \right),$$

$$\nu : Q_1 \longmapsto \left((1,0,0), \begin{pmatrix} 1 & 0 & 0 \\ 0 & 0 & 0 \\ 0 & 0 & 0 \end{pmatrix} \right),$$

$$\nu : Q_2 \longmapsto \left((0,1,0), \begin{pmatrix} 0 & 0 & 0 \\ 0 & 1 & 0 \\ 0 & 0 & 0 \end{pmatrix} \right),$$

$$\nu : Q_3 \longmapsto \left((0,0,1), \begin{pmatrix} 0 & 0 & 0 \\ 0 & 0 & 0 \\ 0 & 0 & 1 \end{pmatrix} \right),$$

where $\gamma_1 = \|Q_1 l_0\|^2$, $\gamma_2 = \|Q_2 l_0\|^2$, $\gamma_3 = 1 - \gamma_1 - \gamma_2$.

PROOF. Introduce the following unitary operator on H:

$$W : I \oplus I \oplus I \oplus \begin{pmatrix} \omega_1 & 0 & 0 \\ 0 & \omega_2 & 0 \\ 0 & 0 & \omega_3 \end{pmatrix},$$

where $\omega_k = \alpha_k / |\alpha_k|$.

Then we have obviously

$$\widetilde{P} = W^{-1} P W = 0 + 0 + 0 + \begin{pmatrix} |\alpha_1|^2 & |\alpha_1\alpha_2| & |\alpha_1\alpha_3| \\ |\alpha_2\alpha_1| & |\alpha_2|^2 & |\alpha_2\alpha_3| \\ |\alpha_3\alpha_1| & |\alpha_3\alpha_2| & |\alpha_3|^2 \end{pmatrix},$$

$$\tilde{Q}_1 = W^{-1}Q_1W = I = 0 + 0 + \begin{pmatrix} 1 & 0 & 0 \\ 0 & 0 & 0 \\ 0 & 0 & 0 \end{pmatrix},$$

$$\tilde{Q}_2 = W^{-1}Q_2W = 0 + I + 0 + \begin{pmatrix} 0 & 0 & 0 \\ 0 & 1 & 0 \\ 0 & 0 & 0 \end{pmatrix},$$

$$\tilde{Q}_3 = W^{-1}Q_3W = 0 + 0 + I + \begin{pmatrix} 0 & 0 & 0 \\ 0 & 0 & 0 \\ 0 & 0 & 1 \end{pmatrix},$$

and the algebra $\mathcal{R} = \mathcal{R}(\mathbb{C}; P, Q_1, Q_2)$ is naturally isomorphic to the algebra $\tilde{\mathcal{R}} = W^{-1}\mathcal{R}W = \mathcal{R}(\mathbb{C}; \tilde{P}, \tilde{Q}_1, \tilde{Q}_2)$ generated by the projections $\tilde{P}$, $\tilde{Q}$, and $\tilde{Q}_2$ (and of course, by I or $\tilde{Q}_3$). The last algebra is naturally isomorphic to $\mathbb{C}^3 \oplus \mathrm{Mat}_3(\mathbb{C})$.

To finish the proof one needs to follow the correspondence of generators and to use new quantities $\gamma_1 = |\alpha_1|^2 = \|Q_1 l_0\|^2$, $\gamma_2 = |\alpha_2|^2 = \|Q_2 l_0\|^2$ and $\gamma_3 = |\alpha_3|^2 = 1 - \gamma_1 - \gamma_2$.

3. Algebra generated by the Bargmann projection and three-valued functions

3.1. Let X_1 and X_2 be two disjoint measurable sets in $\mathbb{R}^n = \mathbb{R}^n_x$, and let

$$X'_k = X_k + i\mathbb{R}^n_y, \quad k = 1, 2$$

be the corresponding sets in $\mathbb{C}^n = \mathbb{R}^n_x + i\mathbb{R}^n_y$.

Introduce the characteristic functions $\chi_{X'_k}$ of the sets X'_k and the orthogonal projections $Q_k = \chi_{X'_k} I$ acting on $L_2(\mathbb{C}^n, d\mu_n)$, here $k = 1, 2$.

Furthermore let

$$Q_3 = I - Q_1 - Q_2.$$

The aim of this section is to describe the C^*–algebra $\mathcal{R} = \mathcal{R}(X_1, X_2)$ with identity I generated by the Bargmann projection $P = P_n$ (1.1) and the projections Q_1, Q_2 acting on $L_2(\mathbb{C}^n, d\mu_n)$.

We have obviously

$$Q_k \cdot Q_j = \delta_{kj} Q_k, \quad k, j = \overline{1,3},$$

and

$$I = Q_1 + Q_2 + Q_3.$$

3.2. Introduce the following operators:

The unitary operator

$$U : L_2(\mathbb{C}^n, d\mu_n) \to L_2(\mathbb{R}^{2n}) = L_2(\mathbb{R}^{2n}, dxdy),$$

defined by

$$(U\varphi)(z) = \pi^{-\frac{n}{2}} e^{-\frac{z \cdot \bar{z}}{2}} \varphi(z),$$

or

$$(U\varphi)(x,y) = \pi^{-\frac{n}{2}} e^{-\frac{x^2+y^2}{2}} \varphi(x+iy);$$

the Fourier transformation $F : L_2(\mathbb{R}^n) \to L_2(\mathbb{R}^n)$, where

$$(Ff)(y) = (2\pi)^{-\frac{n}{2}} \int_{\mathbb{R}^n} e^{-i\eta \cdot y} f(\eta) d\eta;$$

and the isomorphism

$$W = W^* = W^{-1} : L_2(\mathbb{R}^{2n}) \to L_2(\mathbb{R}^{2n}),$$

where

$$(Wf)(x, y) = f(\frac{1}{\sqrt{2}}(x + y), \frac{1}{\sqrt{2}}(x - y)).$$

Then the unitary operator $K = W(I \otimes F)U$ is an isomorphism

$$K : L_2(\mathbb{C}^n, d\mu_n) \to L_2(\mathbb{R}^n, dx) \otimes L_2(\mathbb{R}^n, dy).$$

We have for the generators of the algebra $\mathcal{R}(X_1, X_2)$:

$$KPK^{-1} = I \otimes P_0,$$

where

$$(P_0 f)(y) = \pi^{-\frac{n}{2}} \int_{\mathbb{R}^n} e^{-\frac{1}{2}(y^2 + u^2)} f(u) du$$

is the one-dimensional orthogonal projection of $L_2(\mathbb{R}^n, dy)$ onto the one-dimensional space L_0 generated by $l_0(y) = \pi^{-\frac{n}{4}} e^{-\frac{y^2}{2}}$;

$$\begin{aligned}
KQ_k K^{-1} &= W(I \otimes F)U\chi_{X_k'} U^{-1}(I \otimes F^{-1})W = \\
&= W(I \otimes F)\chi_{X_k}(x)(I \otimes F^{-1})W = \\
&= W\chi_{X_k}(x)W = \chi_{X_k}(\frac{1}{\sqrt{2}}(x + y))I = \\
&= \chi_{\sqrt{2}X_k - x}(y)I, \quad k = \overline{1,3},
\end{aligned}$$

where $X_3 = \mathbb{R}^n \backslash (X_2 \cup X_2)$.

Thus the algebra $K\mathcal{R}K^{-1}$ splits into the direct integral of algebras $\mathcal{R}(x)$, $x \in \mathbb{R}^n$, generated by the orthogonal projections

$$P(x) = P_0, \quad Q_k(x) = \chi_{\sqrt{2}X_k - x}(y)I, \qquad k = 1, 2,$$

acting on $L_2(\mathbb{R}^n, dy)$.

3.3. LEMMA. *The algebra $\mathcal{R}(x)$, $x \in \mathbb{R}^n$, is isomorphic to $\mathbb{C}^3 \oplus \mathrm{Mat}_3(\mathbb{C})$. The isomorphism*

$$\nu_x : \quad \mathcal{R}(x) \to \mathbb{C}^3 \oplus \mathrm{Mat}_3(\mathbb{C})$$

is generated by the following mapping of the generators of the algebra $\mathcal{R}(x)$:

$$\nu_x : \quad P(x) \longmapsto \left((0,0,0), \begin{pmatrix} \gamma_1(x) & \sqrt{\gamma_1(x)\gamma_2(x)} & \sqrt{\gamma_1(x)\gamma_3(x)} \\ \sqrt{\gamma_1(x)\gamma_2(x)} & \gamma_2(x) & \sqrt{\gamma_2(x)\gamma_3(x)} \\ \sqrt{\gamma_1(x)\gamma_3(x)} & \sqrt{\gamma_2(x)\gamma_3(x)} & \gamma_3(x) \end{pmatrix} \right),$$

$$\nu_x : \quad Q_1(x) \longmapsto \left((1,0,0), \begin{pmatrix} 1 & 0 & 0 \\ 0 & 0 & 0 \\ 0 & 0 & 0 \end{pmatrix} \right),$$

$$\nu_x : \quad Q_2(x) \longmapsto \left((0,1,0), \begin{pmatrix} 0 & 0 & 0 \\ 0 & 1 & 0 \\ 0 & 0 & 0 \end{pmatrix} \right),$$

where $\gamma_1(x) = ||Q_1(x)l_0||^2$, $\gamma_2(x) = ||Q_2(x)l_0||^2$, $\gamma_3(x) = 1 - \gamma_1(x) - \gamma_2(x)$.

PROOF. Follows directly from Theorem 2.3.

3.4. Denote by $\mathfrak{S}(X_1, X_2)$ the C^*-algebra with identity generated by the following 3×3 matrix-valued functions

$$p(x) = \begin{pmatrix} \gamma_1(x) & \sqrt{\gamma_1(x)\gamma_2(x)} & \sqrt{\gamma_1(x)\gamma_3(x)} \\ \sqrt{\gamma_1(x)\gamma_2(x)} & \gamma_2(x) & \sqrt{\gamma_2(x)\gamma_3(x)} \\ \sqrt{\gamma_1(x)\gamma_3(x)} & \sqrt{\gamma_2(x)\gamma_3(x)} & \gamma_3(x) \end{pmatrix},$$

$$q_1(x) = \begin{pmatrix} 1 & 0 & 0 \\ 0 & 0 & 0 \\ 0 & 0 & 0 \end{pmatrix},$$

$$q_2(x) = \begin{pmatrix} 0 & 0 & 0 \\ 0 & 1 & 0 \\ 0 & 0 & 0 \end{pmatrix},$$

where $x \in \mathbb{R}^n$ and $\gamma_1(x)$, $\gamma_2(x)$, $\gamma_3(x)$ are as in Lemma 3.3.

3.5. COROLLARY. *The algebra* $\mathcal{R} = \mathcal{R}(X_1, X_2)$ *is isomorphic to the algebra* $\mathbb{C}^3 \oplus \mathfrak{S}(X_1, X_2)$. *The isomorphism*

$$\nu_0 : \mathcal{R} \longmapsto \mathbb{C}^3 \oplus \mathfrak{S}(X_1, X_2)$$

is generated by the following mapping of the generators of the algebra $\mathcal{R}$:

$$\nu_0 : \quad P \longmapsto ((0,0,0), p(x)),$$
$$\nu_0 : \quad Q_1 \longmapsto ((1,0,0), q_1(x)),$$
$$\nu_0 : \quad Q_2 \longmapsto ((0,1,0), q_2(x)),$$

where $x \in \mathbb{R}^n$.

3.6. To describe the algebra $\mathfrak{S}(X_1, X_2)$ we need to introduce some kind of "maximal" algebra. Let

$$\Delta_{\max} = \{t = (t_1, t_2) \in \mathbb{R}^2 : 0 \leq t_1 \leq 1, \ 0 \leq t_2 \leq 1, \ 0 \leq t_1 + t_2 \leq 1\}$$

be the standard two–dimensional simplex.

Denote by $\mathfrak{S}(\Delta_{\max})$ the C^*–algebra with identity generated by the following matrix-valued functions

$$p(t) = \begin{pmatrix} t_1 & \sqrt{t_1 t_2} & \sqrt{t_1 t_3} \\ \sqrt{t_1 t_2} & t_2 & \sqrt{t_2 t_3} \\ \sqrt{t_1 t_3} & \sqrt{t_2 t_3} & t_3 \end{pmatrix},$$

$$q_1(t) = \begin{pmatrix} 1 & 0 & 0 \\ 0 & 0 & 0 \\ 0 & 0 & 0 \end{pmatrix},$$

$$q_2(t) = \begin{pmatrix} 0 & 0 & 0 \\ 0 & 1 & 0 \\ 0 & 0 & 0 \end{pmatrix},$$

where $t = (t_1, t_2) \in \Delta_{\max}$, $t_3 = 1 - t_1 - t_2$.

Now denote by $\mathfrak{S}_{\max}$ the C^*–algebra (with identity) of all 3×3 matrix–valued functions $\sigma(t) = (\sigma_{kj}(t))_{k,j=1}^3$ continuous on $\Delta_{\max}$ and having the following properties:

(i) they are diagonal at the vertices $(0,0)$, $(0,1)$ and $(1,0)$ of $\Delta_{\max}$:

(ii) $\sigma_{12}(t) = \sigma_{13}(t) = \sigma_{21}(t) = \sigma_{31}(t) \equiv 0$, $\sigma_{11}(t) \equiv$ const on the side $t_1 = 0$ of the triangle $\Delta_{\max}$;

(iii) $\sigma_{21}(t) = \sigma_{23}(t) = \sigma_{12}(t) = \sigma_{32}(t) \equiv 0$, $\sigma_{22}(t) \equiv$ const on the side $t_2 = 0$ of the triangle $\Delta_{\max}$;

(iv) $\sigma_{31}(t) = \sigma_{32}(t) = \sigma_{13}(t) = \sigma_{23}(t) \equiv 0$, $\sigma_{33}(t) \equiv$ const on the side $t_1 + t_2 = 1$ $(t_3 = 0)$ of $\Delta_{\max}$.

Standard arguments lead directly to the following lemma.

3.7. LEMMA. *Up to unitary equivalence all irreducible representations of the algebra $\mathfrak{S}_{\max}$ are included in the following list:*

(i) *the family $\{\pi_t\}$ of three-dimensional representations parametrized by the points $t \in \operatorname{Int} \Delta_{\max}$ and defined by the rule*

$$\pi_t : \sigma \in \mathfrak{S}_{\max} \longmapsto \sigma(t) \in \operatorname{Mat}_3(\mathbb{C});$$

(ii) *the three families $\{\pi_t^{(1)}\}$, $\{\pi_t^{(2)}\}$ and $\{\pi_t^{(3)}\}$ of two-dimensional representations parametrized by the interior parts of the sides $t_1 = 0$, $t_2 = 0$ and $t_1 + t_2 = 1$ $(t_3 = 0)$ of the triangle $\Delta_{\max}$ respectively; these representations (for the corresponding points t) are defined by the rule*

$$\pi_t^{(1)} : \sigma \in \mathfrak{S}_{\max} \longmapsto \begin{pmatrix} \sigma_{22}(t) & \sigma_{23}(t) \\ \sigma_{32}(t) & \sigma_{33}(t) \end{pmatrix} \in \operatorname{Mat}_2(\mathbb{C}),$$

$$\pi_t^{(2)} : \sigma \in \mathfrak{S}_{\max} \longmapsto \begin{pmatrix} \sigma_{11}(t) & \sigma_{13}(t) \\ \sigma_{31}(t) & \sigma_{33}(t) \end{pmatrix} \in \operatorname{Mat}_2(\mathbb{C}),$$

$$\pi_t^{(3)} : \sigma \in \mathfrak{S}_{\max} \longmapsto \begin{pmatrix} \sigma_{11}(t) & \sigma_{12}(t) \\ \sigma_{21}(t) & \sigma_{22}(t) \end{pmatrix} \in \operatorname{Mat}_2(\mathbb{C});$$

(iii) *the six one–dimensional representations defined as follows:*

$$\pi_{(1)} : \sigma \in \mathfrak{S}_{\max} \longmapsto \sigma_{11}(1,0),$$

$$\pi_{(2)} : \sigma \in \mathfrak{S}_{\max} \longmapsto \sigma_{11}(0,\tau),$$

$$\pi_{(3)} : \sigma \in \mathfrak{S}_{\max} \longmapsto \sigma_{22}(0,1),$$

$$\pi_{(4)} : \sigma \in \mathfrak{S}_{\max} \longmapsto \sigma_{22}(\tau,0),$$

$$\pi_{(5)} : \sigma \in \mathfrak{S}_{\max} \longmapsto \sigma_{33}(0,0),$$

$$\pi_{(6)} : \sigma \in \mathfrak{S}_{\max} \longmapsto \sigma_{33}(\tau,1-\tau),$$

where $t = (t_1, t_2) \in \partial\Delta_{\max}$.

3.8. THEOREM. *The algebra $\mathfrak{S}(\Delta_{\max})$ coincides with the C^*–algebra $\mathfrak{S}_{\max}$.*

PROOF. The algebra $\mathfrak{S}(\Delta_{\max})$ is obviously a C^*–subalgebra of $\mathfrak{S}_{\max}$. We will prove that $\mathfrak{S}(\Delta_{\max})$ is a massive [4] subalgebra of $\mathfrak{S}_{\max}$, and then by the Stone–Weierstrass theorem for GCR-algebras [4] we will have $\mathfrak{S}(\Delta_{\max}) = \mathfrak{S}_{\max}$. To prove that, we need to check two properties:

(i) for each irreducible representation π of $\mathfrak{S}_{\max}$ the representation $\pi|_{\mathfrak{S}(\Delta_{\max})}$ is irreducible;

(ii) for each two nonequivalent irreducible representations π_1 and π_2 of $\mathfrak{S}_{\max}$ their restrictions $\pi_1|_{\mathfrak{S}(\Delta_{\max})}$ and $\pi_2|_{\mathfrak{S}(\Delta_{\max})}$ are nonequivalent as well.

We check these two statements for three-dimensional representations only.

Fix a point $t_0 \in \Delta_{\max}$ and consider the corresponding representation π_{t_0}. To prove the first statement it is sufficient to find for each matrix $m = (m_{kj})_{k,j=1}^3 \in \mathrm{Mat}_3(\mathbb{C})$ an element $a \in \mathfrak{S}(\Delta_{\max})$ such that

$$\pi_{t_0}(a) = m.$$

It is easy to see that we can take a to be of the form

$$a = (m_{kj} \frac{1}{(p(t_0))_{kj}} (q_k p(t_0) q_j)_{kj})_{k,j=1}^3.$$

To prove the second statement it is sufficient to find for each two fixed different points $t' = (t_1', t_2')$ and $t'' = (t_1'', t_2'')$ of Int $\Delta_{\max}$ an element $a \in \mathfrak{S}(\Delta_{\max})$ such that

$$\pi_{t'}(a) = 0, \quad \pi_{t''}(a) \neq 0.$$

The corresponding element, for example, is

$$a = q_1(p - t_2' e)q_1 + q_2(p - t_2' e)q_2.$$

3.9. Introduce the set

$$\Delta = \Delta(X_1, X_2)$$
$$= \mathrm{clos}\{t = (t_1, t_2) \in \Delta_{\max} : \exists x \in \mathbb{R}^n \text{ with } t_1 = \gamma_1(x), \ t_2 = \gamma_2(x)\},$$

and denote by $\mathfrak{S}(\Delta)$ the restriction of the algebra $\mathfrak{S}_{\max} = \mathfrak{S}(\Delta_{\max})$ to the set Δ; i.e.,

$$\mathfrak{S}(\Delta) = \{\sigma|_\Delta : \sigma \in \mathfrak{S}_{\max}\}.$$

3.10. THEOREM. *The algebra $\mathfrak{S}(X_1, X_2)$ is isomorphic to the algebra $\mathfrak{S}(\Delta)$.*

PROOF. Denote for a moment by $\mathfrak{S}_\Delta$ the C^*–algebra with identity generated by $p(t)$, $q_1(t)$ and $q_2(t)$, $t \in \Delta$.

The set $\Delta = \Delta(X_1, X_2)$ is obviously the set of maximal ideals of the C^*–algebra with identity $\mathcal{R}(\mathbb{C}; \gamma_1, \gamma_2)$ generated by the two functions $\gamma_1(x)$ and $\gamma_2(x)$, $x \in \mathbb{R}^n$. By the Gelfand transformation, the algebra $\mathcal{R}(\mathbb{C}, \gamma_1, \gamma_2)$ is isomorphic to $C(\Delta)$, and t_k is the image of $\gamma_k(x)$, $k = 1, 2$, under this transformation.

Further, the algebras $\mathcal{R}(\mathbb{C}; \gamma_1, \gamma_2) \otimes \mathrm{Mat}_3(\mathbb{C})$ and $C(\Delta) \otimes \mathrm{Mat}_3(\mathbb{C})$ are naturally isomorphic, and their subalgebras $\mathfrak{S}(X_1, X_2)$ and $\mathfrak{S}_\Delta$ are isomorphic as well.

Finally, the Stone–Weierstrass theorem shows, as in the proof of Theorem 3.8, that $\mathfrak{S}_\Delta = \mathfrak{S}(\Delta)$.

3.11. COROLLARY. *The algebra $\mathcal{R}(X_1, X_2)$ is isomorphic to the algebra $\mathbb{C}^3 \oplus \mathfrak{S}(\Delta)$. The isomorphism*

$$\nu : \mathcal{R}(X_1, X_2) \longrightarrow \mathbb{C}^3 \oplus \mathfrak{S}(\Delta)$$

is generated by the following mapping of the projections:

$$\nu : P \longmapsto ((0,0,0), p),$$
$$\nu : Q_1 \longmapsto ((1,0,0), q_1),$$
$$\nu : Q_2 \longmapsto ((0,1,0), q_2),$$

where

$$p = p(t) = \begin{pmatrix} t_1 & \sqrt{t_1 t_2} & \sqrt{t_1 t_3} \\ \sqrt{t_1 t_2} & t_2 & \sqrt{t_2 t_3} \\ \sqrt{t_1 t_3} & \sqrt{t_2 t_3} & t_3 \end{pmatrix},$$

$$q_1 = q_1(t) = \begin{pmatrix} 1 & 0 & 0 \\ 0 & 0 & 0 \\ 0 & 0 & 0 \end{pmatrix},$$

$$q_2 = q_2(t) = \begin{pmatrix} 0 & 0 & 0 \\ 0 & 1 & 0 \\ 0 & 0 & 0 \end{pmatrix},$$

$t = (t_1, t_2) \in \Delta = \Delta(X_1, X_2)$, $t_3 = 1 - t_1 - t_2$.

3.12. Denote by $PC(X_1, X_2)$ the algebra of all piecewise constant functions of the form

$$a(x) = \alpha_1 \chi_1(x) + \alpha_2 \chi_2(x) + \alpha_3 \chi_3(x),$$

where $\alpha_1, \alpha_2, \alpha_3 \in \mathbb{C}$; i.e., the algebra of all three-valued functions constant on the sets X_k, $k = \overline{1,3}$.

Now the algebra $\mathcal{R}(X_1, X_2)$ can be introduced as the algebra generated by the Bargmann projection P and piecewise constant functions from $PC(X_1, X_2)$, i.e., by all operators of the form

$$A = a(x)P + b(x)(I - P),$$

with $a(x) = \sum_1^3 \alpha_k \chi_k(x), b(x) = \sum_1^3 \beta_k \chi_k(x) \in PC(X_1, X_2)$.

It is easy to see that the image of A under the isomorphism

$$\nu : \mathcal{R}(X_1, X_2) \longrightarrow \mathbb{C}^3 \oplus \mathfrak{S}(\Delta)$$

is

$$\nu : A \longmapsto ((\beta_1, \beta_2, \beta_3), a(t)),$$

where

$$a(t) = \begin{pmatrix} \alpha_1 t_1 + \beta_1(1 - t_1) & (\alpha_1 - \beta_1)\sqrt{t_1 t_2} & (\alpha_1 - \beta_1)\sqrt{t_1 t_3} \\ (\alpha_2 - \beta_2)\sqrt{t_1 t_2} & \alpha_2 t_2 + \beta_2(1 - t_2) & (\alpha_2 - \beta_2)\sqrt{t_2 t_3} \\ (\alpha_3 - \beta_3)\sqrt{t_1 t_3} & (\alpha_3 - \beta_3)\sqrt{t_2 t_3} & \alpha_3 t_3 + \beta_3(1 - t_3) \end{pmatrix},$$

with $t = (t_1, t_2) \in \Delta = \Delta(X_1, X_2)$, $t_3 = 1 - t_1 - t_2$.

4. The set $\Delta = \Delta(X_1, X_2)$

4.1. Recall that by definition

$$\Delta = \Delta(X_1, X_2)$$
$$= \text{clos}\{t = (t_1, t_2) \in \Delta_{\max} : \exists x \in \mathbb{R}^n \text{ with } t_1 = \gamma_1(x), \ t_2 = \gamma_2(x)\},$$

where $\gamma_k(x) = ||Q_k(x)l_0||^2 = \int_{\sqrt{2}X_k - x} |l_0(y)|^2 dy$, $k = 1, 2$.

The functions $\gamma_k(x)$, $k = 1, 2$, are continuous on $\mathbb{R}^n$ and thus Δ is a connected compact subset of $\Delta_{\max}$.

4.2. Let us give some examples of sets X_1, X_2 and the corresponding set $\Delta = \Delta(X_1, X_2)$. For simplicity we will consider the two–dimensional case ($X_k \subset \mathbb{R}^2$) only.

EXAMPLE 1. A complete simplex $\Delta_{\max}$.

$$X_k = \{x = re^{i\varphi} \in \mathbb{R}^2 : \ r \in \mathbb{R}_+, \ \varphi \in [\frac{2(k-1)\pi}{3}, \frac{2k\pi}{3})\}, \quad k = 1, 2.$$

It is easy to see that $\Delta = \Delta_{\max}$.

EXAMPLE 2. A polygon.

First recall [10] how to construct for given $0 \le \alpha < \beta \le 1$ the set $X \subset \mathbb{R}$ with the properties

$$\inf_{x \in \mathbb{R}} \int_{\sqrt{2}X - x} |l_0(y)|^2 dy = \alpha,$$

$$\sup_{x \in \mathbb{R}} \int_{\sqrt{2}X - x} |l_0(y)|^2 dy = \beta.$$

It is clear that for each $\gamma \in [0, 1]$ there exists $a_\gamma \in [0, 1]$ such that for the set

$$Q_\gamma = \frac{1}{\sqrt{2}} \bigcup_{n \in \mathbb{Z}} [-a_\gamma + 2n, a_\gamma + 2n]$$

we have

$$\int_{\sqrt{2}Q_\gamma} |l_0(y)|^2 dy = \gamma.$$

For the set $X = ((\mathbb{R} \setminus Q_{1-\alpha}) \bigcap (-\infty, 0)) \bigcup (Q_\beta \bigcap (0, +\infty))$ and

$$\gamma(x) = \int_{\sqrt{2}X - x} |l_0(y)|^2 dy$$

we have

$$\inf_{x \in \mathbb{R}} \gamma(x) = \liminf_{x \to -\infty} \gamma(x) = \alpha,$$

$$\sup_{x \in \mathbb{R}} \gamma(x) = \limsup_{x \to +\infty} \gamma(x) = \beta.$$

Now, for given $\gamma \in (0, 1 - \alpha)$ let $Y \subset \mathbb{R}$ be a set (constructed as above) with

$$\inf_{x \in \mathbb{R}} \int_{\sqrt{2}X - x} |l_0(y)|^2 dy = \liminf_{x \to -\infty} \int_{\sqrt{2}X - x} |l_0(y)|^2 dy = \gamma.$$

Introduce:

$$X_1 = ((\mathbb{R} \setminus X) \times \mathbb{R}_+) \cup (Y \times (\mathbb{R} \setminus \mathbb{R}_+)),$$
$$X_2 = (\mathbb{R} \setminus (X \cup Y)) \times (\mathbb{R} \setminus \mathbb{R}_+).$$

Then, if $\gamma < \alpha$ we have that Δ is the trapezoid bounded by the lines

$$x_1 + x_2 = \alpha, \quad x_1 + x_2 = \beta,$$
$$x_2 = 0, \quad x_1 = \gamma;$$

if $\alpha \le \gamma$ we have that Δ is the triangle bounded by the lines

$$x_1 = \gamma, \quad x_2 = 0, \quad x_1 + x_2 = \beta.$$

Of course, one can modify this example to move this trapezoid or triangle in $\Delta_{\max}$ and to change Δ for a pentagon or a hexagon.

EXAMPLE 3. A closed curve.

Let

$$X_1 = (\bigcup_{k=-\infty}^{\infty} [k, k + \frac{1}{3})) \times \mathbb{R},$$

$$X_2 = (\bigcup_{k=-\infty}^{\infty} [k + \frac{1}{3}, k + \frac{2}{3})) \times \mathbb{R}.$$

Then the set Δ is a simple closed curve passing through the following points of Δ_{max}:

$$(\alpha, \gamma), \quad (\gamma, \alpha), \quad (\delta, \beta), \quad (\beta, \delta),$$

where

$$\alpha = \gamma_1(\frac{2}{3}) = \gamma_2(0), \quad \beta = \gamma_1(\frac{1}{6}) = \gamma_2(\frac{1}{2}),$$

$$\gamma = \gamma_2(\frac{2}{3}) = \gamma_1(0), \quad \delta = \gamma_1(\frac{1}{6}) = \gamma_2(\frac{1}{2}).$$

EXAMPLE 4. A double spiral.

A small change of the X_k, $k = 1, 2$, in the previous Example with

$$X_1 = (\bigcup_{k=-\infty}^{\infty} [k, k + \frac{k(2 + \text{sign } k)}{3(2k + 1)})) \times \mathbb{R},$$

$$X_2 = (\bigcup_{k=-\infty}^{\infty} [k + \frac{k(2 + \text{sign } k)}{3(2k + 1)}, k + \frac{2}{3})) \times \mathbb{R}$$

changes Δ to a double spiral.

5. The algebra of Toeplitz operators

5.1. Denote by $T(PC(X_1, X_2))$ the C^*-algebra generated by all the Toeplitz operators

$$T_a : \varphi \ \longrightarrow \ Pa\varphi$$

with symbols $a(x) = \sum_1^3 \alpha_k \chi_k(x) \in PC(X_1, X_2)$, acting on the Fock space $F^2(\mathbb{C}^n)$

Corollary 3.11 and Subsection 3.12, together with standard considerations, lead directly to the following theorem.

5.2. THEOREM. *The algebra $T(PC(X_1, X_2))$ is isomorphic and isometric to the algebra $C(\Delta(X_1, X_2))$. The isomorphism*

$$\tau : \ T(PC(X_1, X_2)) \ \longrightarrow \ C(\Delta(X_1, X_2))$$

is generated by the following mapping:

$$\tau : \ T_a \ \longmapsto \ \alpha_1 t_1 + \alpha_2 t_2 + \alpha_3 t_3,$$

where $t = (t_1, t_2) \in \Delta = \Delta(X_1, X_2)$, $t_3 = 1 - t_1 - t_2$.

5.3. COROLLARY. *The spectrum of the Toeplitz operator T_a is calculated by the formula*

$$\text{sp} \, T_a = \{\alpha_1 t_1 + \alpha_2 t_2 + \alpha_3 t_3 : \ (t_1, t_2) \in \Delta(X_1, X_2), \ t_3 = 1 - t_1 - t_2\}.$$

REMARK. The examples for the set $\Delta(X_1, X_2)$ given in Section 4 show how complicated $\operatorname{sp} T_a$ can be for general X_1, X_2 and arbitrary $a(x) = \sum_1^3 \alpha_k \chi_k(x) \in PC(X_1, X_2)$. But if we restrict ourselves to two–valued symbols ($\alpha_2 = \alpha_3$), then matters simplify greatly and we come to the known result:

The spectrum of T_a is a line segment

$$\operatorname{sp} T_a = \{\alpha_1 t_1 + \alpha_2(1 - t_1) : \quad t_1 \in [\alpha, \beta]\},$$

where $\alpha = \inf_{x \in \mathbb{R}^n} \gamma_1(x)$, $\beta = \sup_{x \in \mathbb{R}^n} \gamma_1(x)$.

References

[1] V. Bargmann, *On a Hilbert space of analytic functions*, Comm. Pure Appl. Math. **3** (1961), 215–228.

[2] F. A. Berezin, *Wick and anti-Wick symbols of operators*, Math. USSR Sbornik **84** (1971), 578–610.

[3] F. A. Berezin, *Covariant and contravariant symbols of operators*, Math. USSR Izvestia **6** (1972), 1117–1151.

[4] J. Dixmier, *Les C*-algebres et leurs représéntations*, Gauthier-Villars, Paris, 1964.

[5] R. G. Douglas, *Banach algebra techniques in operator theory*, Academic Press, 1972.

[6] V. A. Fock, *Konfigurationsraum und zweite Quantelung*, Z. Phys. **75** (1932), 622–647.

[7] P. Halmos, *Two subspaces*, Trans. Amer. Math. Soc. **144** (1969), 381–389.

[8] V. V. Kisil, E. Ramírez de Arellano, R. Trujillo, and N. L. Vasilevski, *Toeplitz operators with discontinuous presymbols on the Fock space*, Reporte Interno #155, Departamento de Matemáticas, CINVESTAV del I.P.N., Mexico City, 1994.

[9] G. K. Pedersen, *Measure theory for C*-algebras. II*, Mat. Scand. **22** (1968), 63–74.

[10] E. Ramírez de Arellano and N. L. Vasilevski, *Toeplitz operators on the Fock space with presymbols discontinuous on a thick set*, Math. Nachr. **180** (1996), 299–315.

[11] I. E. Segal, *Lectures at the summer seminar on appl. math.*, Boulder, Colorado, 1960.

[12] J. Varela, *Duality of C*-algebras*, Memories Amer. Math. Soc., vol. 148, AMS, Providence, Rhode Island, 1974, pp. 97–108.

[13] N. L. Vasilevski, *Banach algebras generated by two-dimensional integral operators with Bergman kernel and piece-wise continuous coefficients. I*, Soviet Math. (Izv. VUZ) **30** (1986), 14–24.

[14] N. L. Vasilevski and I. M. Spitkovsky, *On the algebra generated by two projectors*, Dokl. Akad. Nauk. UkSSR **A:8** (1981), 10–13.

DEPARTAMENTO DE MATEMÁTICAS, CINVESTAV DEL I.P.N., APARTADO POSTAL 14-740, 07000 MÉXICO, D.F., MEXICO
E-mail address: eramirez@math.cinvestav.mx

DEPARTAMENTO DE MATEMÁTICAS, CINVESTAV DEL I.P.N., APARTADO POSTAL 14-740, 07000 MÉXICO, D.F., MEXICO
E-mail address: nvasilev@math.cinvestav.mx

Contemporary Mathematics
Volume **212**, 1998

Singular Integral Operators in the
$\bar{\partial}$ Theory on Convex Domains in $\mathbb{C}^n$

R. Michael Range

ABSTRACT. In this lecture I shall discuss a class of singular integral operators which arises naturally in the study of the Cauchy-Riemann equations on convex domains in complex Euclidean space of dimension greater than one. These operators have been known for quite a while, yet—except for some special cases—some of their fundamental properties remain unknown. Besides discussing some fairly recent results about these operators, the main purpose of this talk is to suggest to experts in operator theory some problems of major interest in multidimensional complex analysis, with the hope that a fresh approach may perhaps yield some new answers.

1. The one-dimensional case

We briefly review the situation in dimension one. If $z = x + iy \in \mathbb{C}$, the Cauchy-Riemann operator is defined by

$$\frac{\partial}{\partial \bar{z}} = \frac{1}{2}\left(\frac{\partial}{\partial x} - \frac{1}{i}\frac{\partial}{\partial y}\right) .$$

A function $z \mapsto f(z)$ is holomorphic on an open set $D \subset \mathbb{C}$ if and only if $\partial f / \partial \bar{z} = 0$ on D. The associated inhomogeneous equation

$$(1.1) \qquad \frac{\partial f}{\partial \bar{z}} = u$$

is also of great interest. It is classical that the Cauchy transform

$$(1.2) \qquad Tu(z) = -\frac{1}{2\pi i}\int_{\mathbb{C}} u(\zeta)\frac{d\bar{\zeta} \wedge d\zeta}{\zeta - z}$$

solves (1.1) whenever u has compact support. Furthermore, if u is, for example, bounded on the bounded region D, then

$$(1.3) \qquad T_D u(z) = -\frac{1}{2\pi i}\int_D u\frac{d\bar{\zeta} \wedge d\zeta}{\zeta - z}$$

1991 *Mathematics Subject Classification.* Primary 32F20, 47G10; Secondary 32A25.

still solves (1.1)! This extension makes essential use of the fact that the kernel in (1.3) is holomorphic in z for $z \neq \zeta$, as follows. If $W \subset\subset D$, choose $\chi \in C_0^\infty(D)$, such that $\chi \equiv 1$ on W, and decompose

$$(1.4) \qquad T_D u = T_D(1 - \chi)u + T_D(\chi u) \ .$$

On W, the first integral on the right in (1.4) is holomorphic, by the remark just made, and in the second integral χu has compact support. Hence

$$\frac{\partial}{\partial \bar{z}}(T_D u) = 0 + \frac{\partial}{\partial \bar{z}}(T_D(\chi u)) = \chi u = u \text{ on } W \ .$$

Since $W \subset\subset D$ is arbitrary,

$$\frac{\partial}{\partial \bar{z}}(T_D u) = u \text{ on } D \ .$$

The operator T_D is very well understood. In particular, it is "smoothing of order 1" in appropriate function spaces.

2. The Bochner-Martinelli kernel

In several complex variables $z = (z_1, \ldots, z_n)$, one considers the corresponding system of inhomogeneous Cauchy-Riemann equations

$$(2.1) \qquad \frac{\partial f}{\partial \bar{z}_j} = u_j, \quad j = 1, \ldots, n \ .$$

It is evident that solutions (of class C^2) of (2.1) exist only if the given data $(u_1, \ldots u_n)$ satisfies the condition (trivially satisfied for $n = 1$!)

$$(2.2) \qquad \frac{\partial u_j}{\partial \bar{z}_k} = \frac{\partial u_k}{\partial \bar{z}_j}, \quad 1 \leq j, k \leq n \ .$$

At this point it is convenient to introduce the formalism of the $\bar{\partial}$-complex. The given data is represented by the $(0,1)$ form $u = \sum_{j=1}^n u_j d\bar{z}_j$. The operator $\bar{\partial}$ is defined on functions by $\bar{\partial} f = \sum_{j=1}^n \frac{\partial f}{\partial \bar{z}_j} d\bar{z}_j$, and an $(0,1)$ forms by

$$\bar{\partial}\left(\sum_{j=1}^n u_j d\bar{z}_j\right) = \sum_{j=1}^n (\bar{\partial} u_j) \wedge d\bar{z}_j \ .$$

Solving (2.1), given the necessary condition (2.2), is thus equivalent to solving $\bar{\partial} f = u$ for a given $(0,1)$ form u which satisfies $\bar{\partial} u = 0$.

The natural generalization of the Cauchy transform (1.2) to higher dimensions is given by the Bochner-Martinelli transform

$$(2.3) \qquad T_{BM} u(z) = -\int_{\mathbb{C}^n} u \wedge K_{BM}(\cdot, z) \ ,$$

where u is a compactly supported $(0,1)$ form, and

$$(2.4) \qquad K_{BM}(\zeta, z) = \frac{1}{(2\pi i)^n} \frac{\partial_\zeta |\zeta - z|^2 \wedge (\bar{\partial}_\zeta \partial_\zeta |\zeta - z|^2)^{n-1}}{|\zeta - z|^{2n}} \ .$$

Standard techniques (see [7] for a detailed discussion) show that, just as in case $n = 1$,

(2.5) $\bar{\partial}(T_{BM}u) = u$ if u has compact support and $\bar{\partial}u = 0$.

However, if $n > 1$, one cannot eliminate the support condition in (2.4) by the simple procedure used in (1.3) and (1.4). The argument used in dimension 1 breaks down in two critical places: (i) K_{BM} is not holomorphic in $z \neq \zeta$ when $n > 1$, and (ii) the localization χu of u is, in general, no longer $\bar{\partial}$-closed if $n > 1$. Instead, if f is a C^1 function on D, which extends continuously to the boundary bD of D, one obtains the following representation for f, valid at all points $z \in D$:

(2.6) $$f(z) = \int_{bD} f K_{BM}(\cdot, z) - \int_D \bar{\partial}f \wedge K_{BM}(\cdot, z) \ .$$

For f holomorphic on D, i.e. $\bar{\partial}f = 0$, (2.5) reduces to the so-called Bochner-Martinelli integral formula, which agrees with the Cauchy integral formula when $n = 1$.

3. A solution for $\bar{\partial}$ on convex domains

We are thus led to search for integral formulas analogous to (2.5), with a kernel which is holomorphic in $z \in D$ when $\zeta \in bD$. The situation is particularly simple for a (Euclidean) convex domain D, as was recognized first by J. Leray in 1956 [4]. Thus, let us assume that D is convex, with boundary bD of class C^2. Then there exists a convex C^2 function r, such that $D = \{z : r(z) < 0\}$, and $dr \neq 0$ on bD. Since for fixed $\zeta \in bD$ the complex hyperplane $\{z : \sum_{j=1}^n (\partial r/\partial \zeta_j)(\zeta)(\zeta_j - z_j) = 0\}$ is contained in the (real) tangent plane to bD at ζ, the convexity of D implies that $F(\zeta, z) = \sum_{j=1}^n (\partial r/\partial \zeta_j)(\zeta)(\zeta_j - z_j) \neq 0$ for $\zeta \in bD$ and $z \in D$. Hence the coefficients of the $(1,0)$ form $W = \partial r(\zeta)/F(\zeta, z)$ are holomorphic in z on D. For $z \in D$, fixed, we now introduce a homotopy $\hat{W}(\zeta, z, \lambda) = \lambda W + (1 - \lambda)B$ on $bD \times [0, 1]$ between W and the corresponding $(1,0)$ form $B = \partial_\zeta |\zeta - z|^2 / |\zeta - z|^2$ involved in the Bochner-Martinelli kernel K_{BM}. The form

(3.1) $$\Omega_0(\hat{W}) = \left(\frac{1}{2\pi i}\right)^n \hat{W} \wedge (\bar{\partial}_{\zeta,\lambda}\hat{W})^{n-1} \ ,$$

where $\bar{\partial}_{\zeta,\lambda} = \bar{\partial}_\zeta + d_\lambda$, is well defined on $bD \times I$, and its pull back to $bD \times \{0\}$ equals $(1/2\pi i)^n B \wedge (\bar{\partial}_\zeta B)^{n-1} = K_{BM}$, while on $bD \times \{1\}$ one has $\Omega_0(\hat{W})|_{\lambda=1} = \Omega_0(W) = (1/2\pi i)^n (\partial r \wedge (\bar{\partial}\partial r)^{n-1}/F^n)$. Furthermore, the relation $\sum_{j=1}^n \hat{w}_j(\zeta_j - z_j) = 1$ satisfied by the coefficients of $\hat{W}$ implies that

(3.2) $d_{\zeta,\lambda}\Omega_0(\hat{W}) = \bar{\partial}_{\zeta,\lambda}\Omega_0(\hat{W}) = 0$ on $bD \times I$.

Hence, if $f \in C^1$ on bD, an application of Stokes' Theorem to the manifold $bD \times I$ with boundary $-[bD \times \{1\} - bD \times \{0\}]$ yields the formula

(3.3) $$\int_{bD} f K_{BM} = \int_{bD} f \Omega_0(W) + \int_{bD \times I} \bar{\partial}f \wedge \Omega_0(\hat{W}) \ .$$

Given a $(0,1)$ form u on $\overline{D}$, we define

(3.4) $$T_D u = \int_{bD \times I} u \wedge \Omega_0(\hat{W}) - \int_D u \wedge K_{BM} \ .$$

By combining (3.3) and (2.5) one then obtains the following Cauchy-Greene formula for $f \in C^1(\overline{D})$:

$$(3.5) \qquad f(z) = \int_{bD} f\Omega_0(W)(\cdot, z) + T_D(\overline{\partial}f)(z), \quad z \in D .$$

Since in this formula the first term on the right is holomorphic, it follows that

$$(3.6) \qquad \overline{\partial}f = \overline{\partial}(T_D(\overline{\partial}f)) .$$

In fact, one can show that T_D is a solution operator for $\overline{\partial}$, i.e., if u is a continuous $\overline{\partial}$-closed (0,1) form on $\overline{D}$, which is C^1 on D, then $\overline{\partial}(T_D u) = u$. Notice that for $n = 1$, the definition (3.4) reduces to (1.3). The operator T_D coincides with the obvious adaptation to the convex case of the solution operator for $\overline{\partial}$ constructed by Henkin for the case of strictly pseudoconvex domains [3]. Full details may be found in [7].

4. Estimates in dimension 2

The first estimations for a modification $\tilde{T}_D$ of T_D, valid without any additional hypotheses on D, were obtained by J. Polking in 1988 [5], as follows.

THEOREM 1. *Suppose D is a bounded convex domain in $\mathbb{C}^2$ with C^2 boundary. Then the operator $\tilde{T}_D$ satisfies*

$$\|\tilde{T}_D u\|_{L^p(D)} \leq C_p \|u\|_{L^p(D)}, \quad \text{for } 1 < p < \infty .$$

The following analogous result for T_D in Hölder spaces holds as well. Related results were obtained by Chaumat and Chollet [1].

THEOREM 2 [8]. *Suppose D is a bounded convex domain in $\mathbb{C}^2$ with C^∞ boundary. Then*

$$\|T_D u\|_{\wedge_\alpha(D)} \leq C_\alpha \|u\|_{\wedge_\alpha(D)} \quad \text{for all } \alpha > 0$$

and for all $\overline{\partial}$-closed (0,1) forms u in $\wedge_\alpha(D)$.

The most delicate part of the proof of Theorem 2 is based on the following observation. The critical expression in $T_D u$ appears in the boundary integral in 3.4. After integrating in λ over $[0, 1]$, it has the form

$$(4.1) \qquad \int_{bD} u \wedge \frac{\partial r \wedge \partial_\zeta |\zeta - z|^2}{F(\zeta, z)|\zeta - z|^2} .$$

The convexity of D and the definition of $F(\zeta, z)$ imply the estimate

$$(4.2) \qquad |F(\zeta, z)| \geq \frac{1}{2}[|Im F(\zeta, z)| + |r(z)|] \quad \text{for } \zeta \in bD \text{ and } z \in D .$$

Finally, one can show that for (ζ, z) sufficiently close to a point (ζ_0, z_0) with $F(\zeta_0, z_0) = 0$, and for z fixed, $t_2(\zeta) = Im F(\zeta, z)$ is part of a local coordinate system (t_2, t_3, t_4) for $U \cap bD$, where U is a suitable neighborhood of ζ_0. Thus (4.1) and (4.2) show that, qualitatively, the integral (4.1) behaves like the Hilbert transform in the local coordinate t_2, independently of $|r(z)|$, while integration in the remaining coordinates (t_3, t_4) is easily controlled. The details of the proof make this heuristic argument precise.

5. Dimension higher than 2

When $n \geq 3$, the heuristic argument given in the preceding section breaks down. In general, the most critical term which appears in the boundary integral in T_D is of the form

$$(5.1) \qquad \int_{bD} u \wedge \frac{\partial r \wedge \partial_\zeta |\zeta - z|^2 \wedge (\overline{\partial}\partial r)^{n-2}}{F^{n-1}|\zeta - z|^2} .$$

In the worst case, the hyperplane $\{z : Re\ F(\zeta, z) = 0\}$ may be locally contained in bD near ζ, so that, as z approaches bD, the local estimation $2|F| \geq |t_2| + |r(z)|$, with $t_2 = ImF$ a local coordinate for bD, is optimal. Thus, if $n > 2$, the integral (5.1) becomes excessively singular in t_2. On the other hand, in the extreme case just considered, the term $\partial r \wedge (\overline{\partial}\partial r)^{n-2}$, which explicitly involves the Leviform of bD when $n > 2$, will vanish near points ζ where the boundary is flat. This observation opens the door to carry out a delicate analysis of (5.1), which involves a balancing of the order of vanishing in the denominator with the appropriate order of vanishing in the numerator. This approach has been successful for the special class of domains $D_m = \{\sum_{j=1}^n |z_j|^{2m_j} < 1\}$, where the m_j's are positive integers (see [6]), and in slightly more general cases as well (see [2]). However, the general convex case in dimensions ≥ 3, even with the additional hypotheses that bD is real analytic, has remained unsolved for a long time. So we end with the following concrete problem: *Do Theorems 1 and 2 hold in dimension $n \geq 3$, perhaps with other operators T_D and $\hat{T}_D$?*

References

[1] Chaumat, J., and Chollet, A. M., *Estimations hölderiennes pour les équations de Cauchy-Riemann dans les convexes compacts de $\mathbb{C}^n$*, Math. Z. **207** (1991), 501–534.

[2] Diederich, K., Fornaess, J. E., Wiegerinck, J., *Sharp Hölder estimates for $\overline{\partial}$ on ellipsoids*, Manuscr. Math. **56** (1986), 399–413.

[3] Henkin, G. M., *Integral representations in strictly pseudoconvex domains, and application to the $\overline{\partial}$-problem*, Math. Sb. **82** (1970), 300–308; *Math. USSR Sb.* **11** (1970), 273–281.

[4] Leray, J., *Fonction de variable complexe: sa réprésentation comme somme de puissance négatives de fonctions linéaires*, Rend. Accad. Naz. Lincei ser. 8 **20** (1956), 589–590.

[5] Polking, J., *The Cauchy-Riemann equations on convex sets*, Proc. Symp. Pure Math. **52(3)** (1993), 309–322.

[6] Range, R. M., *On Hölder estimates for $\overline{\partial}u = f$ on weakly pseudoconvex domains*, Sc. Norm. Sup. Pisa (1978), 247–267.

[7] Range, R. M., *Holomorphic Functions and Integral Representations in Several Complex Variables*, Springer Verlag, New York, 1986.

[8] Range, R. M., *On Hölder and BMO estimates for $\overline{\partial}$ on convex domains in $\mathbb{C}^2$*, J. Geom. Anal. **2** (1992), 575–584.

DEPARTMENT OF MATHEMATICS, STATE UNIVERSITY OF NEW YORK AT ALBANY, ALBANY, NEW YORK 12222, U.S.A.

E-mail address: `range@math.albany.edu`

Contemporary Mathematics
Volume **212**, 1998

Differentiation and Integration of
Variable Order and the Spaces $L^{p(x)}$

Stefan G. Samko

1. Introduction

Recently, the integration and differentiation of variable fractional order $\alpha(x)$ was considered [16], [18]–[19], [21] in the case of one variable:

$$(1.1) \qquad I^{\alpha(x)}\varphi := \frac{1}{\Gamma[\alpha(x)]} \int_a^x \varphi(t)(x-t)^{\alpha(x)-1}dt,$$

$$(1.2) \qquad D^{\alpha(x)}f := \frac{f(x)}{\Gamma[1-\alpha(x)](x-a)^{\alpha(x)}} + \frac{\alpha(x)}{\Gamma[1-\alpha(x)]} \int_a^x \frac{f(x)-f(t)}{(x-t)^{1+\alpha(x)}}dt,$$

where $x > a \geq -\infty$, $\operatorname{Re}\alpha(x) > 0$ in (1.1) and $0 < \operatorname{Re}\alpha(x) < 1$ in (1.2). It is well known [20] that

$$(1.3) \qquad D^\alpha I^\alpha \varphi \equiv \varphi$$

in the case of constant order $\alpha(x) \equiv \alpha = constant$ and appropriate functions φ, e.g. for $\varphi \in L^p$, $1 \leq p \leq \infty$, if $a > -\infty$ and $1 \leq p < 1/\alpha$, if $a = -\infty$. The relation (1.3) is not valid in case of variable order $\alpha(x)$ and we briefly discuss below what we have instead of (1.3).

The following problems, in particular, arise:

1) How much does $D^{\alpha(x)}$ differs from the real left inverse to the operator $I^{\alpha(x)}$?

2) What are mapping properties of the operator $I^{\alpha(x)}$ in the popular spaces H^λ (of Hölderian functions) and L^p?

3) Can one characterize the intersection $L^p \bigcap I^{\alpha(x)}(L^p)$ with the range

$$I^{\alpha(x)}(L^p) = \{f : f(x) = I^{\alpha(x)}\varphi, \ \varphi \in L^p\}$$

as the space

$$L^{\alpha(x),p} = \{f : f \in L^p, \ D^{\alpha(x)}f \in L^p\}$$

1991 *Mathematics Subject Classification.* Primary 26A33, 42B20.

This research was partially supported by RFFI (Russian Funds of Fundamental Investigations).

The author thanks the referee and Prof. N. K. Karapetiants, who drew the author's attention to the papers [2], [9] and [8], [9], respectively.

which would allow to treat the range of the operator $I^{\alpha(x)}$ as a function space with variable smoothness? We would like to emphasize that the interest we have to the constructions (1)–(2) is mainly caused by the approach they provide for defining fractional Sobolev-Liouville-type function spaces with variable smoothness depending on the point x, see [13]–[14] for Liouville-type spaces in the case of the constant $\alpha(x) \equiv \alpha$. We mention the papers [1], [11], [12], [15], [24], [25] specially devoted to function spaces of smoothness of variable order defined via the PDO approach.

Considering, in particular, the question 2), we naturally arrive at the idea to deal with the spaces H^λ and L^p which have variable orders $\lambda(x)$ and $p(x)$ as well. The first case is not difficult and we mention below the result of the type $I^{\alpha(x)}[H^{\lambda(x)}] \subseteq H^{\alpha(x)+\lambda(x)}$. As for the spaces $L^{p(x)}$, the result of Hardy-Sobolev-type theorem: $I^{\alpha(x)}[L^{p(x)}] \subset L^{q(x)}$, $1/q(x) = 1/p(x) - \alpha(x)$ (or $1/q(x) = 1/p(x) - \alpha(x)/n$ in the multidimensional case, see the multidimensional version of (1.1) in (2.8) below) is much more complicated and remains open. One of the reasons is in the fact that the space $L^{p(x)}$ itself is little studied and many standard methods to investigate operators in L^p do not work in the case of variable $p(x)$.

The paper consists of two parts. The first one contains some discussion of the questions 1)–3) and the statements of results known for the operators (1.1)–(1.2), including the mapping properties of $I^{\alpha(x)}$ in $H^{\lambda(x)}$. In the second part we develop the theory of the spaces $L^{p(x)}$ as a base for future applications to the above questions.

2. Fractional integration and differentiation of variable order

2.1. Preliminaries.
A) Let $-\infty < a < b < \infty$.

DEFINITION 2.1. Let $0 < \lambda(x) \leq 1$. The space $H^{\lambda(x)} = H^{\lambda(x)}([a,b])$ is defined as the space of continuous functions on $[a,b]$, such that

$$|f(x+h) - f(x)| \leq c|h|^{\lambda(x)}$$

for all x, $x+h \in [a,b]$.

The space $H^{\lambda(x)}$ is a normed space with respect to the natural norm $\|f\|_{C([a,b]} + H$ with

$$(2.1) \qquad H = \sup_{x_1,x_2 \in [a,b]} \frac{|f(x_1) - f(x_2)|}{|x_1 - x_2|^{\lambda(x_1)}} = \sup_{x_1,x_2 \in [a,b]} \frac{|f(x_1) - f(x_2)|}{|x_1 - x_2|^{\lambda(x_2)}}.$$

REMARK 2.1. The function $\lambda(x)$ should not necessarily be continuous, at least in the definition of the space. It may be any function defined on $[a,b]$ with values in $(0,1]$.

DEFINITION 2.2. By $H^{\lambda(x),1}$ we denote the space of continuous functions on $[a,b]$ such that

$$|f(x+h) - f(x)| \leq c|h|^{\lambda(x)} \log \frac{1}{|h|}$$

for all $x, x+h \in [a,b]$ and $|h| \leq 1/2$.

B) In the case $a = -\infty$ the operators (1.1)–(1.2) take the form

$$(2.2) \qquad I_+^{\alpha(x)}\varphi = \frac{1}{\Gamma[\alpha(x)]} \int_0^\infty \varphi(x-t)t^{\alpha(x)-1}dt,$$

$$D_+^{\alpha(x)}f = \frac{\alpha(x)}{\Gamma[1-\alpha(x)]} \int_0^\infty \frac{f(x)-f(x-t)}{t^{1+\alpha(x)}}dt = \lim_{\epsilon\to 0} D_{+,\epsilon}^{\alpha(x)}f$$

where

$$D_{+,\epsilon}^{\alpha(x)}f = \frac{\alpha(x)}{\Gamma[1-\alpha(x)]} \int_\epsilon^\infty \frac{f(x)-f(x-t)}{t^{1+\alpha(x)}}dt$$

and it is assumed that $0 < \alpha(x) < 1$.

2.2. Operator $I^{\alpha(x)}$ in the space $H^{\lambda(x)}$. In what follows, $\lambda(x)$ is assumed to be a function on $[a,b]$, $0 < \lambda(x) \le 1$, satisfying the condition

$$(2.3) \qquad \lambda(x) - \lambda(t) \le \frac{c}{\log \dfrac{1}{x-a}}, \qquad a < t < x \ (c > 0),$$

in a neighbourhood of the end point $x = a$ (which is fullfilled automatically if $\lambda(x)$ is either a nonincreasing or Hölderian function in a neighbourhood of this point).

THEOREM 2.1. *Let $\lambda(x)$, $0 < \lambda(x) \le 1$, satisfy the condition (2.3) and let*
1) $0 < m \le \alpha(x) \le 1$;
2) $\alpha(x) \in H^{\mu(x)}([a,b])$ *with* $0 < \delta \le \mu(x) \le 1$,
3) $\alpha(x) + \lambda(x) \le 1$.
If $\varphi(x) \in H^{\lambda(x)}$, then $I^{\alpha(x)}\varphi$ has the form

$$I^{\alpha(x)}\varphi = \frac{\varphi(a)}{\Gamma[1+\alpha(x)]}(x-a)^{\alpha(x)} + \psi(x)$$

where $\psi(x) \in H^{\gamma(x)}$, $\gamma(x) = \min\{\alpha(x)+\lambda(x), \mu(x)\}$, if $\sup_{a\le x\le b}[\alpha(x)+\lambda(x)] < 1$ and $\psi(x) \in H^{\gamma(x),1}$ otherwise; in both cases

$$|\psi(x)| \le c(x-a)^{\lambda(x)+\alpha(x)}$$

with c not depending on x.

PROOF. It follows to what is known for the case when $\alpha(x)$ and $\lambda(x)$ are constant (see [20], pp. 54–55) and can be found in [21].

REMARK 2.2. The statement of Theorem 2.1 can be exactified:

$$(2.4) \qquad |\psi(x+h) - \psi(x)| \le \frac{c}{1-\alpha(x)-\lambda(x)}|h|^{\alpha(x)+\lambda(x)}$$

for all x such that $\alpha(x) + \lambda(x) < 1$, and

$$|f(x+h) - f(x)| \le c|h|^{\gamma(x)}\log\frac{1}{|h|}, \qquad |h| \le 1$$

for those x which give the equality $\alpha(x) + \lambda(x) = 1$, c not depending on x and h.

REMARK 2.3. An assertion similar to that of Theorem 2.1 was proved in [5]–[6] in the case of constant $\alpha(x) = \alpha$; due to this case, in [6] the fractional differentiation D^α was also considered in $H^{\lambda(x)}$.

2.3. "Measure of deviation" of the operator $D_+^{\alpha(x)}$ **from the operator** $[I_+^{\alpha(x)}]^{-1}$. The results given below show that $D_+^{\alpha(x)}$ coincides with the left inverse operator to $I_+^{\alpha(x)}$ up to a composition with an inverible integral operator of Volterra type. To show this we need to calculate the composition $D_{+,\epsilon}^{\alpha(x)} I_+^{\alpha(x)}$ and then pass to a limit as $\epsilon \to 0$. The following auxilliary statement is valid, where $\Omega_b = (-\infty, b)$, $b \leq +\infty$.

LEMMA 2.1. *Let* $0 < \alpha(x) < 1$, $x \in \Omega_b$. *Then, for sufficiently nice functions* φ

$$(2.5) \qquad D_{+,\epsilon}^{\alpha(x)} I_+^{\alpha(x)} = \int_0^\infty A_0(x,t)\varphi(x - \epsilon t)dt + \int_0^\infty C_\epsilon(x,t)\varphi(x - t)dt$$

where

$$A_0(x,t) = \frac{\sin[\pi\alpha(x)]}{\pi t}[t_+^{\alpha(x)} - (t-1)_+^{\alpha(x)}],$$

$$C_\epsilon(x,t) = B_\epsilon(x,t) + \frac{1}{\epsilon}\left[A_\epsilon\left(x, \frac{t}{\epsilon}\right) - A_0\left(x, \frac{t}{\epsilon}\right)\right]$$

with

$$A_\epsilon(x,t) = \frac{\sin[\pi\alpha(x)]}{\pi}[t_+^{\alpha(x)-1} - J_\epsilon(x,t)],$$

$$J_\epsilon(x,t) = \frac{\alpha(x)}{t}\int_1^t \left(\frac{\epsilon t}{y}\right)^{\alpha(x-(t\epsilon/y))-\alpha(x)}(y-1)^{\alpha(x-(t\epsilon/y))-1}dy$$

for $t > 1$ *and* $J_\epsilon(x,t) = 0$ *for* $0 < t < 1$, *and*

$$B_\epsilon(x,t) = \frac{\alpha(x)}{\Gamma[1-\alpha(x)]}\int_\epsilon^t \frac{m(x,y)(t-y)^{\alpha(x-y)-1}}{y^{1+\alpha(x)}}dy, \quad t > \epsilon,$$

$$B_\epsilon(x,t) = 0, \quad 0 < t < \epsilon,$$

where $m(x,y) = 1/\Gamma[\alpha(x)] - 1/\Gamma[\alpha(x-y)]$.

The first term in (2.5) is the identity approximation, while the second one will generate a certain integral operator as $\epsilon \to 0$.

By means of Lemma 2.1 the following theorem is proved.

THEOREM 2.2. *Let* $\alpha(x) \in C^1(\Omega_b)$, $0 < \alpha(x) < 1$, *and* $\max_{x\in\Omega_b} |\alpha'(x)| < \infty$. *Then*

$$(2.6) \qquad D_+^{\alpha(x)} I_+^{\alpha(x)}\varphi = (I + K)\varphi := \varphi(x) + \int_{-\infty}^x \mathcal{K}(x, x-t)\varphi(t)dt$$

where

$$\mathcal{K}(x,\xi) = \frac{\alpha(x)\sin[\pi\alpha(x)]}{\pi}\int_0^\xi \frac{(\xi-y)^{\alpha(x)-1} - \dfrac{\Gamma[\alpha(x)]}{\Gamma[\alpha(x-y)]}(\xi-y)^{\alpha(x-y)-1}}{y^{1+\alpha(x)}}dy.$$

Under some additional assumptions on $\alpha(x)$ the following result on compactness of the integral operator in (2.6) is valid.

THEOREM 2.3. *Let* $0 < m \leq \alpha(x) \leq M < 1$, $\alpha(x) \in C^1(\Omega_b)$, $|\alpha'(x)| \leq c(1+|x|)^{-\lambda}$, $\lambda > 1$; *let also* $|\alpha(x) - \alpha(x-h)| \leq ch(1+|x|)^{-1}(1+|x-h|)^{-1}$, $h \geq 0$. *Then the integral operator in* (2.6) *is compact in* $L_p(\Omega_b)$ *if* $1 \leq p < 1/\alpha(-\infty)$.

Proofs of Lemma 2.1 and Theorems 2.2 and 2.3 can be found in [19].

REMARK 2.4. In Theorem 2.3 the Lipschits-type condition on $\alpha(x)$ is satisfied automatically if $\lambda \geq 2$.

From Theorems 2.2 and 2.3 it can be derived that *the integration operator $I_+^{\alpha(x)}$ of variable order $\alpha(x)$, considered on functions in $L_p(\Omega_b)$, $b < \infty$, $1 < p < 1/\alpha(-\infty)$, is left invertible. The differentiation operator $D_+^{\alpha(x)}$ of the same variable order coincides with the operator $[I_+^{\alpha(x)}]^{-1}$, left inverse to $I_+^{\alpha(x)}$, up to the composition with the Volterra integral operator $I + K$ (2.6) bilaterraly invertible in $L_p(\Omega_b)$:*

$$(2.7) \qquad D_+^{\alpha(x)} = (I + K)[I_+^{\alpha(x)}]^{-1}.$$

We should emphasize, however, that the representation (2.7) has been proved on the range $I^{\alpha(x)}[L_p(\Omega_b)]$, $1 \leq p < 1/\alpha(-\infty)$. Still, we do not know whether the domain $\mathcal{D}$ of the operator $D_+^{\alpha(x)}$ in the space $L^p(\Omega_b)$ coincides with this range. Anyhow, the relation (2.7) allows to assert that

$$\mathcal{D} \subseteq I^{\alpha(x)}(L^p)$$

which is half an answer to the question 3) in the introduction. The invertibility of the operator $I + K$ should, in fact, mean that $\mathcal{D}$ coincides with $I^{\alpha(x)}(L^p)$. However, this does not follow from (2.6) directly and an independent proof is required (compare with a similar situation in the case of constant order [20]).

2.4. In the multidimensional case it is natural to introduce the Riesz-type potential operator of variable order:

$$(2.8) \qquad I^{\alpha(x)}\varphi = \int_{R^n} \frac{\varphi(y)dy}{|x - y|^{n - \alpha(x)}}$$

(we omit the normalizing factor before the integral), although the partial fractional integrals of variable order may be also considered. Similarly to the one-dimensional fractional differentiation operator $D_+^{\alpha(x)}$ we may also consider the hypersingular integral of variable order:

$$D^{\alpha(x)}f = \int_{R^n} \frac{f(x) - f(x - y)}{|y|^{n + \alpha(x)}}dy.$$

The question of validity of the relation of the type (2.6) in the multidimensional case remains open.

Another, even more important question, is the problem of validity of Sobolev-type theorem for potential type operators (2.8) within the framework of $L^{p(x)}$-spaces (which is is close to the problem of embedding of function spaces with different variable smoothness and different variable orders of sumability). This will be the topic of our next paper where , in particular, we show that the Sobolev-type theorem in the form

$$I^{\alpha(x)} : L^{p(x)} \to L^{q(x)},$$

$$\frac{1}{q(x)} = \frac{1}{p(x)} - \frac{\alpha(x)}{n},$$

holds in some cases for bounded domains.

3. Spaces $L^{p(x)}(\Omega)$

3.1. Definitions and Preliminaries. Let Ω be a measurable set in R^n and $p(x)$ be a non-negative measurable function on Ω. Let

$$E_a = E_a(p) := \{x \in \Omega : p(x) = a\}$$

where we shall be interested in the cases $a = 0$, $a = 1$ and $a = \infty$. Everywhere below it is assumed that

$$|E_0| = 0$$

(By $|E|$ we denote the Lebesgue measure of a measurable set $E \subset R^n$.)

DEFINITION 3.1. By $L^{p(x)}(\Omega)$ we denote the set of measurable functions $f(x)$ on Ω such that

$$(3.1) \qquad I_p(f) := \int_{\Omega \setminus E_\infty} |f(x)|^{p(x)} dx < \infty$$

and

$$(3.2) \qquad \sup_{x \in E_\infty} |f(x)| < \infty.$$

REMARK 3.1. Evidently, allowing the exponent $p(x)$ to vanish at some points, we admit functions $f(x)$ with a fast increase near these points.

The idea of consideration of the spaces $L^{p(x)}$ was firstly given, to all appearance, in [23] in the case $n = 1$, $\Omega = [0,1]$ and continuous functions $p(x)$ and $f(x)$ in connection with some problems of the approximation theory. Some elements of the theory of the spaces $L^{p(x)}[0,1]$ were in fact firstly presented in [22], in connection with the approximation theory as well. It should be noted that the space $L^{p(x)}$ is an Orlicz space if and only if $p(x) = p = const$.

Afterwards, the papers [2], [8], [9] appeared, where the further development of the theory of the spaces $L^{p(x)}$ was undertaken, including some results for the Sobolev-type spaces $W^{k,p(x)}$. Unfortunately, the author of this paper was unaware of the above mentioned publications before submitting this paper for publication. Some of our results in this section are similar to those obtained in [8], [9], being given under a little different assumptions, others are new. We also remark that in [8] the case $n = 1$, $|E_\infty| = 0$, was considered and in [9] the version (3.18) of the norm was used, not (3.10). (We show the equivalence of the norms (3.10) and (3.18) in Theorem 3.2 below.)

We present here a development of some topics of the theory of the spaces $L^{p(x)}$. The main attention is paid to the Riesz-type norm and its equivalence to the Kolmogorov-Minkowsky-type norm. For completeness of the presentation we give also some results from [22] with certain exactifications or developments.

Following [22] we introduce a natural topology in $L^{p(x)}$ defined by the convergence

$$(3.3) \qquad \int_{\Omega \setminus E_\infty} |f_m(x) - f(x)|^{p(x)} dx + \sup_{x \in E_\infty} |f_m(x) - f(x)| < \epsilon.$$

We shall essentially use the numbers

$$(3.4) \qquad P = \sup_{x \in \Omega \setminus E_\infty} p(x), \quad p_0 = \inf_{x \in \Omega} p(x)$$

so that $0 \leq p_0 \leq P \leq \infty$. In the case $P < \infty$ we set

$$d(f,g) = \left\{ \int_{\Omega \setminus E_\infty} |f(x) - g(x)|^{p(x)} \right\}^{1/P_1} + \sup_{x \in E_\infty} |f(x) - g(x)|$$

with $P_1 = \max\{P, 1\}$. We shall also use the notation

(3.5) $$\rho(f) = \rho_p(f) = d(f, 0).$$

LEMMA 3.1. [22]. *The topological space defined by (3.1)–(3.3), is linear if and only if $P < \infty$ and then $d(f,g)$ is a metric on this space.*

Let $S = S(\Omega)$ be the set of simple functions (step functions) of the form $\sum_{k=1}^{N} c_k \chi_{\Omega_k}(x)$ where Ω_k are arbitrary measurable sets in Ω and $\chi_{\Omega_k}(x)$ are their characteristic functions.

LEMMA 3.2. *Let $P < \infty$. Then $S \subset L^{p(x)}$.*

Proof is straightforward.

3.2. Kolmogorov-Minkowski-type norm in $L^{p(x)}$, $1 \leq p(x) \leq \infty$. Theorem 3.1 below introduces a norm inspired by the Kolmogorov's theorem on norming topological spaces [7], [4, Ch. 4, p. 122].

LEMMA 3.3. *Let $f(x) \in L^{p(x)}(\Omega)$, $0 \leq p(x) \leq \infty$. The function*

(3.6) $$F(\lambda) := I_p\left(\frac{f}{\lambda}\right), \quad \lambda > 0,$$

takes finite values for all $\lambda \geq 1$, is continuous and decreases and $\lim_{\lambda \to \infty} F(\lambda) = 0$. If $P < \infty$, the same is true for all $\lambda > 0$.

Proof is straightforward.

THEOREM 3.1. *Let $0 \leq p(x) \leq \infty$. For any $f(x) \in L^{p(x)}(\Omega)$ the functional*

(3.7) $$\|f\|_{(p)} = \inf\left\{ \lambda : \lambda > 0, \int_{\Omega \setminus E_\infty} \left|\frac{f(x)}{\lambda}\right|^{p(x)} dx \leq 1 \right\}$$

takes a finite value and

(3.8) $$I_p\left(\frac{f}{\|f\|_{(p)}}\right) \leq 1, \quad \|f\|_{(p)} \neq 0.$$

If either $P < \infty$ or $\|f\|_{(p)} \geq 1$, then

(3.9) $$I_p\left(\frac{f}{\|f\|_{(p)}}\right) = 1, \quad \|f\|_{(p)} \neq 0.$$

Finally, if $1 \leq p(x) \leq P < \infty$, $x \in \Omega \setminus E_\infty$, then

(3.10) $$\|f\|_p = \|f\|_{(p)} + \sup_{x \in E_\infty} |f(x)|$$

is a norm in $L^{p(x)}(\Omega)$.

PROOF. By Lemma 3.3, $\|f\|_{(p)}$ is finite for $f(x) \in L^{p(x)}(\Omega)$ and (3.8)–(3.9) are derived directly from the definition (3.7). To see that (3.10) is a norm, we should only check the triangle inequality for $\|f\|_{(p)}$, which is straightforward if one takes into account the inequality $|\lambda y_1 + (1-\lambda)y_2|^p \leq \lambda|y_1|^p + (1-\lambda)|y_2|^p$, $0 \leq \lambda \leq 1$, $p \geq 1$ (the convexity property of the power function).

COROLLARY 1. *The functional* (3.7) *satisfies the estimates*

(3.11)
$$\left(\frac{\|f\|_{(p)}}{\lambda}\right)^P \leq I_p\left(\frac{f}{\lambda}\right) \leq \left(\frac{\|f\|_{(p)}}{\lambda}\right)^{p_0}, \quad \lambda \geq \|f\|_{(p)},$$

(3.12)
$$\left(\frac{\|f\|_{(p)}}{\lambda}\right)^{p_0} \leq I_p\left(\frac{f}{\lambda}\right) \leq \left(\frac{\|f\|_{(p)}}{\lambda}\right)^P, \quad 0 < \lambda \leq \|f\|_{(p)}.$$

COROLLARY 2. *Let E be a measurable set in $\Omega \backslash E_\infty$ and let $\chi_E(x)$ be its characteristic function. If $0 < p_0 \leq P < \infty$, we have*

(3.13)
$$|E|^{1/p_0} \leq \|\chi_E\|_{(p)} \leq |E|^{1/P}, \quad |E| \geq 1,$$

signs of the inequalities being opposite if $|E| \leq 1$, so that the equality $\|\chi_E\|_{(p)} = 1$ is equivalent to the equality $|E| = 1$.

REMARK 3.2. In the case $P = \infty$, the functional $\|f\|_{(p)}$ does exist for any $f \in L^{p(x)}$ according to Theorem 3.1. However, if $\|f\|_{(p)} < \infty$, it does not necessarily imply that $f \in L^{p(x)}(\Omega \backslash E_\infty)$, but

$$f(x) \in \mathcal{L}L^{p(x)}(\Omega \backslash E_\infty)$$

where $\mathcal{L}L^{p(x)}$ denotes the linear envelope of the class $L^{p(x)}(\Omega \backslash E_\infty)$.

REMARK 3.3. A realization of Kolmogorov-Minkowsky norm for the Orlicz spaces, similar to (3.7), is known in the theory of Orlicz spaces [10] as the Luxemburg norm.

LEMMA 3.4. *Let $0 < p_0 \leq P \leq \infty$. If*

(3.14)
$$I_p\left(\frac{f}{a}\right) < b, \quad a > 0, \ b > 0,$$

then $\|f\|_{(p)} \leq ab^\nu$ with $\nu = 1/p_0$ if $b \geq 1$ and $\nu = 1/P$ if $b \leq 1$.

PROOF. Since (3.14) yields the inequality $I_p(f/(ab^\nu)) \leq 1$, by Definition (3.7) we have $\|f\|_{(p)} \leq ab^\nu$.

LEMMA 3.5. *Let $0 < \gamma(x) \leq p(x) \leq P < \infty$, $x \in \Omega \backslash E_\infty$. Then*

(3.15)
$$\|f\|_{(p)}^{\gamma_0} \leq \|f^\gamma\|_{(\frac{p}{\gamma})} \leq \|f\|_{(p)}^\Gamma, \quad \|f\|_{(p)} \geq 1,$$

where $f^\gamma = |f(x)|^{\gamma(x)}$ and

$$\gamma_0 = \inf_{x \in \Omega \backslash E_\infty} \gamma(x), \quad \Gamma = \sup_{x \in \Omega \backslash E_\infty} \gamma(x),$$

the signs of the inequalities being opposite if $\|f\|_{(p)} \leq 1$. *If* $p(x)$ *and* $\gamma(x)$ *are continuous, there exists a point* $x_0 \in \Omega \backslash E_\infty$ *such that*

$$\|f^\gamma\|_{(p/\gamma)} = \|f\|_{(p)}^{\gamma(x_0)}.$$

PROOF. Let $\lambda = \|f\|_{(p)}$, $\mu = \|f^\gamma\|_{(p/\gamma)}$. Since $E_\infty(p/\gamma) = E_\infty(p)$, by means of (3.9) we obtain

$$\int_{\Omega \backslash E_\infty} |f(x)|^{p(x)} \frac{\lambda^{p(x)} - \mu^{p(x)/\gamma(x)}}{\lambda^{p(x)} \mu^{p(x)/\gamma(x)}} dx = 0.$$

Hence, by easy arguments we arrive at lemma's assertions.

LEMMA 3.6. *Let* $0 < p_0 \leq p(x) \leq P \leq \infty$, $x \in \Omega$, $|E_\infty| = 0$. *Then*

$$(3.16) \qquad L^{p_0}(\Omega) \bigcap L^P(\Omega) \subseteq L^{p(x)}(\Omega) \subseteq L^{p_0}(\Omega) + L^P(\Omega)$$

where the algebraic sum of the spaces stands in the right hand side. Besides, $\|f\|_p \leq \max\{\|f\|_{p_0}, \|f\|_P\}$.

Proof is straightforward.

The property of semiadditivity of the norm:

$$(3.17) \quad \max\{\|f\|_{L^{p(x)}(\Omega_1)}, \|f\|_{L^{p(x)}(\Omega_2)}\} \leq \|f\|_{L^{p(x)}(\Omega)} \leq \|f\|_{L^{p(x)}(\Omega_1)} + \|f\|_{L^{p(x)}(\Omega_2)}$$

with $\Omega_1 \bigcup \Omega_2 = \Omega$, well known for the case of constant exponents, is covered by the following lemma.

LEMMA 3.7. *Let* $\Omega = \Omega_1 \bigcup \Omega_2$ *and let* $p(x)$ *be a function on* Ω, $p(x) \geq 1$ *and* $P < \infty$. *Then* (3.17) *holds for any* $f(x) \in L^{p(x)}(\Omega)$.

Proof is straightforward.

The Kolmogorov-Minkowski-type norm can be also introduced directly with respect to the whole set Ω:

$$(3.18) \qquad \|f\|_p^1 = \inf \left\{ \lambda > 0 : I_p\left(\frac{f}{\lambda}\right) + \sup_{x \in E_\infty} \left|\frac{f(x)}{\lambda}\right| \leq 1 \right\}$$

which is well defined for $f(x) \in L^{p(x)}(\Omega)$, whatever measurable function $p(x)$, $0 \leq p(x) \leq \infty$, is used; it is a norm, if $1 \leq p(x) \leq \infty$. This can be proved similarly to Theorem 3.1. Analogously to (3.9) it can be proved that

$$(3.19) \qquad \int_{\Omega \backslash E_\infty} \left|\frac{f(x)}{\|f\|_p^1}\right|^{p(x)} dx + \frac{\|f\|_{L^\infty(E_\infty)}}{\|f\|_p^1} = 1$$

if $P < \infty$ or $P = \infty$, but $\|f\|_p^1 \geq 1$.

THEOREM 3.2. *The norms* (3.10) *and* (3.18) *are equivalent:*

$$(3.20) \qquad \frac{1}{2} \|f\|_p \leq \|f\|_p^1 \leq \|f\|_p$$

where $f(x) \in L^{p(x)}(\Omega)$, $1 \leq p(x) \leq \infty$, $P < \infty$.

PROOF. The right-hand side inequality is equivalent to the inequality

$$\inf\left\{\lambda > 0 : F(\lambda) + \frac{c}{\lambda} \leq 1\right\} \leq \lambda_0 + c,$$

where $F(\lambda)$ is defined in (3.6) and

(3.21) $c = \|f\|_{L^\infty(E_\infty)}, \quad \lambda_0 = \|f\|_{(p)}.$

So, it is sufficient to verify that $F(\lambda_0 + c) + c/(\lambda_0 + c) \leq 1$, that is $F(\lambda_0 + c) \leq \lambda_0/(\lambda_0 + c)$. Since $F(\lambda_0 + c) = I_p\left(f/(\|f\|_{(p)} + c)\right)$, by (3.11) we obtain $F(\lambda_0 + c) \leq \|f\|_{(p)}/(\|f\|_{(p)} + c) = \lambda_0/(\lambda_0 + c)$ which was required.

The left-hand side inequality in (3.20) follows from the inequalities

$$\inf\left\{\lambda : F(\lambda) + \frac{c}{\lambda} \leq 1\right\} \geq \inf\left\{\lambda : F(\lambda) \leq 1\right\} = \lambda_0,$$

$$\inf\left\{\lambda : F(\lambda) + \frac{c}{\lambda} \leq 1\right\} \geq \inf\left\{\lambda : \frac{c}{\lambda} \leq 1\right\} = c,$$

because then the left-hand side of the last inequalities is not less than $(\lambda_0 + c)/2$.

3.3. Hölder inequality.

THEOREM 3.3. *Let* $f(x) \in L^{p(x)}(\Omega)$, $1 \leq p(x) \leq \infty$, *and* $\varphi(x) \in L^{q(x)}(\Omega)$, $(1/p(x)) + (1/q(x)) \equiv 1$, $x \in \Omega$. *Then*

(3.22) $\int_\Omega |f(x)\varphi(x)|dx \leq k\|f\|_p \|\varphi\|_q$

with $k = (1/p_0) + (1/q_0) = \sup(1/p(x)) + \sup(1/q(x))$.

PROOF. First we note that under the assumptions of the theorem, the functionals $\|f\|_p$ and $\|\varphi\|_q$ may not be norms and the classes $L^{p(x)}$ and $L^{q(x)}$ are not necessarily linear, but these functionals always do exist by Theorem 3.1.

To prove (3.22), we make use of the standard inequality

(3.23) $ab \leq \dfrac{a^p}{p} + \dfrac{b^q}{q}$

with $a > 0$, $b > 0$, $(1/p) + (1/q) = 1$, $1 < p < \infty$. The inequality (3.23) is valid for $p = 1$ in the form $ab \leq a^p/p$ if $b \leq 1$ and for $p = \infty$ in the form $ab \leq b^q/q$ if $a \leq 1$. So we have

$$\left|\frac{f(x)\varphi(x)}{\|f\|_p\|\varphi\|_q}\right| \leq \frac{1}{p(x)}\left|\frac{f(x)}{\|f\|_p}\right|^{p(x)} + \frac{1}{q(x)}\left|\frac{\varphi(x)}{\|\varphi\|_q}\right|^{q(x)}, \quad x \in \Omega\backslash E_\infty(p)\bigcup E_\infty(q),$$

while for $x \in E_\infty(p)$ and $x \in E_\infty(q)$ we should omit the first and the second term, respectively, in the right-hand side, since $|f(x)/\|f\|_p| \leq 1$ for $x \in E_\infty$ and $|\varphi(x)/\|\varphi\|_q| \leq 1$ for $x \in E_\infty(q)$. Integrating over the whole Ω and estimating $p(x)$ and $q(x)$, we arrive at (3.22).

REMARK 3.4. Hölder inequality holds also in the form

(3.24) $$\int_\Omega |f_1(x) \cdots f_m(x)| dx \le c\|f_1\|_{p^1} \cdots \|f_m\|_{p^m}$$

where $p^1(x) \ge 1, \ldots, p^m(x) \ge 1$ and $\sum_{k=1}^m 1/p^k(x) \equiv 1$, $x \in \Omega$, and $c = \sum_{k=1}^m 1/p_0^k$, $p_0^k = \min_{x\in\Omega} p^k(x)$. To obtain (3.24), instead of (3.23) one should use the inequality $a_1 \cdots a_m \le \frac{a_1^{p^1}}{p^1} + \cdots + \frac{a_m^{p^m}}{p^m}$ ([3]).

LEMMA 3.8. *Let* $(1/p(x)) + (1/q(x)) \equiv 1/r(x)$, $p(x) \ge 1$, $q(x) \ge 1$, $r(x) \ge 1$ *and let* $\sup_{x\in\Omega\backslash E_\infty(r)} r(x) < \infty$. *Then*

(3.25) $$\|uv\|_r \le c\|u\|_p\|v\|_q$$

for all $u \in L^{p(x)}$ *and* $v \in L^{q(x)}$ *with*

$$c = \sup_{x\in\Omega\backslash E_\infty(r)} (r(x)/p(x)) + \sup_{x\in\Omega\backslash E_\infty(r)} (r(x)/q(x)).$$

PROOF. We note first that the inequality (3.25) is not an immediate consequence of the Hölder inequality itself, as in the case of constant p and q, since $\||u|^r\|_{\frac{p}{r}} \ne \|u\|_p^r$ (we have only the estimates of Lemma 3.5). To prove (3.25) we should use the inequality

(3.26) $$(AB)^r \le \frac{r}{p}A^p + \frac{r}{q}B^q$$

with $A > 0$, $B > 0$, $p > 0$, $q > 0$ and $(1/p) + (1/q) = 1/r$ and follow then the proof of Theorem 3.3, taking into account that $E_\infty(r) = E_\infty(p) \bigcap E_\infty(q)$.

REMARK 3.5. The inequality (3.25) is also valid in the form

(3.27) $$I_r(uv) \le c\|u\|_p\|v\|_q$$

if $\|u\|_p \le 1$ and $\|v\|_q \le 1$, which follows from the Hölder inequality (3.22) and the estimate of the type (3.15) corresponding to the case $\|u\|_p \le 1$.

3.4. On the imbedding $L^{p(x)} \subseteq L^{r(x)}$.

THEOREM 3.4. *Let* $0 \le r(x) \le p(x) \le \infty$ *and let* $|\Omega\backslash E_\infty(r)| < \infty$. *If* $E_\infty(r) \subseteq E_\infty(p)$ *and*

$$R := \sup_{x\in E_\infty(p)\backslash E_\infty(r)} r(x),$$

then $L^{p(x)}(\Omega) \subseteq L^{r(x)}(\Omega)$ *and*

(3.28) $$I_r(f) \le I_p(f) + |E_\infty(p)\backslash E_\infty(r)|\|f\|_{L^\infty(E_\infty(p)\backslash E_\infty(r))}^R + |\Omega\backslash E_\infty(r)|$$

for any $f \in L^{p(x)}$. *(In the case* $E_\infty(p) = E_\infty(r)$*, the second term in the right hand side should be omitted and* R *is allowed to be infinite.) If, moreover,* $1 \le r(x) \le p(x)$ *and* $E_\infty(p) = E_\infty(r)$*, the inequality for norms also holds:*

(3.29) $$\|f\|_{(r)} \le c_0^\nu\|f\|_{(p)}$$

where $c_0 = c_2 + (1 - c_1)|\Omega\backslash E_\infty(p)|$, $c_1 = \inf_{x\in\Omega\backslash E_\infty(p)}(r(x)/p(x))$, $c_2 = \sup_{x\in\Omega\backslash E_\infty(p)}(r(x)/p(x))$, *and* $\nu = 1/r_0$ *if* $c_0 \ge 1$ *and* $\nu = 1/R$ *if* $c_0 \le 1$.

PROOF. The estimate (3.28) is derived by direct estimations from the equality $I_r(f) = \int_{\Omega_1} + \int_{\Omega_2} + \int_{\Omega_3}$ with $\Omega_1 = \{x \in \Omega \backslash E_\infty(p) : |f(x)| \geq 1\}$, $\Omega_2 = \{x \in E_\infty(p) \backslash E_\infty(r) : |f(x)| \geq 1\}$, $\Omega_3 = \{x \in \Omega \backslash E_\infty(r) : |f(x)| \leq 1\}$.

The familiar way to prove the inequality (3.29) for norms by means of the Hölder inequality with the exponents $p_1(x) = p(x)/r(x)$ and $p_2 = r(x)/(p(x) - r(x))$ is not appropriate because we may have $p(x) = r(x)$ at an arbitrary set. So, again, we make use of the inequality (3.26). Taking $A = |f(x)|/\|f\|_{(r)}$ and $B = 1$, we arrive at

$$\int_{\Omega \backslash E_\infty} \left| \frac{f(x)}{\|f\|_{(r)}} \right|^{r(x)} dx \leq c_0$$

in view of (3.8). Hence, by Lemma 3.4 we obtain (3.29).

3.5. Riesz-type norm in $L^{p(x)}(\Omega)$. We introduce now the norm inspired by the Riesz theorem on the representation of a linear functional in L^p. We introduce first the space

$$(3.30) \qquad \tilde{L}^{p(x)}(\Omega) = \left\{ f(x) : \left| \int_\Omega f(x)\varphi(x)dx \right| < \infty \quad \forall \varphi(x) \in L^{q(x)}(\Omega) \right\}$$

where $1 \leq p(x) \leq \infty$ and $(1/p(x)) + (1/q(x)) \equiv 1$. This space will in fact coincide with $L^{p(x)}(\Omega)$ under some natural assumptions on $p(x)$ as it reaches the values 1 and ∞. The imbedding

$$(3.31) \qquad L^{p(x)} \subseteq \tilde{L}^{p(x)}(\Omega), \quad 1 \leq p(x) \leq \infty$$

is an immediate consequence of the Hölder inequality (3.22).

We note that the space (3.30) is always linear, so that, by Lemma 3.1, it cannot coincide with $L^{p(x)}$ á priori if $P = \infty$.

Besides the notations p_0 and P for $p(x)$, see (3.4), and q_0 and Q for $q(x)$, from now on we shall also use

$$p_0^1 = \inf_{x \in \Omega \backslash E_1(p)} p(x), \quad q_0^1 = \inf_{x \in \Omega \backslash E_1(q)} q(x).$$

Evidently,

$$(3.32) \qquad E_1(p) = E_\infty(q), \quad E_1(q) = E_\infty(p), \quad Q = \frac{p_0^1}{p_0^1 - 1}, \quad q_0^1 = \frac{P}{P - 1}.$$

The space (3.30) can be equipped with the natural norms

$$(3.33) \qquad \|f\|_p^* = \sup_{\rho_q(\varphi) \leq 1} \left| \int_\Omega f(x)\varphi(x)dx \right|,$$

$$(3.34) \qquad \|f\|_p^{**} = \sup_{\|\varphi\|_q \leq 1} \left| \int_\Omega f(x)\varphi(x)dx \right|,$$

where $\rho_q(\varphi)$ is the distance (3.5) taken with respect to the variable exponent $q(x)$ and it is assumed that $Q < \infty$ (that is $p_0^1 > 1$) in (3.33), while $p(x)$ may be arbitrary ($1 \leq p(x) \leq \infty$) in case of (3.34).

LEMMA 3.9. *Let* $f(x) \in \tilde{L}^{p(x)}(\Omega)$, $p_1^0 > 1$. *Then* $\|f\|_p^* < \infty$ *and*

$$(3.35) \qquad \int_\Omega |f(x)\varphi(x)|dx \le \|f\|_p^*\|\varphi\|_q^1 \le \|f\|_p^*\|\varphi\|_q$$

for all $\varphi(x) \in L^{q(x)}(\Omega)$, $1/p(x) + 1/q(x) \equiv 1$, *where* $\|\varphi\|_q^1$ *is the norm* (3.18). *Besides, the functional* (3.33) *is a norm in* $\tilde{L}^{p(x)}(\Omega)$.

PROOF. (We follow some ideas from [10, p. 84].) Suppose that $\|f\|_p^* < \infty$. Then there exists a function $f_0(x) \in \tilde{L}^{p(x)}(\Omega)$ and a sequence $\varphi_k(x) \in L^{q(x)}$ such that $\rho_q(\varphi_k) \le 1$ and

$$\int_\Omega f_0(x)\varphi_k(x)dx \ge 2^{Qk}, \quad k = 1, 2, \ldots$$

($f_0 \ge 0$, $\varphi_k \ge 0$). Then $j_m = \sum_{k=1}^m 2^{-Qk}\varphi_k(x)$ is an increasing sequence. Direct calculations show that $\rho_q(j_m) \le 1$ and

$$(3.36) \qquad \int_\Omega f_0(x)j_m(x)dx = \sum_{k=1}^m 2^{-Qk}\int_\Omega f_0(x)\varphi_k(x)dx \ge m.$$

The sequence $j_m(x)$ converges monotonically to the function

$$j(x) = \sum_{k=1}^\infty 2^{-Qk}\varphi_k(x).$$

So,

$$\int_{\Omega \setminus E_\infty(q)} |j(x)|^{q(x)}dx = \lim_{m \to \infty} \int_{\Omega \setminus E_\infty(q)} |j_m(x)|^{q(x)}dx \le 1,$$

by Levi's theorem and, evidently, $\sup_{x \in E_\infty(q)} j(x) = \sum_{k=1}^\infty 2^{-Qk} < \infty$. Hence, $j(x) \in L^{q(x)}$. By Levi's theorem again and in view of (3.36), $\int_\Omega f_0(x) j(x)dx = \infty$ which contradicts to the fact that $f_0(x) \in \tilde{L}^{p(x)}$.

Then $\|f\|_p^* < \infty$ and from the definition (3.33) we have

$$(3.37) \qquad \left| \int_\Omega f(x)\varphi(x)dx \right| \le A\|f\|_p^*$$

where $A > 0$ and $\rho_q(\varphi/A) \le 1$. Taking *infinum* with respect to A, we arrive at the left-hand side inequality in (3.35) in accordance with the definition (3.18). The right-hand side inequality follows from (3.20).

It remains to check axioms for norms. Homogenity and the triangle inequality are obvious. Let $\|f\|_p^* = 0$. Then $\int_\Omega f(x)\varphi(x)dx = 0$ for all $\varphi(x) \in L^{q(x)}$ and then for all $\varphi(x) \in S$ by Lemma 3.2. Then $f(x) \equiv 0$ as is well known.

LEMMA 3.10. *Let* $1 \le p(x) \le \infty$, $p_0^1 > 1$ *and* $P < \infty$. *The norms* (3.33) *and* (3.34) *are equivalent on functions* $f(x) \in \tilde{L}^{p(x)}(\Omega)$:

$$(3.38) \qquad 2^{1-Q/q_0^1}\|f\|_p^{**} \le \|f\|_p^* \le \|f\|_p^{**}.$$

They coincide with each other in the cases: (1) $|E_1(p)| = 0$, (2) $p(x) = const$ *for* $x \in \Omega \setminus (E_\infty \bigcup E_1)$.

PROOF. To arrive at the right-hand side inequality, we shall prove that

(3.39) $$\{\varphi : \rho_q(\varphi) \le 1\} \subseteq \{\varphi : \|\varphi\|_q \le 1\}$$

for $\varphi(x) \in L^{q(x)}(\Omega)$. Let $\rho_q(\varphi) \le 1$. Then $I_q(\varphi) \le 1$ so that $\|\varphi\|_{(q)} \le 1$ by (3.11)–(3.12). Therefore, by (3.11) $\|\varphi\|_{(q)} \le [I_q(\varphi)]^{1/Q} \le 1$ and then $\|\varphi\|_q \le [I_q(\varphi)]^{1/Q} + \sup_{x \in E_\infty(q)} |\varphi(x)| = \rho_q(\varphi) \le 1$ so that (3.39) is proved.

Further, let $c = 2^{1-Q/q_0} \le 1$. We shall prove the inequality

(3.40) $$\{\varphi : \|\varphi\|_q \le 1\} \subseteq \{\varphi : \rho_q(c\varphi) \le 1\}$$

which will give the left-hand side inequality in (3.38). We have $\|\varphi\|_q \le 1 \Longrightarrow \|c\varphi\|_{(q)} \le 1 \Longrightarrow [I_q(c\varphi)]^{1/Q} \le \|c\varphi\|_{(q)}^{q_0^1/Q}$ by (3.11). Hence, $\rho_q(c\varphi) \le \|c\varphi\|_{(q)}^{q_0^1/Q} + \|c\varphi\|_{L^\infty[E_\infty(q)]}$. Since $A^\lambda + B \le 2^{1-\lambda}(A + B)^\lambda$, $0 \le \lambda \le 1$, $A \ge 0$, $0 \le B \le 1$, we easily obtain that $\rho_q(c\varphi) \le 1$ and (3.40) is proved together with the left-hand side inequality in (3.38).

Finally, if $|E_1(p)| = 0$ or $p(x) = const$ for $x \in \Omega \backslash (E_\infty \bigcup E_1)$, then $\|\varphi\|_{L^\infty[E_\infty(q)]} = 0$ or $q_0^1/Q = 1$, respectively, and we have (3.40) with $c = 1$, which implies the coincidence of the norms.

THEOREM 3.5. *Let $p_0^1 > 1$ and $P < \infty$. The spaces $L^{p(x)}(\Omega)$ and $\tilde{L}^{p(x)}(\Omega)$ coincide up to the equivalence of norms:*

(3.41) $$\frac{1}{3}\|f\|_p \le \|f\|_p^* \le \left(\frac{1}{p_0} + \frac{1}{q_0}\right)\|f\|_p$$

where $1/3$ may be replaced by 1 if $|E_1| = |E_\infty| = 0$.

PROOF. In view of (3.31) it is sufficient to prove the imbedding

(3.42) $$\tilde{L}^{p(x)}(\Omega) \subseteq L^{p(x)}(\Omega).$$

Let $f(x) \in \tilde{L}^{p(x)}(\Omega)$ and let $\|f\|_p^* \le 1$ first. We set $\varphi_0(x) = |f(x)|^{p(x)-1}$ if $x \in \Omega \backslash (E_1 \bigcup E_\infty)$ and $\varphi_0(x) = 0$ otherwise. We will show that

(3.43) $$\varphi_0(x) \in L^{q(x)}(\Omega) \quad and \quad I_q(\varphi_0) \le 1.$$

Suppose that $I_q(\varphi_0) > 1$. Then

(3.44) $$I_p(f) \ge \int_{\Omega \backslash E_\infty(q)} |\varphi_0(x)|^{q(x)} dx > 1.$$

Let $f_{N,k}(x) = f(x)$ if both $|x| \le k$ and $|f(x)| \le N$ and $f_{N,k} \equiv 0$ otherwise. Then $\varphi_{N,k}(x) = |f_{N,k}|^{p(x)-1} \in L^{q(x)}(\Omega)$. From (3.44) we derive that there exist $N_0 \to \infty$ and $k_0 \to \infty$ such that

(3.45) $$\int_{\Omega \backslash E_\infty(p)} |f_{N_0,k_0}|^{p(x)} dx > 1.$$

Then, by (3.35)

$$1 < I_p(f_{N_0,k_0}) \le \|f_{N_0,k_0}\|_p^* \|f_{N_0,k_0}^{p(x)-1}\|_q.$$

So, in view of (3.11)–(3.12)

(3.46) $$1 < \|f_{N_0,k_0}\|_p^* \max\left\{[I_p(f_{N_0,k_0})]^{1/Q}, [I_p(f_{N_0,k_0})]^{1/q_0'}\right\}.$$

Hence,

$$\min\left\{[I_p(f_{N_0,k_0})]^{1-(1/Q)}, [I_p(f_{N_0,k_0})]^{1-1/q_0'}\right\} \leq \|f_{N_0,k_0}\|_p^*$$

and so, by (3.45), $1 < \|f_{N_0,k_0}\|_p^*$. This means that

$$\sup_{\rho_q(\varphi)\leq 1}\left|\int_\Omega f(x)\varphi^{N,k}(x)dx\right| > 1$$

where $\varphi^{N,k}(x) = \varphi(x)$ if both $|x| \leq k$ and $|f(x)| \leq N$ and $\varphi^{N,k} = 0$ otherwise. However, since $\rho_q(\varphi^{N,K}) \leq \rho_q(\varphi)$, this contradicts to the supposition that $\|f\|_p^* \leq 1$. Hence, (3.43) is obtained.

Then $\int_{\Omega\setminus(E_1(p)\bigcup E_\infty(p))} |f(x)|^{p(x)}dx \leq 1$ and to have the imbedding (3.42), it remains to show that $\int_{E_1(p)} |f(x)|dx < \infty$ and $\sup_{x\in E_\infty(p)} |f(x)| < \infty$, which follows from the inequalities

$$\int_{\Omega_i} |f(x)\varphi(x)|dx \leq c\|\varphi\|_{L^q(\Omega_i)}, \quad i = 1, 2,$$

derived from (3.35), where $\Omega_1 = E_1(p)$, $\Omega_2 = E_\infty(p)$ and $f \in L^1$, $\varphi \in L^\infty$ ($q = 1$) in the first case and $f \in L^\infty$, $\varphi \in L^1$ ($q = \infty$) in the second one.

Let now $\|f\|_p^* > 1$. Then $f(x)/\|f\|_p^* \in L^{p(x)}(\Omega)$ as was just proved. Then $f(x) \in L^{p(x)}(\Omega)$ by the linearity of the space $L^{p(x)}(\Omega)$ under the condition $P < \infty$. The imbedding (3.42) is proved.

It remains to prove the inequalities (3.41) for the norms. The right-hand side inequality is a consequence of the Hölder inequality (3.22) and the definition of the norm (3.34). To prove the left-hand side inequality we put $f(x) = f_1(x) + f_2(x) + f_3(x)$ with $f_2(x) = f(x)$, $x \in E_1$ and $f_2(x) = 0$, $x \in \Omega\setminus E_1$ and $f_3(x) = f(x)$, $x \in E_\infty$, and $f_3(x) = 0$, $x \in \Omega\setminus E_\infty$. We shall prove that

$$(3.47) \qquad \|f_1\|_p \leq \|f\|_{L^{p(x)}[\Omega\setminus(E_1\bigcup E_\infty)]}^*.$$

We have

$$(3.48) \qquad I_p\left(\frac{f_1}{\lambda}\right) = \frac{1}{\lambda}\int_{\Omega\setminus E_\infty} |f_1(x)|\varphi_\lambda(x)dx, \quad \lambda > 0,$$

with $\varphi_\lambda(x) = |f_1(x)/\lambda|^{p(x)-1}$. We choose $\lambda = \|f_1\|_{(p)}$ and from (3.35) and (3.48) have

$$(3.49) \qquad 1 = \frac{1}{\|f_1\|_{(p)}}\int_{\Omega\setminus E_\infty} |f_1(x)|\,\varphi_\lambda(x)dx \leq \frac{\|f_1\|_p^*}{\|f_1\|_p}\|\varphi_\lambda\|_q.$$

Since $\rho_q(\varphi_\lambda) \leq I_p(f_1/\lambda) = 1$, we also have $\|\varphi_\lambda\|_q \leq 1$ by (3.11)–(3.12) and the coincidence $\|\varphi_\lambda\|_q = \|\varphi_\lambda\|_{(q)}$ follows. Then (3.49) yields (3.47). Since $\|f_2\|_p = \|f\|_{L^1(E_1)}^*$ and $\|f_3\|_p = \|f\|_{L^\infty(E_\infty)}^*$, we arrive at the left-hand side inequality.

We conclude this section by the following

THEOREM 3.6. *Let* $1 \leq p(x) \leq \infty$, $x \in \Omega$, *and* $P < \infty$. *The space* $L^{p(x)}(\Omega)$ *is complete.*

PROOF. Evidently, $L^{p(x)}(\Omega)$ is an algebraic sum $L^{p(x)}(E) + L^{\infty}(E_{\infty})$ where $E = \Omega \backslash E_{\infty}$ and each of the spaces is understood as the space of functions equal to zero beyond the sets E and E_{∞}, respectively. So, the completeness of the space $L^{p(x)}(E)$ is only to be proved. This is done in the standard way, taking use of the completeness of the space $L^1(E)$, which is known, and the Hölder inequality for the spaces $L^{p(x)}(E)$ which was proved in Theorem 3.3.

COROLLARY. *Let* $1 \le p(x) \le \infty$, $x \in \Omega$, $P < \infty$, *and* $p_0^1 > 1$. *The space* $\tilde{L}^{p(x)}(\Omega)$ *is complete.*

3.6. Minkowski inequality.

THEOREM 3.7. *Let* $1 \le p(x) \le \infty$, $P < \infty$ *and* $p_0^1 > 1$. *Then*

$$(3.50) \qquad \left\| \int_{\Omega} f(\cdot, y) dy \right\|_p^{**} \le \int_{\Omega} \| f(\cdot, y) \|_p^{**} dy.$$

PROOF. Let J be the expression in the left-hand side. We have

$$J \le \sup_{\|\varphi\|_q \le 1} \int_{\Omega} dy \int_{\Omega} |\varphi(x) f(x, y)| dx.$$

Hence, by the definition of the norm (3.34), (3.50) follows.

COROLLARY. *Let* $1 \le p(x) \le \infty$, $P < \infty$ *and* $p_0^1 > 1$. *Then*

$$(3.51) \qquad \left\| \int_{\Omega} f(\cdot, y) dy \right\|_p^{*} \le c_1 \int_{\Omega} \| f(\cdot, y) \|_p^{*} dy$$

where $c_1 = 1$ *if* $|E_1| = 0$ *and* $c_1 = 2^{-1+Q/q_0^1}$ *otherwise. The inequality (3.51) holds also with respect to the Kolmogorov-Minkowsky norm (3.10) with* c_1 *replaced by* $c_2 = kc_1$ *if* $|E_{\infty}| = |E_1| = 0$ *and* $c_2 = 3kc_1$ *otherwise;* $k = (1/p_0) + (1/q_0)$.

References

[1]   R. Beals, *Weighted distribution spaces and pseudodifferential operators*, J. Anal. Math. **39** (1981), 131–187.

[2]   D. E. Edmunds and J. Rákosník, *Density of smooth functions in* $W^{k,p(x)}(\Omega)$, Proc. Roy. Soc. London A. **437** (1992), 229–236.

[3]   G. H. Hardy, J. E. Littlewood, and G. Pólya, *Inequalities*, Cambridge Univ. Press, 1952.

[4]   L. V. Kantorovich and G. P. Akilov, *Functional Analysis* (Russian), Nauka, Moscow, 1977.

[5]   N. K. Karapetiants and A. I. Ginzburg, *Fractional integrodifferentiation in Hölder classes of variable order* (Russian), Dokl. Akad. nauk Rossii **339** (1994), 439–441.

[6]   N. K. Karapetiants and A. I. Ginzburg, *Fractional integrals and singular integrals in the Hölder classes of variable order* (Russian), Integr. Transf. and Spec. Funct. **2** (1994), 91–96.

[7]   A. N. Kolmogorov, *Zur Normierbarkeit eines allgemeinen topologischen linearen Räumes*, Studia Math. **5** (1934), 29–33.

[8]   O. Kováčik, *Some generalization of Lebesgue and Sobolev spaces*, Constructive Theory of Functionso 1984, Sofia, 487-492.

[9]   O. Kováčik and J. Rákosník, *On spaces* $L^{p(x)}$ *and* $W^{k,p(x)}$, Czechoslovak Math. J. **41** (1991), 592–618.

[10] M. A. Krasnoselskii and Ya. B. Rutitskii, *Convex functions and Orlicz spaces* (Russian), Moscow, 1958.

[11] H. G. Leopold, *On Besov spaces of variable order of differentiation*, J. Anal. Anw. **8** (1989), 69–82.

[12] H. G. Leopold, *On function spaces of variable order of differentiation*, Forum Math. **3** (1991), 1–21.

[13] P. I. Lizorkin, *Generalized Liouville differentiation and the functional spaces $L_p^r(E_n)$. Imbedding theorems*, Matem. sb. (N.S.) **60** (1963), 325–353.

[14] S. M. Nikolskii, *The approximation of functions of several variables and the imbedding theorems* (Russian), Nauka, Moscow, 1977.

[15] L. S. Novozhilova (Urazhdina), *Spaces of variable order in nonbounded domains*, Teor. funkts., funkts. analiz i ikh priloz., Kharkov **20** (1974), 122–133.

[16] B. Ross and S. G. Samko, *Fractional integration operator of variable order in the Hölder spaces $H^{\lambda(x)}$*, Intern. J. Math. and Math. Sci. **18** (1995)), 777–788.

[17] S. G. Samko, *Hypersingular integrals and their applications*, Izdat. Rostov. Univ., Rostov-on-Don, 1984.

[18] S. G. Samko, *Differentiation and integration of variable (fractional) order*, Dokl. Akad. Nauk Rossii **342** (1995).

[19] S. G. Samko, *Fractional integration and differentiation of variable order*, Analysis Mathem. **21** (1995), 213–236.

[20] S. G Samko, A. A. Kilbas, and O. I. Marichev, *Fractional Integrals and Derivatives. Theory and Applications*, Gordon & Breach Sci. Publ., 1993.

[21] S. G. Samko and B. Ross, B, *Integration and differentiation to a variable fractional order*, Integr. Transf. and Spec. Funct. **1** (1993), 277–300.

[22] I. I. Sharapudinov, *On a topology of the space $L^{p(t)}([0,1])$*, Matem. Zametki **26** (1979), 613–632.

[23] I. V. Tsenov, *A generalization of the problem of the best approximation in the space L^s*, Uchen. Zap. Daghestan. Gos. Univ. **7** (1961), 25–37.

[24] A. Unterberger, *Résolution d'équation aux derivées partielles dans dès espaces de distributions d'ordre de régularité variable*, Ann. Inst. Fourier, Grenoble **21** (1971), 85–128.

[25] L. R. Volevich and V. M. Kagan, *Pseudodifferential hypoelliptic operators in the theory of function spaces* (Russian), Trudy Mosk. Matem. obsch. **20** (1969), 241–275.

MATHEMATICS DEPARTMENT, ROSTOV UNIVERSITY, ROSTOV-ON-DON, UL. ZORGE 5, ROSTOV-ON-DON, 344104 RUSSIA

Current address: Unidade de Ciencias Exactas e Humanas, Universidade do Algarve, Campus de Gam belas, Faro 8000, Portugal

E-mail address: ssamko@ualg.pt

Contemporary Mathematics
Volume 212, 1998

Twistor Quantization of Loop Spaces
and General Kähler Manifolds

A. G. Sergeev

ABSTRACT. A twistor reformulation of the geometric quantization scheme for
loop spaces of Lie groups and general (finite- or infinite-dimensional) Kähler
manifolds is presented.

The geometric quantization method for the loop space ΩG of a Lie group G
has a natural reformulation in terms of twistors (cf. [1–2]). Namely, one can define
a symplectic twistor bundle $Z \to \Omega G$ of invariant (w.r. to the natural action of the
mapping group $LG = \mathrm{Map}(S^1, G)$ on ΩG) complex structures on ΩG compatible
with symplectic structure of ΩG. Structures of that type in a given point of ΩG
are parametrized by points of an infinite-dimensional manifold $\mathcal{S} = \mathrm{Diff}(S^1)/S^1$
where $\mathrm{Diff}(S^1)$ is the group of orientation-preserving diffeomorphisms of S^1. The
twistor space Z can be provided with an almost complex structure so that the
natural projection $p : Z \to \mathcal{S}$ will be a holomorphic mapping. In twistor terms the
geometric quantization means the construction of a quantization bundle $\mathcal{H} \to \mathcal{S}$ over
$\mathcal{S}$ with a flat unitary connection on it. Such a bundle can be defined as the tensor
product of the Fock bundle H (with the connection provided by Kostant-Souriau
prequantization) and the square root $K^{-1/2} \to \mathcal{S}$ of the anticanonical bundle of $\mathcal{S}$
(provided with a natural spinorial connection). Then the tensor-product connection
on $\mathcal{H}$ will be flat under the critical dimension condition (making sense for simple
compact Lie groups G or the group of translations of a vector space). It means
that the twistor quantization problem (and also the original geometric quantization
problem) may be solved under the critical dimension condition. We shortly present
this quantization scheme here referring for details to [1–2].

In case of general (finite- or infinite-dimensional) Kähler manifolds the construction of Fock bundle represents a (well-known) problem. It's not at all clear,
for example, what should be the base of this bundle. (In the loop space case we
have used the invariance of the involved complex structures on ΩG in order to define the Fock bundle over $\mathcal{S}$). We propose here a definition of the Fock bundle for a

1991 *Mathematics Subject Classification.* Primary 32L25, 58F06; Secondary 58D15, 81D27,
81E30.
    Research partly supported by the International Science Foundation (grant NFT 000) and
Russian Foundation for Fundamental Research (grant 95-011-00027)

general Kähler manifold M as a bundle over the space of all Kähler structures on M (such structures are identified with sections of a symplectic twistor bundle $Z \to M$ with vanishing Nijenhuis tensor). This Fock bundle has a natural almost complex structure and the twistor quantization problem for a general Kähler manifold can be formulated as before: to construct a flat connection on the Fock bundle. Note that the Fock bundle, so defined, may be thought of as the "biggest possible" Fock bundle for M. One can define "smaller" Fock bundles by restricting this "big" bundle to subspaces of Kähler structures on M satisfying to some extra conditions (e.g. invariance with respect to group actions on M).

1. Geometric quantization

We recall here basic facts from the geometric quantization theory (cf. [3–4] for a detailed exposition).

1.1. Classical system. We start with a *classical (mechanical) system* defined by a pair $(M, \mathcal{A})$ consisting of a manifold M and a Lie algebra $\mathcal{A}$. Here M is the *phase space* of the system, i.e., a $2n$-dimensional symplectic manifold with symplectic form ω. Locally, M is isomorphic to $\mathbb{R}^{2n}$ with the standard symplectic form $\omega_0 = \sum_{i=1}^{n} dp_i \wedge dq_i$ where (p_i, q_i) are local coordinates on M. The Lie algebra $\mathcal{A}$ is the *algebra of observables* of the system, i.e., a Lie algebra of smooth real functions on M with Poisson bracket. A standard example of $\mathcal{A}$ for the local model $M = \mathbb{R}^{2n}$, $\omega = \omega_0$ is given by the Heisenberg (or oscillator) algebra $\mathrm{Heis}(M)$ generated by coordinate and momenta functions q_i, p_i. If the phase space M is provided with a group Γ of symplectic diffeomorphisms acting on M one can take for $\mathcal{A}$ the Hamiltonian algebra $\mathrm{Ham}(\Gamma)$ consisting of Hamiltonians generating transformations from Γ.

1.2. Formal quantization. By (formal) *quantization of a classical system* $(M, \mathcal{A})$ we mean an irreducible Lie-algebra representation $r : \mathcal{A} \to \mathrm{End}^* H$ of the algebra of observables $\mathcal{A}$ in the algebra of linear self-adjoint operators in a complex (pre)Hilbert space H called *quantization* or *Fock space* of the system. We assume that the algebra $\mathrm{End}^* H$ is provided with the Lie bracket given by the usual commutator of linear operators multiplied by imaginary unit i. We also add the normalization condition requiring that the function $f \equiv 1$ corresponds to the identity operator on H under r. If r satisfies to all above conditions apart from irreducibility it is called a *prequantization*.

1.3. Kostant–Souriau prequantization. The most interesting case for us is that of Kähler phase manifolds, i.e., symplectic manifolds M which can be endowed with a complex structure J compatible with symplectic structure ω. Recall that *symplectic structure ω is compatible with complex structure J* if: (i) $\omega(J\xi, J\eta) = \omega(\xi, \eta)$ for any tangent vectors $\xi, \eta \in T_x M$, $x \in M$; (ii) symmetric form $g(\xi, \eta) = \omega(\xi, J\eta)$ on $T_x M$ is positive definite for any $x \in M$, i.e., g defines a Riemannian metric on M. Complex structures compatible with symplectic structure of M are called *Kähler structures* on M. In Kähler case we have a natural candidate for quantization space H.

Consider first the topologically trivial situation, i.e., assume that symplectic structure ω is exact: $\omega = d\theta$. Then we define H to be the space $\mathcal{L}^2(M; \omega)$ of square integrable functions on M with respect to volume form given by the nth power ω^n of symplectic form. Take for the algebra of observables $\mathcal{A}$ the algebra $C^\infty(M)$ of smooth real functions on M and define a representation $r : \mathcal{A} \to \text{End}^* H$ by the formula

$$r : f \longmapsto r(f) = f - iX_f - \theta(X_f)$$

where X_f is the Hamiltonian vector field corresponding to f and $f - \theta(X_f)$ acts on H as multiplication operator. Easy to check that this representation defines a prequantization of $(M, C^\infty(M))$.

In general case ω is only locally exact so it has only local potentials θ_α with respect to some open covering U_α of M. However, if M satisfies to a certain topological condition then these local potentials can be glued together to a connection on a line bundle over M. The topological condition is that of integrality of symplectic form, more precisely, we call a symplectic manifold (M, ω) *quantizable* if the 2-form $\omega/2\pi$ defines an integral cohomology class in $H^2(M, \mathbb{C})$. Under this condition there exists a Hermitian complex line bundle $L \to M$ with a Hermitian connection ∇ having the curvature $F_\nabla = \omega$. This line bundle is called the *prequantum bundle* on M.

We can take for the Hilbert space H in this case the space $\mathcal{L}^2(M, L; \omega)$ of square integrable sections of L with scalar product given by

$$(s_1, s_2) = \int_M \langle s_1, s_2 \rangle \omega^n$$

where $\langle \cdot, \cdot \rangle$ is the Hermitian structure on L. A prequantization $r : C^\infty(M) \to \text{End}^* H$ in this quantization space can be defined by the formula

$$r : f \longmapsto r(f) = f - i\nabla_{X_f}.$$

It is called the *Kostant–Souriau prequantization*. Note that the Kostant–Souriau representation is reducible. It is clear from "physical considerations" because, according to Heisenberg uncertainty principle, a "physical" quantization space should involve only "half" of wave functions. Mathematically, it means that, in order to get an irreducible representation, one should restrict the above quantization space $H = \mathcal{L}^2(M, L; \omega)$ to a subspace containing only a "half" of sections from H. On a Kähler phase manifold it is reasonable to take for such a subspace the space of sections of L holomorphic with respect to the complex structure J. In other words, we define the quantization space to be the Fock space $H = F(M, L, ; J) := \mathcal{L}^2_{\mathcal{O}}(M, L; \omega)$ of holomorphic square integrable sections of $L \to M$. We restrict next the Kostant–Souriau representation to this subspace in order to get an irreducible representation $r : \mathcal{A} \to F(M, L; J)$. In the next Section we shall discuss the difficulties one may encounter proceeding this way.

2. Loop groups

We want to apply the ideas of the first Section to the quantization of the open bosonic string theory. The phase space of this theory, describing open bosonic strings moving in a Lie group G, may be identified with the space of loops of G (cf. [5]). We start with some general properties of such spaces (cf. [1–2, 6] for a detailed exposition).

2.1. Phase space of string theory. Define the *loop space* ΩG of a Lie group G as

$$\Omega G = LG/G$$

where $LG = \mathrm{Map}(S^1, G)$ is the space of smooth mappings $S^1 \to G$ of the unit circle $S^1 \subset \mathbb{C}$ into the group G, the denominator being identified with the group G of constant mappings $S^1 \to g_0 \in G$. The quantization of ΩG (identified with the phase space of string theory) may be carried out for two different classes of Lie groups G. The mathematically interesting case corresponds to compact Lie groups G (provided with the Killing metric). The group $G = \mathbb{R}^d$ of translations of the d-dimensional vector space $\mathbb{R}^d$ is another important example. In the physically interesting situation it is provided with the Lorentz metric so that G is identified with the d-dimensional Minkowski space $\mathbb{R}^{d-1,1}$. We shall restrict here mainly to this second case in order to simplify our construction, with natural modifications it is applied also in the compact group case (cf. [1–2]).

According to a Birkhoff's theorem (cf. [6]), the loop space ΩG can be also represented in the form

$$\Omega G = LG^{\mathbb{C}}/L_+ G^{\mathbb{C}}$$

where $G^{\mathbb{C}}$ is the complexification of G (so that $LG^{\mathbb{C}} = \mathrm{Map}(S^1, G^{\mathbb{C}})$) and $L_+ G^{\mathbb{C}} = \mathrm{Hol}(\Delta, G^{\mathbb{C}})$ is the subspace of $LG^{\mathbb{C}}$ consisting of maps which can be extended to holomorphic maps of the unit disc $\Delta \subset \mathbb{C}$ into $G^{\mathbb{C}}$.

It is well known (cf. [1–2, 6]) that ΩG for a compact Lie group G is an infinite-dimensional Kähler manifold, i.e., it can be provided with a symplectic structure ω and a complex structure J^0 which are invariant under the natural left action of the group LG and compatible with each other. For $G = \mathbb{R}^{d-1,1}$ the loop space ΩG is an infinite-dimensional pseudoKähler manifold, the only difference with the Kähler case is that in the definition of compatibility of symplectic and complex structures (cf. Section 1.3) one should require the symmetric form g to be non-degenerate (so that g defines a pseudoRiemannian metric).

2.2. Quantization: statement of the problem. For the group Γ of symplectic diffeomorphisms of ΩG (cf. Section 1.1) we take now the group $\mathrm{Diff}(S^1)$ of diffeomorphisms of the unit circle S^1 preserving its orientation. Its action on loops from $LG = \mathrm{Map}(S^1, G)$ is given by the reparametrization of a loop and preserves the symplectic form on ΩG. The Lie algebra $\mathrm{Vect}(S^1)$ of $\mathrm{Diff}(S^1)$ consists of tangent vector fields on S^1 and is identified with the Lie algebra $\mathrm{Ham}(\Gamma)$ of Hamiltonians generating reparametrizations of strings of Section 1.1.

We take for the algebra of observables $\mathcal{A}$ on ΩG the Lie algebra generated by the Heisenberg algebra of ΩG and the Lie algebra $\mathrm{Vect}(S^1)$. In order to quantize the bosonic string theory, we need to construct an irreducible representation $r : \mathcal{A} \to \mathrm{End}^* H$. Following the idea proposed at the end of Section 1.3, we should take for H the Fock space $F(\Omega G; J^0)$ of functions holomorphic with respect to a complex structure generated by J^0. (Note that for $G = \mathbb{R}^{d-1,1}$ the symplectic form of ΩG is exact, so we are in a topologically trivial situation. The prequantum bundle L in this case is a trivial line bundle provided with a flat connection ∇. The Fock space of its sections can be identified with the space of functions holomorphic with respect to a complex structure generated by J^0 and ∇). In order to get an irreducible representation of $\mathcal{A}$ in $F(\Omega G; J^0)$, we should restrict the Kostant–Souriau representation operators to $F(\Omega G; J^0)$. Unfortunately, this idea does not work because these operators, in general, do not preserve the Fock space $F(\Omega G; J^0)$. The reason for that is that the group $\mathrm{Diff}(S^1)$ does not preserve the complex structure J^0.

2.3. Space of admissible complex structures. The last statement of the previous subsection can be made more precise. The complex structure J on ΩG, obtained from the original structure J^0 by the action of a diffeomorphism $f \in \text{Diff}(S^1)$, coincides with J^0 if and only if f is a rotation, i.e., f belongs to the subgroup S^1 of rotations in $\text{Diff}(S^1)$. All complex structures J, obtained in this way from J^0, are invariant with respect to LG-action on ΩG and compatible with ω, i.e., they also define pseudoKähler structures on ΩG. We call these complex structures on ΩG *admissible*. Admissible complex structures on ΩG are parametrized by points of the space

$$\mathcal{S} = \text{Diff}(S^1)/S^1.$$

This space is an infinite-dimensional Kähler manifold having a family of Kähler metrics parametrized by two real numbers (cf. [7]).

3. Twistor quantization of string theory

We present here the twistor approach to the geometric quantization of bosonic string theory (cf. [1–2] for details). In order to simplify our construction, we assume that $G = \mathbb{R}^{d-1,1}$ and denote the loop space ΩG in this case simply by Ω.

3.1. Twistor bundle. As we have seen in Section 2.3, the Kostant-Souriau representation operators do not preserve Fock spaces of functions on Ω holomorphic with respect to a fixed admissible complex structure J on Ω. It means, in other words, that Ω has no distinguished admissible complex structure. The idea of twistor approach (proposed in [1–2]) is that, instead of fixing one admissible complex structure on Ω, one should rather consider them all together simultaneously and define the quantization in terms of the space of all admissible complex structures on Ω. This space <u>has</u> a distinguished almost complex structure (this structure is integrable in case of $G = \mathbb{R}^{d-1,1}$) so that one can define the quantization in terms of this natural almost complex structure.

The *space Z of all admissible complex structures* on Ω is, by definition, a bundle $\pi : Z \to \Omega$ over Ω with fibre $\pi^{-1}(\gamma)$ at $\gamma \in \Omega$ given by restrictions J_γ of admissible complex structures J on Ω to the tangent space $T_\gamma \Omega$; points $z \in Z$ are the pairs $z = (\gamma, J_\gamma)$ where $\gamma \in \Omega$, J_γ is a complex structure on $T_\gamma \Omega$.

Admissible complex structures are invariant with respect to LG-action on Ω, so there is a natural action of LG on Z such that the quotient of this action coincides with the manifold $\mathcal{S} = \text{Diff}(S^1)/S^1$. Thus we have a double fibration

$$
\begin{array}{ccc}
Z & \xrightarrow{\ p\ } & \mathcal{S} \\
{\scriptstyle \pi}\downarrow & & \\
\Omega & &
\end{array}
$$

where p is the natural projection of Z to $\mathcal{S}$. The fibre $p^{-1}(J)$ at $J \in \mathcal{S}$ can be identified with the complex manifold $\Omega_J = (\Omega, J)$, i.e., with the space Ω provided with the complex structure corresponding to $J \in \mathcal{S}$ (we denote a point $J \in \mathcal{S}$ and corresponding complex structure on Ω by the same letter J).

As in conventional twistor theory, the space Z of admissible complex structures on Ω has a *natural almost complex structure* which is defined as follows. The tangent bundle TZ of Z can be split into the direct sum $TZ = V \oplus H$ of vertical and horizontal subbundles. The vertical subbundle $V \subset Z$ is identified, as usual, with the kernel of differential π_* of projection $\pi : Z \to \Omega$. Respectively, the horizontal subbundle $H \subset Z$ is given by the kernel of p_* for projection $p : Z \to \mathcal{S}$. The fibre V_z at $z \in Z$ is identified by p_* with the tangent space $T_J\mathcal{S}$ at $J = p(z) \in \mathcal{S}$, and so has a natural complex structure $\mathcal{J}^v$ induced by the Kähler complex structure of $\mathcal{S}$. We define now an almost complex structure $\mathcal{J}$ on Z in terms of splitting $TZ = V \oplus H$ as

$$\mathcal{J} = \mathcal{J}^v \oplus \mathcal{J}^h$$

where the complex structure $\mathcal{J}^h$ at $z \in Z$ is equal to the complex structure J_γ on $T_zH \approx T_{\pi(z)}\Omega$ defined by the point $z = (\gamma, J_\gamma)$. This structure is integrable in our case and projection $p : Z \to \mathcal{S}$ is holomorphic with respect to $\mathcal{J}$.

3.2. Twistor quantization. The quantization problem may be reformulated in terms of the twistor bundle, constructed in the last subsection, in the following way (cf. [1–2]). The role of Fock space H is played by a Fock bundle $\mathcal{H} \to \mathcal{S}$ which should be a holomorphic vector $\mathrm{Diff}(S^1)$-homogeneous bundle over $\mathcal{S}$. To quantize Ω in twistor terms means: *to construct a Fock bundle $\mathcal{H} \to \mathcal{S}$ and a flat unitary $\mathrm{Diff}(S^1)$-invariant connection on it.* Using such a connection one can (in principle) identify different fibres of the Fock bundle (corresponding to different complex structures $J \in \mathcal{S}$) and define a quantization space which will not depend on the choice of a complex structure on ΩG. That gives a connection between the original and twistor approaches to the quantization of ΩG (cf. [1–2] for details).

In case of bosonic string theory, the Fock bundle $\mathcal{H} \to \mathcal{S}$, which gives a solution to the above twistor quantization problem, has the form (cf. [1–2])

$$\mathcal{H} = H \otimes K^{-1/2}.$$

Here the fibre of H_J at $J \in \mathcal{S}$ coincides with the Fock space $F(\Omega_J)$ of functions on $p^{-1}(J) = \Omega_J$ (we recall that Z has a natural complex structure). A connection D on $H \to \mathcal{S}$ is provided by the Kostant–Souriau construction and its curvature F_D is computed explicitly. On the other hand, $K \to \mathcal{S}$ is the canonical line bundle of the Kähler manifold $\mathcal{S}$, so that $K^{-1/2}$ is similar to holomorphic spinor bundles in finite dimensions. The Riemannian connection on $\mathcal{S}$ (corresponding to the Kähler metric on $\mathcal{S}$) generates a spinorial connection ∇ on $K^{-1/2}$, having the curvature equal to $-1/2\,\mathrm{Ric}(\mathcal{S})$ where $\mathrm{Ric}(\mathcal{S})$ is the Ricci curvature of $\mathcal{S}$. This curvature was computed explicitly in [8, 9].

The Fock bundle $\mathcal{H} = H \otimes K^{-1/2}$, provided with connection $\mathcal{D}$ given by tensor product of the above connections D and ∇, yields a solution to twistor quantization problem under the well-known "critical dimension condition". In other words, the curvature $F_\mathcal{D}$ is zero if and only if the dimension d of the target space $\mathbb{R}^{d-1,1}$ is equal to 26. In case of a compact simple Lie group G the "critical dimension condition" takes the form $c(\mathfrak{g}) = 26$. Here $\mathfrak{g}$ is the Lie algebra of G and $c(\mathfrak{g})$ depends on an integer k (parametrizing different symplectic structures on ΩG), the dimension of $\mathfrak{g}$ and a certain invariant of $\mathfrak{g}$ (the dual Coxeter number).

4. Twistor quantization of Kähler phase manifolds

The method of twistor quantization from the last Section used heavily the holomorphic projection

$$p : Z \to \mathcal{S} = \mathrm{Diff}(S^1)/S^1.$$

The existence of this projection is due to specific properties of ΩG, namely, it is essential that ΩG is a homogeneous space and the twistor space Z may be defined in terms of LG-invariant complex structures on ΩG. For a general Kähler manifold there is no hope to find a twistor bundle with a holomorphic projection of above type. We use another approach.

4.1. Twistor bundle. For a general (finite- or infinite-dimensional) Kähler manifold M we can introduce its *twistor bundle* as follows. For a finite-dimensional Kähler manifold of real dimension $2n$, we define $\pi : Z \to M$ to be the bundle of Kähler structures on M, i.e., complex structures on M compatible with symplectic structure. In other words, it is a vector bundle, associated with the Lagrangian $\mathrm{Sp}(2n, \mathbb{R})$-bundle of symplectic frames on M, with fibre $\pi^{-1}(x)$ at $x \in M$ identified with the space $\mathrm{Sp}(2n, \mathbb{R})/\mathrm{U}(n)$ of Kähler structures on T_xM.

If M is a Hilbert-space Kähler manifold modelled on a Hilbert space, we define $\pi : Z \to M$ to be the bundle of <u>restricted</u> Kähler structures on M. Its fibre at $x \in M$ is identified now with the space $\mathrm{Sp}_{res}(T_xM)/\mathrm{U}(T_xM)$ of restricted Kähler structures on T_xM where $\mathrm{Sp}_{res}(T_xM)$ is the space of linear symplectic transformations of T_xM differing from a unitary one by a Hilbert–Schmidt operator (cf. [6] for details).

In contrast with the case of ΩG, the twistor space Z for a general Kähler manifold M has no distinguished complex (or almost complex) structure. In order to define it as in Section 3.1, we need a vertical-horizontal decomposition $TZ = V \oplus H$ of the tangent bundle TZ, i.e., a symplectic connection on M. But (opposite to Riemannian case where we have a canonical Levi-Civita connection) we have no canonical choice of symplectic connection on a general Kähler manifold.

Though the twistor space Z, in general, has no distinguished almost complex structure, we still can associate with a general Kähler manifold canonically some almost complex space. Namely, we consider the *space $\mathcal{Z}$ of Kähler structures* on M which is identified with the space $\Gamma(M, Z)$ of smooth sections J of $\pi : Z \to M$ satisfying to the Nijenhuis integrability condition. A tangent vector X to $\mathcal{Z}$ at J is represented by a tangent vector field $X = \{X_z \in T_zZ : z \in J(M)\}$ to Z along the image $J(M)$ of $J : M \to Z$. Any such vector X has a vertical-horizontal decomposition $X = X^v \oplus X^h$ where X_z^v, $z \in J(M)$, is tangent to the fibre of $\pi : Z \to M$ at z and X_z^h is tangent to $J(M)$ at z.

We define a *canonical almost complex structure $\mathcal{J}$* on $\mathcal{Z}$ by setting

$$\mathcal{J}_J X = \{\mathcal{J}_z^v X_z^v \oplus z X_z^h : z \in J(M)\}$$

at $J \in \mathcal{Z}$. Here $\mathcal{J}_z^v$ is the Kähler structure on the Kähler manifold $\pi^{-1}(\pi(z))$ and z in the second summand is considered as the complex structure (defined by z) on $T_zJ(M)$ identified with $T_{\pi(z)}M$ by π_*.

4.2. Twistor quantization problem. Suppose that a Kähler manifold M is quantizable, i.e., there exists a Hermitian prequantum bundle $L \to M$ with a Hermitian connection ∇ on it. Denote by $\tilde{L} \to Z$ the pull-back of L to Z provided with the pulled-back connection $\tilde{\nabla}$.

We define a *prequantum bundle $\mathcal{L} \to \mathcal{Z}$* over $\mathcal{Z}$ with fibre at $J \in \mathcal{Z}$ given by $\mathcal{L}_J = \Gamma(J(M), \tilde{L})$. A local section σ of $\mathcal{L}$ at J is a collection $\sigma = \{\sigma_z : z \in J(M)\}$. The prequantum bundle $\mathcal{L}$ can be provided with a natural *structure of holomorphic vector bundle* over $\mathcal{Z}$. This structure is determined by $\bar{\partial}$-operator acting on local sections of $\mathcal{L} \to \mathcal{Z}$ according to the following rule. If $\sigma = \{\sigma_z : z \in J(M)\}$ is a section of $\mathcal{L}$ at J then

$$\bar{\partial}_J \sigma = \{\tilde{\nabla}_z^{(0,1)} \sigma_z : z \in J(M)\}$$

where $\tilde{\nabla}_z^{(0,1)}$ is the pull-back to Z of the $(0,1)$-component of $\nabla_{\pi(z)}$ with respect to the complex structure on $T_{\pi(z)}M$ given by z.

We define a *Fock bundle* $\mathcal{H} \to \mathcal{Z}$ over $\mathcal{Z}$ with fibre at $J \in \mathcal{Z}$ given by the Fock space $F(J(M), \tilde{L})$ of holomorphic sections σ of $\mathcal{L}$ at J. They satisfy to the equation: $\tilde{\nabla}_z^{(0,1)} \sigma_z = 0$ for $z \in J(M)$.

With this definition, we can formulate the twistor quantization problem for Kähler phase manifolds as follows: *construct a (projectively) flat unitary connection on the Fock bundle* $\mathcal{H} \to \mathcal{Z}$.

References

[1] A. Popov and A. Sergeev, *Symplectic twistors and geometric quantization of strings*, Algebraic Geometry and Its Applications, ed. by A. Tikhomirov, A. Tyurin, Vieweg Verlag, Wiesbaden, 1994, pp. 137–157.

[2] A. Popov and A. Sergeev, *Geometric quantization of string theory using twistor approach*, Quantization and Infinite-Dimensional Systems, ed. by J.-P. Antoine et al., Plenum Press, New York, 1994, pp. 43–51.

[3] J. Sniatycki, *Geometric Quantization and Quantum Mechanics*, Springer, New York, 1980.

[4] N. J. M. Woodhouse, *Geometric Quantization,* 2nd ed., Clarendon Press, Oxford, 1992.

[5] J. Scherk, Rev. Mod. Phys. **47** (1975), 123.

[6] A. Pressley and G. Segal, *Loop Groups*, Clarendon Press, Oxford, 1986.

[7] G. Segal, *Unitary representations of some infinite dimensional groups*, Commun. Math. Phys. **80** (1981), 301–342.

[8] M. J. Bowick and S. G. Rajeev, *The holomorphic geometry of closed bosonic string theory and Diff(S^1)/S^1*, Nucl. Phys. **B293** (1987), 348–384.

[9] A. A. Kirillov and D. V. Juriev, *Kähler geometry of the infinite-dimensional homogeneous space M = Diff(S^1)/Rot(S^1)*, Funct. Anal. Appl. **21** (1987), 284–294.

STEKLOV MATHEMATICAL INSTITUTE, GUBKINA 8, 117966, GSP-1, MOSCOW, RUSSIA
E-mail address: `armen@sergeev.mian.su`

Contemporary Mathematics
Volume **212**, 1998

On a Class of Integral Representations Related to the Two-Dimensional Helmholtz Operator

Michael Shapiro and Luis Manuel Tovar

ABSTRACT. For the theory of quaternion-valued functions of two real variables which is in the same relation to the 2-dimensional Helmholtz equation as usual holomorphic functions of one variable are to the Laplace equation, We establish analogues of the basic integral formulas of the complex analysis: Borel-Pompeiu's, Cauchy's, etc.

Introduction

In [17] we began constructing a theory of quaternion-valued functions (called hyperholomorphic) of two real variables which is in the same relation to the 2-dimensional Helmholtz equation as the usual one-dimensional complex analysis is to the Laplace equation in $\mathbb{R}^2$. Basically we followed the ideas of a series of work [10], [11], [12], [13] by V. V. Kravchenko and M. V. Shapiro in which a corresponding theory for functions of three real variables has been constructed. There are at least two reasons for developing the theory for two variables. First of all, phenomena described by such hyperholomorphic functions are of sufficient interest for the plane situation which requires exact formulas, theorems, and definitions. Secondly, although the general line of study of the 2-dimensional situation is quite clear, the rigorous treatment of it contains enough obstacles to justify such a work. In fact, the mere existence of a good multiplicative structure (that of complex numbers) in the plane separate the 3-dimensional case from the 2-dimensional one making the latter interesting in itself. Some speculations on this point can be found in [17].

It is worth mentioning also that until now there have been published quite a lot of articles devoted to the 2-dimensional Helmholtz operator, hence it can be expected that our theory will find its own applications and usage (some recent work on the two-dimensional Helmholtz equation are [4], [5], [7], [16].

One more motivating factor is that one can consider the 2-dimensional case as a good model for constructing the Clifford analysis related to the Helmholtz equation

1991 *Mathematics Subject Classification.* Primary 30G35, 30J05; Secondary 35J67.

This work was partially supported by the Project 21085-5-4069E of the CONACyT (México) and by the Comisión de Operación y Fomento a las Actividades Académicas del I. P. N. (México). We are grateful to the referee whose criticism helped us to amend the work.

with a complex or Clifford wave number. For the pioneering work in that direction we refer the reader to [2], [3], [9], [14], [21].

We have chosen to make both [17] and the present article self-contained, i.e., to give all the proofs directly, without referring to the 3-dimensional case. In our opinion, although this would be possible but in most cases it is not convenient nor easier to extract from the 3-dimensional theory what we need for the 2-dimensional situation. We have also taken into account that the 3-dimensional theory itself is not too well known.

The article is organized as follows. In Section 1 some basic facts about quaternions are described. A brief account of what has been done in [17] is contained in Section 2. We give there the definition of a hyperholomorphic function of two real variables, which necessarily satisfies the Helmholtz equation. In particular, when the latter equation degenerates to the Laplace equation, the notion of a hyperholomorphic function reduces to that of a common complex holomorphic function. One more fine point arises here: a fundamental solution θ_α of the Helmholtz operator Δ_{α^2} is expressed in terms of Hankel's functions, which differs essentially from the 3-dimensional case, where a corresponding fundamental solution contains the exponential function. That is why our hyperholomorphic Cauchy kernel, a fundamental solution to the hyperholomorphic Cauchy-Riemann operator (differential operator defining a class of hyperholomorphic functions), contains also Hankel's functions of different kinds.

Sections 3, 4, and 5 are central in the work. In Section 3 we give, first, a version of the usual Green's formula which includes the hyperholomorphic Cauchy-Riemann operator and its conjugate. This implies immediately various forms of the Cauchy integral theorem. It should be noted that, in contrast to the complex holomorphic situation, this theorem includes both a curvilinear and an area integral.

Sections 4 and 5 present analogues, for our theory, of the principal integral formulas of complex analysis based on the Cauchy kernel.

Section 4 describes the case of complex values of the parameter α. To do the same for a complex-quaternionic parameter we are forced to consider quite complicated defining expressions (see Subsection 5.1) which include integrals with kernels generated by Hankel's functions introduced in Subsection 3.6.

1. Algebras of real and complex quaternions

1.1. We shall denote by $\mathbb{H}(\mathbb{R})$ and by $\mathbb{H}(\mathbb{C})$ the sets of real and complex quaternions (=biquaternions) correspondingly. This means that each quaternion a is represented in the form $a = \sum_{k=0}^{3} a_k i_k$ where $\{a_k\} \subset \mathbb{R}$ for real quaternions and $\{a_k\} \subset \mathbb{C}$ for complex quaternions, i_0 is the multiplicative unit and $\{i_k \mid k \in \mathbb{N}_3\}$ are the quaternionic imaginary units, that is, the standard basis elements possessing the following properties:

$$i_0^2 = i_0 = -i_k^2, \quad i_0 i_k = i_k i_0 = i_k, \quad k \in \mathbb{N}_3,$$

(1.1) $$i_1 i_2 = -i_2 i_1 = i_3; \quad i_2 i_3 = -i_3 i_2 = i_1; \quad i_3 i_1 = -i_1 i_3 = i_2,$$

where $\mathbb{N}_p := \{1, 2, \dots, p\}$, $\mathbb{N}_p^0 := \mathbb{N}_p \cup \{0\}$.

We denote the imaginary unit in $\mathbb{C}$ by i as usual. By definition

(1.2) $$i \cdot i_k = i_k \cdot i, \quad k \in \mathbb{N}_3^0.$$

Mostly we will be in need of the set $\mathbb{H}(\mathbb{C})$ of complex quaternions which by the multiplication laws (1.1), (1.2) and by the natural component-wise operation of addition is a complex non-commutative, associative algebra with zero divisors. The set $\mathbb{H}(\mathbb{R})$ is a real, non-commutative, associative algebra without zero divisors.

1.2. Quaternions have the following vector representation: $a = a_0 + \vec{a}$, where $\vec{a} := \sum_{k=1}^{3} a_k i_k$. a_0 will be called the scalar part and $\vec{a}$ the vector part of a complex quaternion $a : a_0 =: \mathrm{Sc}(a)$, $\vec{a} =: \mathrm{Vec}(a)$. Complex quaternion for which $a = \vec{a}$ will be called purely vectorial.

1.3. We will be in need of two different conjugations on $\mathbb{H}(\mathbb{C})$. For $a \in \mathbb{H}(\mathbb{C})$, the usual complex conjugation is defined by

$$Z_{\mathbb{C}}(a) := a^* = \mathrm{Re}(a) - i\mathrm{Im}(a) := \sum_{k=0}^{3} \mathrm{Re}(a_k)i_k - i\sum_{k=0}^{3}\mathrm{Im}(a_k)i_k$$

and the quaternionic conjugation by

$$Z_{\mathbb{H}}(a) := \bar{a} = a_0 - \vec{a}.$$

Further properties of quaternions can be found in [8] for instance.

1.4. Let us denote by $\mathfrak{G}$ the set of zero divisors from $\mathbb{H}(\mathbb{C})$, and let $G\mathbb{H}(\mathbb{C})$ be the subset of invertible elements from $\mathbb{H}(\mathbb{C})$.

As was proved in [10], the following conditions are equivalent:
1. $a \in \mathfrak{G}$.
2. $a\bar{a} = 0$.
3. $a_0^2 = \vec{a}^2$.
4. $a^2 = 2a_0 a = 2\vec{a}a$.

2. Two-dimensional Helmholtz operator and hyperholomorphic functions of two real variables

2.1. We give in this section a brief account of [17]. All proofs and details can be found there.

We consider $\mathbb{H}(\mathbb{C})$-valued functions defined in a domain $\Omega \subset \mathbb{R}^2$. On the left $\mathbb{H}(\mathbb{C})$-module $C^2(\Omega; \mathbb{H}(\mathbb{C}))$ we introduce the two-dimensional Helmholtz operator with a quaternionic wave number:

(2.1) $$\Delta_\lambda := \Delta_{\mathbb{R}^2} + M^\lambda,$$

where $\Delta_{\mathbb{R}^2} := \partial_1^2 + \partial_2^2$, $\partial_k := \partial/\partial x_k$, $M^\lambda[f] := f\lambda, \lambda \in \mathbb{H}(\mathbb{C})$.

2.2. The well-known operators

(2.2) $$\bar{\partial} := \partial_1 + i\partial_2$$

and

(2.3) $$\partial := \partial_1 - i\partial_2$$

define very important classes in the plane: respectively, these of holomorphic and anti-holomorphic functions. As is known, the following factorization holds:

(2.4) $$\partial \cdot \bar{\partial} = \Delta_{\mathbb{R}^2},$$

which is, in fact, the base of holomorphic function theory. We will obtain the analogue of (2.4) for the operator (2.1). This requires introducing some analogues of (2.2) and (2.3) taking into account the peculiarity of the quaternionic situation.

Let $\psi := \{\psi^1, \psi^2\} \subset \mathbb{H}(\mathbb{R}) \times \mathbb{H}(\mathbb{R})$, denote $\bar{\psi} := \{\overline{\psi^1}, \overline{\psi^2}\}$.

Let

$$(2.5) \qquad {}_\psi\partial := \psi^1 \partial_1 + \psi^2 \partial_2, \quad \partial_\psi := \partial_1 \psi^1 + \partial_2 \psi^2.$$

Then it can be easily verified (compare with [15], [19]) that the equalities

$$(2.6) \qquad \partial_\psi \cdot \partial_{\bar{\psi}} = \partial_{\bar{\psi}} \cdot \partial_\psi = \Delta_{\mathbb{R}^2}, \quad {}_\psi\partial \cdot_{\bar{\psi}} \partial =_{\bar{\psi}} \partial \cdot_\psi \partial = \Delta_{\mathbb{R}^2}$$

hold if and only if,

$$(2.7) \qquad \psi^j \cdot \overline{\psi^k} + \psi^k \cdot \overline{\psi^j} = 2\delta_{jk},$$

for any j, k from $\mathbb{N}_2$, δ_{jk} being the Kroneker delta. Following [15], [19] we shall call ψ with the property (2.7) a "structural set". It is evident that ψ and $\bar{\psi}$ are structural sets simultaneously.

Let us note that for purely vectorial ψ the factorization (2.6) reduces to

$$(2.8) \qquad {}_\psi\partial^2 = \partial_\psi^2 = -\Delta_{\mathbb{R}^2},$$

which is paradoxically different from that of (2.4) for many purposes.

2.3. Let $\alpha^2 = \lambda$, $\alpha \in \mathbb{H}(\mathbb{C})$. Let also ψ consist of purely vectorial quaternions: $\mathrm{Sc}(\psi^1) = \mathrm{Sc}(\psi^2) = 0$. We introduce a "metaharmonic generalization" of the operators (2.2) and (2.3).

Let

$$_\alpha\partial_\psi := \partial_\psi +^\alpha M \quad \text{and} \quad {}_\psi\partial_\alpha :=_\psi \partial + M^\alpha.$$

Then (2.6) implies the following factorizations of the Helmholtz operator:

$$(2.9) \qquad \Delta_\lambda = -_\psi\partial_\alpha \cdot_\psi \partial_{-\alpha} = -_\psi\partial_{-\alpha} \cdot_\psi \partial_\alpha.$$

We will call λ-metaharmonic functions, the elements belonging to $\ker \Delta_\lambda$.

It is of interest to mention explicitly that for $\lambda = \lambda_0 \in \mathbb{C}$, apart from the complex square roots, we have also a family of purely imaginary complex quaternions $\vec{\alpha}$ satisfying the condition $-\alpha_1^2 - \alpha_2^2 - \alpha_3^2 = \lambda_0$. In particular for $\lambda = 0$ we have a factorization of the Laplace operator other than (2.4): when $\alpha \in \mathfrak{G}$ and $\alpha_0 = 0$ then

$$\Delta_{\mathbb{R}^2} = -_\psi\partial_{\vec{\alpha}} \cdot_\psi \partial_{-\vec{\alpha}} = -_\psi\partial_{-\vec{\alpha}} \cdot_\psi \partial_{\vec{\alpha}}.$$

2.4. It is a well known result (see e.g. [20]) that if $\lambda = \alpha^2 \in \mathbb{C}$, a fundamental solution θ_α of Δ_λ is given by

$$\theta_\alpha(a) = \begin{cases} \dfrac{-i}{4} H_0^1(\alpha|a|), & \alpha \neq 0, \\[2mm] \dfrac{1}{2\pi} \ln |a|, & \alpha = 0, \end{cases}$$

where $a = (x, y)$ and H_0^1 is the Hankel function of the first kind of order zero. For our purposes we just need that H_0^1 satisfies the relation

$$\frac{d}{dt} H_0^1(t) = -H_1^1(t),$$

with H_1^1 the Hankel function of the first kind of order one. We recall that for $s = 0, 1$

$$H_s^1 := I_s + iN_s$$

where I_s is the Bessel function of the first kind

$$I_s(r) = \frac{r^s}{2^s \Gamma(1/2)\Gamma(s+1/2)} \int_0^\pi e^{ir\cos\theta} \sin^{2s}\theta d\theta,$$

Γ is the gamma function, and N_s is the Bessel function of the second kind

$$N_s(r) = \frac{I_s(r)\cos s\pi - I_{-s}(r)}{\sin s\pi}$$

For general properties of Hankel functions see [6], [20].

2.5. THEOREM. *Let $0 \neq \alpha \in \mathbb{H}(\mathbb{C})$. Then a fundamental solution θ_α to the operator $\Delta_{\alpha^2} = \Delta + M^{\alpha^2}$ evaluated at $a \in \mathbb{R}^2 \backslash \{0\}$ is given by:*

1. *If $\alpha^2 \notin \mathfrak{G}$, $\vec{\alpha}^2 \neq 0$, let $\gamma := \sqrt{\vec{\alpha}^2}$ and $\xi_\pm := \alpha_0 \pm \gamma$. Then,*

$$\theta_\alpha(a) = \frac{1}{2\gamma}(\theta_{\xi_+}(a)(\gamma + \vec{\alpha}) + \theta_{\xi_-}(a)(\gamma - \vec{\alpha}))$$

$$= \begin{cases} \dfrac{-i}{8\gamma}[H_0^1((\alpha_0 + \gamma)|a|)(\gamma + \vec{\alpha}) + H_0^1((\alpha_0 - \gamma)|a|)(\gamma - \vec{\alpha})], & \xi_\pm \neq 0, \\[3mm] \dfrac{1}{4\pi\gamma}\ln|a|(\gamma + \alpha) - \dfrac{i}{8\gamma}H_0^1((\alpha_0 - \gamma)|a|)(\gamma - \vec{a}), & \xi_+ = 0, \\[3mm] -\dfrac{i}{8\gamma}H_0^1(\alpha_0 + \gamma)|a|)(\gamma + \vec{\alpha}) + \dfrac{1}{4\pi\gamma}\ln|a|(\gamma - \vec{\alpha}), & \xi_- = 0. \end{cases}$$

2. *If $\alpha^2 \notin \mathfrak{G}$, $\vec{\alpha}^2 = 0$ then*

$$\theta_\alpha(a) = \theta_{\alpha_0}(a) + \frac{\partial}{\partial\alpha_0}[\theta_{\alpha_0}(a)]\vec{\alpha}$$

$$= \frac{-i}{4}[H_0^1(\alpha_0|a|) - H_1^1(\alpha_0|a|)|a|\vec{\alpha}].$$

3. *If $\alpha^2 \in \mathfrak{G}$ then*

$$\theta_\alpha(a) = \frac{1}{2\alpha_0}(\theta_{2\alpha_0}(a) \cdot \alpha + \theta_0(a) \cdot \vec{\alpha})$$

$$= \frac{-i}{8\alpha_0}H_0^1(2\alpha_0|a|)\alpha + \frac{1}{4\alpha_0\pi}\ln|a|\vec{\alpha}.$$

2.6. Abbreviate

$$\psi_{st} := \psi_{i_1, i_2}, \qquad {}_{st}\partial := \psi_{st}\partial, \qquad \partial_\alpha := {}_{st}\partial_\alpha,$$

and denote the set of (ψ, α)-hyperholomorphic functions by ${}^\psi\mathfrak{M}_\alpha$:

$$^\psi\mathfrak{M}_\alpha := {}^\psi\mathfrak{M}_\alpha(\Omega, \mathbb{H}(\mathbb{C})) := \ker {}_\psi\partial_\alpha.$$

Note that relations (2.11) mean that each hyperholomorphic function in $C^2(\Omega, \mathbb{H}(\mathbb{C}))$ is λ-metaharmonic and it follows immediately that if $\alpha \in \mathbb{H}(\mathbb{R})$ then

$$^\psi\mathfrak{M}_\alpha(\Omega, \mathbb{H}(\mathbb{C})) = {}^\psi\mathfrak{M}_\alpha(\Omega, \mathbb{H}(\mathbb{R})) \pm i^\psi\mathfrak{M}_\alpha(\Omega, \mathbb{H}(\mathbb{R})).$$

We define in a similar way $_\alpha\mathfrak{M}^\psi$ as

$$_\alpha\mathfrak{M}^\psi := \ker_\alpha \partial_\psi.$$

2.7. For the variable $a = (x, y)$ we define the differential form $d_\psi a$ as:

$$d_\psi a := \psi^1 dy - \psi^2 dx.$$

Let $d\ell$ be the length element in $\mathbb{R}^2$, thus $|d_\psi a| = d\ell$. If Γ is a smooth curve then

$$d_\psi a = \psi^2 n_1(a)d\ell + \psi^1 n_2(a)d\ell = n_\psi(a)d\ell$$

where $n = (n_1, n_2)$ is a unit vector of the outward normal to Γ at the point $a \in \Gamma$, $n_\psi := \psi^1 n_1 + \psi^2 n_2$.

Consider $\{f, g\} \subset C^1(\bar{\Omega}, \mathbb{H}(\mathbb{C}))$. If we apply the operator d to the differential form $g \cdot d_\psi a \cdot f$ we obtain

$$d(g \cdot d_\psi a \cdot f) = dg \wedge d_\psi a \cdot f - g d_\psi a \wedge df$$
$$= (\partial_\psi[g] \cdot f + g_\psi \partial[f])dx \wedge dy.$$

It follows that

$$(2.10) \qquad d(g \cdot d_\psi a \cdot f) = \begin{cases} (_\alpha\partial_\psi[g] \cdot f + g_\psi \partial_\alpha[f] - (\alpha g f + g f \alpha))dx \wedge dy, \\ (_\alpha\partial_\psi[g] \cdot f - g_{\bar\psi}\partial_\alpha[f] - (\alpha g f - g f \alpha))dx \wedge dy, \\ (g_\psi \partial_\alpha[f] -_\alpha \partial_\psi[g]f - (g f \alpha - \alpha g f)dx \wedge dy. \end{cases}$$

In particular if $\alpha \in \mathbb{C}$ these formulas are simpler:

$$(2.11) \qquad d(g \cdot d_\psi a \cdot f) = \begin{cases} (_\alpha\partial_\psi[g] \cdot f + g_\psi \partial_\alpha[f] - 2\alpha g f)dx \wedge dy, \\ (_\alpha\partial_\psi[g] \cdot f - g_{\bar\psi}\partial_\alpha[f])dx \wedge dy, \\ (g_\psi \partial_\alpha[f] -_\alpha \partial_{\bar\psi}[g]f)dx \wedge dy. \end{cases}$$

2.8. The following hyperholomorphic function plays an important role as an analogue of the "Cauchy kernel" in one-dimensional complex analysis:

$$(2.12) \qquad \mathcal{K}_{\psi,\alpha} := \begin{cases} -_\psi\partial_{-\alpha}[\theta_\alpha], & \alpha \in G\mathbb{H}(\mathbb{C}) \text{ or } \alpha \in \mathfrak{G} \text{ but } \alpha_0 \neq 0, \\ -_\psi\partial_{-\alpha}[\theta_0], & \alpha \in \mathfrak{G} \text{ and } \alpha_0 = 0, \\ -_\psi\partial[\theta_0], & \alpha = 0. \end{cases}$$

2.9. THEOREM. (*Fundamental solution to the operator* $_\psi\partial_\alpha$). *Let* $\alpha \in \mathbb{H}(\mathbb{C})$, $a = (x, y) \in \mathbb{R}^2\backslash\{0\}$, *and* $P^\pm := (1/2\gamma)M^{\gamma\pm\bar\alpha}$. *Then a fundamental solution* $\mathcal{K}_{\psi,\alpha}$ *to the operator* $_\psi\partial_\alpha$ *is given by the formulas:*
   1. *If* $0 \neq \alpha = \alpha_0 \in \mathbb{C}$, *then*

$$(2.13) \qquad \mathcal{K}_{\psi,\alpha}(a) = \begin{cases} \dfrac{-i}{4}\alpha[H_1^1(\alpha|a|)\dfrac{a_\psi}{|a|} - H_0^1(\alpha|a|)], & \alpha \neq 0, \\ \dfrac{-a_\psi}{2\pi|a|^2}, & \alpha = 0, \end{cases}$$

where $a_\psi = \psi^1 x + \psi^2 y$.

2. *If $\alpha \notin \mathfrak{G}$, $\vec{\alpha}^2 \neq 0$, then*

(2.14)

$$\mathcal{K}_{\psi,\alpha}(a)$$

$$= P^+[\mathcal{K}_{\psi,\xi_+}](a) + P^-[\mathcal{K}_{\psi,\xi_-}](a)$$

$$= \begin{cases} \dfrac{-i}{8\gamma}\{\xi_+[H_1^1(\xi_+|a|)\dfrac{a_\psi}{|a|} - H_0^1(\xi_+|a|)](\gamma + \vec{\alpha}) + \xi_-[H_1^1(\xi_-|a|)\dfrac{a_\psi}{|a|} \\ \qquad - H_0^1(\xi_-|a|)](\gamma - \vec{\alpha})\}, & \xi_\pm \neq 0, \\[2ex] \dfrac{-a_\psi}{4\pi\gamma|a|^2}(\gamma + \vec{\alpha}) - \dfrac{i\xi_-}{8\gamma}[H_1^1(\xi_-|a|)\dfrac{a_\psi}{|a|} - H_0^1(\xi_-|a|)](\gamma - \vec{\alpha}), & \xi_+ = 0, \\[2ex] \dfrac{-i\xi_+}{8\gamma}[H_1^1(\xi_+|a|)\dfrac{a_\psi}{|a|} - H_0^1(\xi_+|a|)](\gamma + \vec{\alpha}) - \dfrac{a_\psi}{4\pi\gamma|a|^2}(\gamma - \vec{\alpha}), & \xi_- = 0 \end{cases}$$

where $\mathcal{K}_{\psi,\xi_\pm}(a)$ are the fundamental solutions to the operators ${}_\psi\partial_{\xi_\pm}$ with the complex parameters $\xi_\pm$ defined as before.

3. *If $\alpha \notin \mathfrak{G}$, $\vec{\alpha}^2 = 0$, then*

(2.15)

$$\mathcal{K}_{\psi,\alpha} = \mathcal{K}_{\psi,\alpha_0} + \frac{\partial}{\partial \alpha_0}[\mathcal{K}_{\psi,\alpha_0}]\vec{\alpha}$$

$$= \frac{-i}{4}\left\{\alpha\left[H_1^1(\alpha_0|a|)\frac{a_\psi}{|a|} - H_0^1(\alpha_0|a|)\right] + \alpha_0\frac{\partial}{\partial \alpha_0}\left[H_1^1(\alpha_0|a|)\frac{a_\psi}{|a|}\right.\right.$$

$$\left.\left. - H_0^1(\alpha_0|a|)\right]\vec{\alpha}\right\}.$$

4. *If $\alpha \in \mathfrak{G}$, $\alpha_0 \neq 0$, then*

(2.16)

$$\mathcal{K}_{\psi,\alpha} = P^+[\mathcal{K}_{\psi,2\alpha_0}] + P^-[\mathcal{K}_{\psi,0}]$$

$$= \frac{-i}{4\gamma}\alpha_0\left[H_1^1(2\alpha_0|a|)\frac{a_\psi}{|a|} - H_0^1(2\alpha_0|a|)\right](\gamma + \vec{\alpha}) - \frac{a_\psi}{2\pi\gamma|a|^2}(\gamma - \vec{\alpha}).$$

5. *If $\alpha \in \mathfrak{G}$, $\alpha_0 = 0$, then*

(2.17)

$$\mathcal{K}_{\psi,\alpha} = \mathcal{K}_{\psi,0} + \theta_0\alpha = \frac{1}{2\pi}\left[\ln|a| - \frac{a_\psi}{|a|}\alpha\right].$$

2.10. Let us remark finally that through the operator ${}_\psi\partial_{\xi_\pm}$ we get a very practical representation for ${}_\psi\partial_\alpha$ if $\vec{\alpha}^2 \neq 0$:

(2.18)

$$_\psi\partial_\alpha = P^+ \cdot {}_\psi\partial_{\xi_+} + P^- \cdot {}_\psi\partial_{\xi_-}.$$

Now if $\alpha \in \mathfrak{G}$, we have $\vec{\alpha}^2 \neq 0 \iff \alpha_0 \neq 0$, so $\xi_+ = 2\alpha_0$, $\xi_- = 0$ therefore ${}_\psi\partial_\alpha$ takes the form

$$_\psi\partial_\alpha = P^+ \cdot {}_\psi\partial_{2\alpha_0} + P^- \cdot {}_{st}\partial.$$

3. Two-dimensional Helmoltz Operator and integral formulas for hyperholomorphic functions

3.1. We are going to give the quaternionic version of Green's formula.

THEOREM. (Green's Formula): *Let* $f, g \in C^1(\bar{\Omega}, \mathbb{H}(\mathbb{C}))$, ψ *a structural set and* $\gamma_0, \gamma_1, \ldots, \gamma_n$ *a system of* $(n+1)$ *closed rectifiable Jordan curves in* $\bar{\Omega}$ *satisfying the following conditions:*

a) $\gamma_i \subset E(\gamma_j)$ *if* $i \neq j$, $i, j = 1, \ldots, n$,

b) $\gamma_i \subset I(\gamma_0)$,

c) Ω *contains the multiply connected domain*

$$D := I(\gamma_0) \backslash (\bigcup_{i=1}^{n} \overline{I(\gamma_i)}),$$

with boundary $\gamma_0 \cup \gamma_1 \cup \cdots \cup \gamma_n$, *then:*

$$(3.1) \quad \int_{\gamma_0} g \cdot d_\psi \cdot f - \sum_{i=1}^{n} \int_{\gamma_i} g \cdot d_\psi \cdot f$$
$$= \begin{cases} \int_D \,_\alpha\partial_\psi[g] \cdot f + g_\psi\partial_\alpha[f] - (\alpha gf + gf\alpha)dx \wedge dy, \\ \int_D \,_\alpha\partial_\psi[g]f - g \,_{\bar{\psi}}\partial_\alpha[f] - (\alpha gf - gf\alpha)dx \wedge dy, \\ \int_D g_\psi\partial_\alpha[f] - \alpha\partial_{\bar{\psi}}[g]f - (gf\alpha - \alpha gf)dx \wedge dy, \end{cases}$$

where $I(\gamma_j)$ *and* $E(\gamma_j)$ *denote the interior and exterior regions given by the Jordan curve theorem.*

PROOF. Is a direct consequence of (2.10) and (2.11).

3.2. COROLLARY. *Under the same conditions of the preceeding theorem if* $\alpha = 0$ *and* ψ *is the standard structural set we obtain*

$$\int_{\gamma_0} g \cdot d_{st}a \cdot f - \sum_{i=1}^{n} \int_{\gamma_i} g \cdot d_{st}a \cdot f = \int_D (\partial_{st}[g]f + g_{st}\partial[f])dx \wedge dy.$$

3.3. THEOREM. (Quaternionic Cauchy integral theorem for conjugate classes of hyperholomorphy). *If* $g \in_\alpha \mathfrak{M}^{\bar{\psi}}(\bar{\Omega}, \mathbb{H}(\mathbb{C}))$, $f \in^\psi \mathfrak{M}_\alpha(\bar{\Omega}, \mathbb{H}(\mathbb{C}))$, ψ *a structural set and* $\gamma_0, \gamma_1, \ldots, \gamma_n$ *is a system of* $(n+1)$ *closed rectifiable Jordan curves satisfying conditions* a), b) c) *of Green's Theorem, then*

$$\int_{\gamma_0} g \cdot d_\psi a \cdot f - \sum_{i=1}^{n} \int_{\gamma_j} g \cdot d_\psi a \cdot f = \int_D (\alpha gf - gf\alpha)dx \wedge dy.$$

In particular if $\alpha \in \mathbb{C}$,

$$\int_\gamma g \cdot d_\psi a \cdot f = \sum_{i=1}^{n} \int_{\gamma_j} g \cdot d_\psi a \cdot f.$$

PROOF. Follows directly from Green's Theorem.

3.4. THEOREM. (Quaternionic Cauchy integral theorem for a pair of (ψ, α)-hyperholomorphic functions). *Let* $g \in_\alpha \mathfrak{M}^\psi(\bar{\Omega}, \mathbb{H}(\mathbb{C}))$, $f \in^\psi \mathfrak{M}_\alpha(\bar{\Omega}, \mathbb{H}(\mathbb{C}))$, ψ *a structural set and* $\gamma_0, \gamma_1, \ldots, \gamma_n$, *a system of* $(n+1)$ *curves as in the preceeding theorems; then*

$$\int_{\gamma_0} g \cdot d_\psi a \cdot f - \sum_{i=1}^{n} \int_{\gamma_i} g \cdot d_\psi a \cdot f = -\int_D (\alpha gf + gf\alpha)dx \wedge dy,$$

if $\alpha \in \mathbb{C}$ then this formula reduces to

$$\int_{\gamma_0} g \cdot d_\psi a \cdot f - \sum_{i=1}^n \int_{\gamma_i} g \cdot d_\psi a \cdot f = -2\alpha \int_D g \cdot f dx \wedge dy.$$

3.5. Observe that if $\alpha = 0$, then $\mathfrak{M}^\psi = \mathfrak{M}^{\bar\psi}$, so Theorems 3.3 and 3.4 give the same result. Although constants are not (ψ, α)-hyperholomorphic functions so Cauchy integral theorem for one function is not an immediate consequence of the preceeding theorems, this theorem follows from the first equality in Green's Theorem:

THEOREM. (Quaternionic Cauchy integral theorem for a hyperholomorphic function). *If $f \in^\psi \mathfrak{M}_\alpha(\bar\Omega, \mathbb{H}(\mathbb{C}))$ and $g \in_\alpha \mathfrak{M}^\psi(\bar\Omega, \mathbb{H}(\mathbb{C}))$ and $\gamma_0, \gamma_1, \ldots, \gamma_n$ is a system of $(n+1)$ curves satisfying the conditions of the Green Theorem then*

$$\int_{\gamma_0} d_\psi a \cdot f = \sum_{i=1}^n \int_{\gamma_i} d_\psi a \cdot f - \int_D f \cdot \alpha dx \wedge dy,$$

$$\int_{\gamma_0} g \cdot d_\psi a = \sum_{i=1}^n \int_{\gamma_i} g \cdot d_\psi a - \alpha \int_D g dx \wedge dy.$$

3.6. We generalize three classical one-dimensional complex analysis operators: the Cauchy-type operator, the T-operator and the operator of singular integration.

Consider the Cauchy kernel $\mathcal{K}_{\psi,\alpha_0}$ with $\alpha_0 \in \mathbb{C}$ given by the formula (2.13). Let $^\psi T_{\alpha_0}$, $^\psi K_{\alpha_0}$ and $^\psi S_{\alpha_0}$ be the operators defined by the formulas:

$$^\psi T_{\alpha_0}[f](x,y) := \int_\Omega \mathcal{K}_{\psi,\alpha_0}(x-u, y-v)f(u,v)du \wedge dv, \quad (x,y) \in \mathbb{R}^2,$$

$$^\psi K_{\alpha_0}[f](x,y) := -\int_\Gamma \mathcal{K}_{\psi,\alpha_0}(x-u, y-v)n_\psi(u,v)f(u,v)d\Gamma_{(u,v)}, \quad (x,y) \in \mathbb{R}^2 \backslash \Gamma,$$

$$^\psi S_{\alpha_0}[f](t_1,t_2) := -2\int_\Gamma \mathcal{K}_{\psi,\alpha_0}(t_1-\tau_1, t_2-\tau_2)n_\psi(\tau_1,\tau_2)f(\tau_1,\tau_2)d\Gamma_{(\tau_1,\tau_2)},$$
$$(t_1,t_2) \in \Gamma,$$

where Ω is a domain in $\mathbb{R}^2$ with boundary given by a closed rectifiable Jordan curve Γ. The last integral exists in the sense of the Cauchy principal value.

Using the explicit formulas for $\mathcal{K}_{\psi,\alpha_0}$, we get: for $(x,y) \in \mathbb{R}^2$:

$$^\psi T_{\alpha_0}[f](x,y) = \begin{cases} \dfrac{-i}{4}\displaystyle\int_\Omega \alpha_0[H_1^1(\alpha_0|(x-u, y-v)|)\dfrac{(x-u, y-v)_\psi}{|(x-u, y-v)|} \\ \qquad -H_0^1(\alpha_0|(x-u, y-v)|)]f(u,v)du \wedge dv, & \alpha_0 \neq 0, \\ \dfrac{1}{2\pi}\displaystyle\int_\Omega \dfrac{(x-u, y-v)_\psi}{|(x-u, y-v)|^2}f(u,v)du \wedge dv, & \alpha_0 = 0, \end{cases}$$

for $(x,y) \in \mathbb{R}^2 \backslash \Gamma$:

$$^\psi K_{\alpha_0}[f](x,y) = \begin{cases} \dfrac{i}{4}\displaystyle\int_\Gamma \alpha_0[H^1(\alpha_0|(x-u, y-v)|)\dfrac{(x-u, y-v)_\psi}{|(x-u, y-v)|} \\ \qquad -H_0^1(\alpha_0|(x-u, y-v)|)]n_\psi(u,v)f(u,v)d\Gamma_{(u,v)}, & \alpha_0 \neq 0, \\ \dfrac{1}{2\pi}\displaystyle\int_\Gamma \dfrac{(x-u, y-v)_\psi}{|(x-u, y-v)|^2}n_\psi(u,v)f(u,v)d\Gamma_{(u,v)}, & \alpha_0 = 0, \end{cases}$$

for $(t_1, t_2) \in \Gamma$:

$$
{}^{\psi}S_{\alpha_0}[f](t_1, t_2)
$$

$$
=
\begin{cases}
\dfrac{i}{2} \displaystyle\int_{\Gamma} \alpha_0 [H^1(\alpha_0|(t_1 - \tau_1, t_2 - \tau_2)|) \dfrac{(t_1 - \tau_1, t_2 - \tau_2)_{\psi}}{|(t_1 - \tau_1, t_2 - \tau_2)|} \\
\qquad - H_0^1(\alpha_0|(t_1 - \tau_1, t_2 - \tau_2)|)] n_{\psi}(\tau_1, \tau_2) f(\tau_1, \tau_2) d\Gamma_{(\tau_1, \tau_2)}, \quad \alpha_0 \neq 0, \\[2mm]
\dfrac{1}{\pi} \displaystyle\int_{\Gamma} \dfrac{(t_1 - \tau_1, t_2 - \tau_2)_{\psi}}{|(t_1 - \tau_1, t_2 - \tau_2)|^2} n_{\psi}(\tau_1, \tau_2) f(\tau_1, \tau_2) d\Gamma_{(\tau_1, \tau_2)}, \quad\quad\quad \alpha_0 = 0.
\end{cases}
$$

4. Main integral formulas for a complex parameter α

4.1. THEOREM. (Quaternionic Borel-Pompeiu formula for a complex parameter α). *Let Ω be a domain in $\mathbb{R}^2$ and Γ a closed rectifiable Jordan curve such that $\partial\Omega = \Gamma$. Let $\alpha_0 \in \mathbb{C}$ and $f \in C^1(\Omega, \mathbb{H}(\mathbb{C})) \cap C(\bar{\Omega}, \mathbb{H}(\mathbb{C}))$. Then*

$$(4.1) \qquad f(x, y) = {}^{\psi}K_{\alpha_0}[f](x, y) + {}^{\psi}T_{\alpha_0} \cdot {}_{\psi}\partial_{\alpha_0}[f](x, y), \quad (x, y) \in I(\Gamma).$$

PROOF. By definition of ${}^{\psi}T_{\alpha_0}$, we have: let $w = (u, v)$ and $z = (x, y)$ then

$$
{}^{\psi}T_{\alpha_0} \cdot {}_{\psi}\partial_{\alpha_0}[f](x, y) = \int_{I(\Gamma)} \mathcal{K}_{\psi,\alpha_0}(x - u, y - v) {}_{\psi}\partial_{\alpha_0,w}[f](u, v) du \wedge dv
$$

$$
= \lim_{\epsilon \to 0} \int_{\theta_\epsilon} \mathcal{K}_{\psi,\alpha_0}(x - u, y - v) {}_{\psi}\partial_{\alpha_0,w}[f](u, v) du \wedge dv.
$$

The subindex w in ${}_{\psi}\partial_{\alpha,w}$ means differentiation with respect to w and $\theta_\epsilon :=$ $I(\Gamma) \backslash \{(u, v) \| (x - u, y - v)| \leq \epsilon\}$. Now

$$
\int_{\theta_\epsilon} \mathcal{K}_{\psi,\alpha_0}(x - u, y - v) {}_{\psi}\partial_{\alpha_0,w}[f](u, v) du \wedge dv
$$

$$
= \int_{\theta_\epsilon} \mathcal{K}_{\psi,\alpha_0}(x - u, y - v) {}_{\psi}\partial_w[f](u, v) du \wedge dv
$$

$$
+ \alpha_0 \int_{\theta_\epsilon} \mathcal{K}_{\psi,\alpha_0}(x - u, y - v) f(u, v) du \wedge dv
$$

$$
= \int_{\theta_\epsilon} \mathcal{K}_{\psi,\alpha_0}(x - u, y - v) {}_{\psi}\partial_w[f](u, v) du \wedge dv
$$

$$
+ \alpha_0 \int_{\theta_\epsilon} \mathcal{K}_{\psi,\alpha_0}(x - u, y - v) f(u, v) du \wedge dv
$$

$$
- \int_{\theta_\epsilon} \partial_{\psi,w}[\mathcal{K}_{\psi,\alpha_0}(x - u, y - v)] f(u, v)
$$

$$
+ \int_{\theta_\epsilon} \partial_{\psi,w}[\mathcal{K}_{\psi,\alpha_0}(x - u, y - v) f(u, v)] du \wedge dv.
$$

Using

$$
-\partial_{\psi,w}[\mathcal{K}_{\psi,\alpha_0}(x - u, y - v)] = -{}_{\psi}\partial_w[\mathcal{K}_{\psi,\alpha_0}(x - u, y - v)]
$$

$$
= {}_{\psi}\partial_z[\mathcal{K}_{\psi,\alpha_0}(x - u, y - v)]
$$

and applying Green's formula (3.1) we have

$$\int_{\theta_\epsilon} \mathcal{K}_{\psi,\alpha_0}(x - u, y - v)_\psi \partial_{\alpha_0,w}[f](u,v) du \wedge dv$$

$$= \int_{\theta_\epsilon} \mathcal{K}_{\psi,\alpha_0}(x - u, y - v)_\psi \partial_w[f(u,v)]$$

$$+ \partial_{\psi,w}[\mathcal{K}_{\psi,\alpha_0}(x - u, y - v)] f(u,v) du \wedge dv$$

$$+ \int_{\theta_\epsilon} (\alpha_0 \mathcal{K}_{\psi,\alpha_0}(x - u, y - v) +_\psi \partial_z[\mathcal{K}_{\psi,\alpha_0}(x - u, y - v)] f(u,v) du \wedge dv$$

$$= \int_{\gamma_0^\epsilon} \mathcal{K}_{\psi,\alpha_0}(x - \tau_1, y - \tau_2) n_\psi(\tau_1, \tau_2) f(\tau_1, \tau_2) d\gamma_0^\epsilon$$

$$+ \int_{\theta_\epsilon} {}_\psi \partial_{\alpha_0}[\mathcal{K}_{\psi,\alpha_0}(x - u, y - v)] f(u,v) du \wedge dv,$$

where $\gamma_0^\epsilon = \partial \theta_\epsilon$. Passing to the limit we get the result.

An inmediate consequence of this theorem is the quaternionic version of Cauchy's integral formula.

4.2. THEOREM. (Quaternionic Cauchy integral formula for a complex parameter α_0).

Let Ω be a domain in $\mathbb{R}^2$ and Γ a closed rectifiable Jordan curve with $\overline{I(\Gamma)} \subset \bar{\Omega}$. Let $f \in^\psi \mathfrak{M}_{\alpha_0}(\Omega) \cap C(\bar{\Omega})$ and $\alpha_0 \in \mathbb{C}$. Then

$$f(x,y) =^\psi K_{\alpha_0}[f](x,y), \quad (x,y) \in I(\Omega).$$

We are now in a position to give the quaternionic version of Morera's Theorem.

4.3. THEOREM. (Quaternionic Morera Theorem for a complex parameter α_0)

Let $\alpha_0 \in \mathbb{C}$, $f \in C^1(\Omega, \mathbb{H}(\mathbb{C}))$, $_\psi \partial_{\alpha_0}[f] \in L_p(\Omega, \mathbb{H}(\mathbb{C}))$ for some $p > 1$. If for any closed rectifiable Jordan curve Γ such that $\overline{I(\Gamma)} \subset \Omega$ we have that

$$\int_\Gamma n_\psi \cdot f d\Gamma = -\int_\Omega f \alpha_0 dx \wedge dy$$

then f is (ψ, α_0)-hyperholomorphic in Ω.

PROOF. Let $\{\Omega_k\}_{k \in \mathbb{N}}$ be a regular sequence of domains converging to the point $(x_0, y_0) \in \Omega$. Let Γ_k be the boundary of Ω_k.

Then by Lebesgue's Theorem (see [18, p.218]) for any $\mathbb{H}(\mathbb{C})$-valued function $g \in L_p(\Omega)$, $p > 1$, there exists $\lim_{k \to \infty}(1/|\Omega_k|) \int_{\Omega_k} g dx \wedge dy := \bar{g}(x_0, y_0)$ and $g = \bar{g}$ in $L_p(\Omega)$. If we choose $g :=_\psi \partial[f]$, then by (3.1) with $\alpha = 0$, we have

$$\int_{\Omega_k} {}_\psi \partial[f] dx \wedge dy = \int_{\Gamma_k} n_\psi \cdot f d\Gamma_k.$$

By hypothesis, it follows that

$$\frac{1}{|\Omega_k|} \int_{\Omega_k} {}_\psi \partial_{\alpha_0}[f] dx \wedge dy = 0, \quad k \in \mathbb{N} \cup \{0\},$$

therefore $_\psi \partial_{\alpha_0}[f](x,y) = 0$ almost everywhere in Ω.

4.4. THEOREM. (Right inverse for the quaternionic Cauchy-Riemann operator in the case of a complex parameter). *Let* $\alpha_0 \in \mathbb{C}$, $f \in C^1(\Omega, \mathbb{H}(\mathbb{C})) \cap C(\bar{\Omega}, \mathbb{H}(\mathbb{C}))$. *Then the following representation for f holds:*

$$(4.2) \qquad f(x,y) =_\psi \partial_{\alpha_0} \cdot {}^\psi T_{\alpha_0}[f](x,y), \quad \forall (x,y) \in \Omega.$$

PROOF. It follows from quaternionic Green's formula (3.1) that if $\Gamma = \partial\Omega$ is a closed rectifiable Jordan curve and $\tau = (\tau_1, \tau_2)$, then

$$\int_\Gamma \theta_{\alpha_0}(x - \tau_1, y - \tau_2) n_\psi(\tau_1, \tau_2) f(\tau_1, \tau_2) d\Gamma_\tau$$

$$= \int_\Omega \{\mathcal{K}_{\psi,\alpha_0}(x - u, y - v) f(u,v) - \theta_{\alpha_0}(x - u, y - v)_{\bar\psi}\partial_{\alpha_0}[f(u,v)]\} dx \wedge dy,$$

so it follows that

$${}^\psi T_{\alpha_0} = \int_\Omega \theta_{\alpha_0}(x - u, y - v)_{\bar\psi}\partial_{\alpha_0}[f](u,v) dx \wedge dy$$

$$+ \int_\Gamma \theta_{\alpha_0}(x - \tau_1, y - \tau_2) n_\psi(\tau_1, \tau_2) f(\tau_1, \tau_2) d\Gamma_\tau.$$

If we apply ${}_\psi \partial_{\alpha_0}$ and the Borel-Pompeiu formula we get:

$${}_\psi \partial_{\alpha_0} \cdot {}^\psi T_{\alpha_0}[f](x,y)$$

$$= -\int_\Omega {}_\psi\partial_{\alpha_0}[\theta_{\alpha_0}(x - u, y - v)]_\psi\partial_{-\alpha_0}[f](u,v) du \wedge dv$$

$$+ \int_\Gamma {}_\psi\partial_{\alpha_0}[\theta_{\alpha_0}(x - \tau_1, y - \tau_2)] n_\psi(\tau_1, \tau_2) f(\tau_1, \tau_2) d\Gamma_\tau$$

$$= \int_\Omega \mathcal{K}_{\psi,-\alpha_0}(x - u, y - v)_\psi\partial_{-\alpha_0}[f](u,v) du \wedge dv$$

$$- \int_\Gamma \mathcal{K}_{\psi,-\alpha_0}(x - \tau_1, y - \tau_2) n_\psi(\tau_1, \tau_2) f(\tau_1, \tau_2) d\Gamma_\tau$$

$$= ({}^\psi T_{-\alpha_0} \cdot {}_\psi\partial_{-\alpha_0} + {}^\psi K_{-\alpha_0})[f](x,y) = f(x,y)$$

5. Main integral formulas for a complex-quaternionic parameter α

5.1. Now we are going to generalize for a complex quaternionic parameter, the quaternionic Borel-Pompeiu formula, the Cauchy integral formula, the Morera theorem and the Theorem on the right inverse for the operator ${}_\psi\partial_\alpha$. For this purpose we need first, to extend the definition of the above integral operators ${}^\psi T_\alpha$, ${}^\psi K_\alpha$, $\alpha \in \mathbb{H}(\mathbb{C})$ as follows:

$${}^\psi T_\alpha := \begin{cases} P^+ \cdot {}^\psi T_{\xi_+} + P^- \, {}^\psi T_{\xi_-}, & \alpha \notin \mathfrak{G}, \vec{\alpha}^2 \neq 0, \\ {}^\psi T_{\alpha_0} + M^{\vec\alpha}\dfrac{\partial}{\partial\alpha_0}[{}^\psi T_{\alpha_0}], \alpha \notin \mathfrak{G}, & \vec{\alpha}^2 = 0, \\ P^+ \cdot {}^\psi T_{2\alpha_0} + P^- \, {}^\psi T_0, & \alpha \in \mathfrak{G}, \alpha_0 \neq 0, \\ {}^\psi T_0 + M^\alpha \cdot W_0, & \alpha \in \mathfrak{G}, \qquad \alpha_0 = 0, \end{cases}$$

where $W_\mu[f](x,y) := \int_\Omega \theta_\mu(x-u, y-v)f(u,v)du \wedge dv, \quad \mu \in \mathbb{C}, \quad (x,y) \in \mathbb{R}^2$.

$$
{}^\psi K_\alpha := \begin{cases} P^+ \cdot {}^\psi K_{\xi_+} + P^- \; {}^\psi K_{\xi_-}, & \alpha \notin \mathfrak{G}, \; \vec{\alpha}^2 \neq 0, \\[2mm] {}^\psi K_{\alpha_0} + M^{\vec{\alpha}} \dfrac{\partial}{\partial \alpha_0}[{}^\psi K_{\alpha_0}], & \alpha \notin \mathfrak{G}, \; \vec{\alpha}^2 = 0, \\[2mm] P^+ \cdot {}^\psi K_{2\alpha_0} + P^- \; {}^\psi K_0, & \alpha \in \mathfrak{G}, \; \alpha_0 \neq 0, \\[2mm] {}^\psi K_0 - M^\alpha \cdot {}^\psi V_0, & \alpha \in \mathfrak{G}, \alpha_0 = 0, \end{cases}
$$

where ${}^\psi V_\mu[f](x,y) := \int_\Omega \theta_\mu(x-\tau_1, y-\tau_2)n_\psi(\tau_1, \tau_2)f(\tau_1, \tau_2)d\Gamma_\tau$, with $\mu \in \mathbb{C}$, $(x,y) \in \mathbb{R}^2 \backslash \Gamma$. Observe that when $\alpha = \alpha_0 \in \mathbb{C}$ both operators ${}^\psi T_\alpha$, ${}^\psi K_\alpha$ coincide with those introduced in Subsection **3.6**. The explicit formulas here are too large, and thus are not shown.

5.2. THEOREM. (Main integral theorems for hyperholomorphic functions with arbitrary complex quaternionic parameter α). *Let α be an arbitrary complex quaternion, Ω be a domain in $\mathbb{R}^2$ whose boundary is given by a closed rectifiable Jordan curve Γ and let ${}^\psi T_\alpha, {}^\psi K_\alpha$ be the operators defined by **5.1**. Then the following assertions are true:*

1. (Quaternionic Borel-Pompeiu formula for a complex quaternionic parameter α). *If $f \in C^1(\Omega, \mathbb{H}(\mathbb{C})) \cap C(\bar{\Omega}, \mathbb{H}(\mathbb{C}))$ then $\forall (x,y) \in \Omega$*

$$
{}^\psi K_\alpha[f](x,y) + {}^\psi T_\alpha \cdot_\psi \partial_\alpha[f](x,y) = f(x,y).
$$

2. (Quaternionic Cauchy integral formula for a complex-quaternionic parameter α). *If $f \in {}^\psi \mathfrak{M}_\alpha(\Omega, \mathbb{H}(\mathbb{C})) \cap C(\bar{\Omega}, \mathbb{H}(\mathbb{C}))$, then $\forall (x,y) \in \Omega$*

$$
f(x,y) = {}^\psi K_\alpha[f](x,y).
$$

3. (Quaternionic Morera theorem for a complex quaternionic parameter α). *If $f \in C^1(\Omega; \mathbb{H}(\mathbb{C}))$, ${}^\psi \partial_\alpha[f] \in L_p(\Omega, \mathbb{H}(\mathbb{C})), p > 1$, and if for any closed rectifiable Jordan curve Γ^* with $\overline{I(\Gamma^*)} \subset \Omega$ we have that*

$$
\int_{\Gamma^*} n_\psi f d\Gamma = -\int_{I(\Gamma^*)} f\alpha dx \wedge dy
$$

then f is $(\psi, \alpha)-$hyperholomorphic in Ω.

4. (Right inverse for the quaternionic Cauchy-Riemann operator for a complex quaternionic parameter α). *If $f \in C^1(\Omega, \mathbb{H}(\mathbb{C})) \cap C(\bar{\Omega}, \mathbb{H}(\mathbb{C}))$ then $\forall (x,y) \in \Omega$*

$$
{}_\psi \partial_\alpha \cdot {}^\psi T_\alpha[f](x,y) = f(x,y).
$$

PROOF. It is essential in the proof, that the corresponding results for $\alpha = \alpha_0 \in \mathbb{C}$ have already been obtained and that for that case, definitions of ${}^\psi T_\alpha$, ${}^\psi K_\alpha$ coincide with those given in Subsection 3.6.

1) We separate the proof in four cases.

If $\alpha \notin \mathfrak{G}$, $\vec{\alpha}^2 \neq 0$, then we apply the Borel-Pompeiu formula for the complex parameter:

$$
\begin{aligned}
(5.1) \quad {}^\psi K_\alpha[f] &= P^+[f - {}^\psi T_{\xi_+}[f]] + P^-[f - {}^\psi T_{\xi_-} \; {}_\psi \partial_{\xi_-}[f]] \\
&= f - {}^\psi T_{\xi_+} P^+ \; {}_\psi \partial_{\xi_+}[f] - {}^\psi T_{\xi_-} P^- \cdot {}_\psi \partial_{\xi_-}[f].
\end{aligned}
$$

By applying the commutativity of the operators $P^{\pm}$ with $\partial_{\xi_{\pm}}$ and formula (2.28) we obtain:

$$
(5.2) \qquad P^{+} \,_{\psi}\partial_{\xi_{+}}[f] = P^{+} \,_{\psi}\partial_{\alpha}[f],
$$

$$
(5.3) \qquad P^{-} \,_{\psi}\partial_{\xi_{-}}[f] = P^{-} \,_{\psi}\partial_{\alpha}[f].
$$

If we substitute (5.2) and (5.3) in (5.1) we get:

$$
\begin{aligned}
{}^{\psi}K_{\alpha}[f] &= f - (P^{+} \,^{\psi}T_{\xi_{+}} + P^{-} \,^{\psi}T_{\xi_{-}}) \,_{\psi}\partial_{\alpha}[f] \\
&= f - {}^{\psi}T_{\alpha} \cdot_{\psi} \partial_{\alpha}[f].
\end{aligned}
$$

For the cases $\alpha \notin \mathfrak{G}$, $\vec{\alpha}^{2} \neq 0$ and $\alpha \in \mathfrak{G}$, $\alpha_{0} \neq 0$ the proof is similar.

For the case $\alpha \in \mathfrak{G}$, $\alpha_{0} = 0$ observe first that from the simplest version of the quaternionic Green's formula (3.1) we obtain directly

$$
{}^{\psi}V_{0}[f](x,y) = \int_{\Omega} \,_{\psi}\partial[\theta_{0}(x-u, y-v)f(u,v)]du \wedge dv.
$$

But also we have that

$$
\,_{\psi}\partial[\theta_{0}(x-u, y-v)] = {}^{\psi}\mathcal{K}_{0}(x-u, y-v).
$$

Hence after an easy calculation we arrive at

$$
{}^{\psi}V_{0}[f](x,y) = {}^{\psi}T_{0}[f](x,y) + W_{0} \cdot_{\psi} \partial[f](x,y),
$$

therefore

$$
{}^{\psi}K_{\alpha}[f] = f - {}^{\psi}T_{0} \cdot_{\psi} \partial[f] - W_{0} \,_{\psi}\partial_{\alpha}[f]\alpha - {}^{\psi}T_{0}[f]\alpha = f - {}^{\psi}T_{\alpha} \cdot {}^{\psi}\partial_{\alpha}[f],
$$

i.e., the quaternionic Borel-Pompeiu formula is proved.

With the same kind of arguments, the proofs of 2) and 3) follow.

4) Let us consider first, the case $\alpha \in \mathfrak{G}$, $\vec{\alpha}^{2} \neq 0$:

From equality (2.18) it follows that

$$
\begin{aligned}
\,_{\psi}\partial_{\alpha} \,^{\psi}T_{\alpha}[f] &= (\,_{\psi}\partial_{\xi_{+}}P^{+} +_{\psi} \partial_{\xi_{-}}P^{-}) (P^{+} \,^{\psi}T_{\xi_{+}} + P^{-} \,^{\psi}T_{\xi_{-}})[f] \\
&= (P^{+} \,_{\psi}\partial_{\xi_{+}} \cdot^{\psi} T_{\xi_{+}} + P^{-} \,_{\psi}\partial_{\xi_{-}} \,^{\psi}T_{\xi_{-}})[f] \\
&= (P^{+} + P^{-})[f] = f
\end{aligned}
$$

where we have made use of formula (4.2) (right inverse for a complex parameter).

The proof for the case $\alpha \in \mathfrak{G}$, $\alpha_{0} \neq 0$ is similar.

Consider now the case $\alpha \in \mathfrak{G}$, $\alpha_{0} = 0$:

$$
\begin{aligned}
\,_{\psi}\partial_{\alpha} \,^{\psi}T_{\alpha}[f] &= (\,_{\psi}\partial_{\alpha} \,^{\psi}T_{0} +_{\psi} \partial_{\alpha}M^{\alpha}W_{0})[f] \\
&= f + M^{\alpha} \,^{\psi}T_{0}[f] + M^{\alpha} \,_{\psi}\partial W_{0}[f].
\end{aligned}
$$

But $_{\psi}\partial W_{0} = -^{\psi}T_{0}$, hence for $\alpha \in \mathfrak{G}$, $\alpha_{0} = 0$, the formula for the right inverse for the quaternionic Cauchy-Riemann operator with complex parameter is valid too.

Finally consider the case $\alpha \notin \mathfrak{G}$, $\vec{\alpha}^2 = 0$

$$
\begin{aligned}
{}_\psi \partial_\alpha \, {}^\psi T_\alpha[f] &= {}_\psi \, \partial_\alpha \, {}^\psi T_{\alpha_0}[f] + {}_\psi \partial_\alpha \frac{\partial}{\partial \alpha_0}[{}^\psi T_{\alpha_0}[f]]\vec{\alpha} \\
&= f + {}^\psi T_{\alpha_0}[f]\vec{\alpha} + {}_{\alpha_0} \partial_\psi \cdot \frac{\partial}{\partial \alpha_0}[{}^\psi T_{\alpha_0}[f]]\vec{\alpha} \\
&= f + ({}^\psi T_{\alpha_0}[f])\vec{\alpha} + \frac{\partial}{\partial \alpha_0}({}_\psi \partial_{\alpha_0} \, {}^\psi T_{\alpha_0}[f])\vec{\alpha} \\
&\quad - {}^\psi T_{\alpha_0}[f]\vec{\alpha} = f + \frac{\partial}{\partial \alpha_0}(f)\vec{\alpha} = f.
\end{aligned}
$$

Thus we conclude the proof of the theorem.

References

[1] S. Bergman, *Integral Operators in the Theory of Linear Partial Differential Equations*, Springer-Verlag, New York, 1969.

[2] S. Bernstein, *The Left-linear Riemann Problem in Clifford Analysis as an Integral Equation*, Preprint 95-04, Fakultat für Mathematik und Informatik, Technische Universität Bergakademie Freiberg, 25 pp.

[3] F. Brackx and N. Van Acker, *Boundary Value Theory for Eigenfunctions of the Dirac Operator*, Bull. Soc. Math. Belgium, ser. B **45** (1993), 113–123.

[4] L. Bragg and J. Dettman, *Function Theories for the Yukawa and Helmholtz Equation*, Rocky Mountain J. Math. **25** (1995), 887–917.

[5] S. Chumakov and K. B. Wolf, *Super Symmetry in Helmholtz Optics*, Phys. Lett. A. **193** (1994), 51–53.

[6] R. Dautray and J. L. Lions, *Mathematical Analysis and Numerical Methods for Science and Technology, Vol. 1*, Springer-Verlag, 1985.

[7] P. González-Casanova and K. B. Wolf, *Interpolation for Solutions of the Helmholtz Equation*, Numerical Methods for P.D.E. **11** (1995), 77–91.

[8] Gürlebeck by W. Sprössig, *Quaternionic Analysis and Elliptic Boundary Value Problems*, Akademic-Verlag, Berlin, 1989.

[9] V. G. Kravchenko and M. V. Shapiro, *Hypercomplex Factorization of the Multidimensional Helmholtz Operator and Some of its Applications*, Doklady rasshirennyh zasedaniy seminara instituta prikl. mat. imeni I. N. Vekua, Tbilisi **5** (1990), 106–109 (in Russian).

[10] V. V. Kravchenko and M. V. Shapiro, *Helmholtz Operator with a Quaternionic Wave Number and Associated Function Theory*, Deformations of Mathematical Structures. II, rm (ed.: J. Lawrynowicz), Kluwer Academic Publishers, 1994, pp. 101–128.

[11] V. V. Kravchenko and M. V. Shapiro, *Helmholtz Operator with a Quaternionic Wave Number and Associated Function Theory. II: Integral Representations*, Acta Applicandae Mathematicae **32** (1993), 243–265.

[12] V. V. Kravchenko and M. V. Shapiro, *On the Generalized System of Cauchy-Riemann Equations with a Quaternion Parameter*, Doklady Akademii Nauk, Russia **329** (1993), , 547–549 (in Russian); English transl.: Russian Acad. Sci. Dokl. Math., **47** (1993), 315–319.

[13] V. Kravchenko and M. Shapiro, *Integral Representations for Spatial Models of Mathematical Physics*, Reporte Interno # 172, Enero, 1995, Departamento de Matematicas, CINVESTAV del IPN, Mexico City, 171 pp.

[14] M. Mitrea, *Boundary Value Problems and Hardy Spaces Associated to the Helmholtz Equation in Lipschitz Domains*, Research Report 1992:02, Dept. of Math., University of South Carolina.

[15] K. Nôno, *On the Quaternion Linearization of Laplacian Δ*, Bull Fukuoka Univ. Educ. Nat. Sci. **35** (1985), 5–10.

[16] S. Prössdorf and J. Saranen, *A Fully Discrete Approximation Method for the Exterior Neumann Problem of the Helmholtz Equation*, Zeitschrift Anal. Anwend **13** (1994), 683–695.

[17] M. Shapiro and L. M. Tovar, *Two-dimensional Helmholtz Operator and its Hyperholomorphic Solutions*, Journal of Natural Geometry (to appear).

[18] G. E. Shilov and B. L. Gurevich, *Integral Measure and Derivative: a Unified Approach*, Prentice Hall, 1966.

[19] N. L. Vasilevski and M. V. Shapiro, *On an Analogue of Monogenicity in the Sense of Moisil-Teodoresco and Some Applications in the Theory of Boundary Value Problems*, Reports of Enlarged Session of Seminars of the I. N. Vekua Institute of App. Math., Tbilisi **1** (1985), 63–66 (in Russian).

[20] V. S. Vladimirov, *Equations of Mathematical Physics*, Nauka, Moscow, 1984 (in Russian); Engl. transl. of the first edition: Marcel Dekker, New York, 1971.

[21] Xu Zhenyuan, *A Function Theory for the Operator $D - \lambda$*, Complex Variables, Theory and Appl **16** (1991), 27–42.

DEPARTAMENTO DE MATEMÁTICAS, ESCUELA SUPERIOR DE FÍSICA Y MATEMÁTICAS, INSTITUTO POLITÉCNICO NACIONAL, MEXICO CITY, MEXICO
E-mail address: shapiro@esfm.ipn.mx

DEPARTAMENTO DE MATEMÁTICAS, ESCUELA SUPERIOR DE FÍSICA Y MATEMÁTICAS, INSTITUTO POLITÉCNICO NACIONAL, MEXICO CITY, MEXICO
E-mail address: tovar@esfm.ipn.mx

Contemporary Mathematics
Volume **212**, 1998

Cocycles on the Gauge Group and
the Algebra of Chern-Simons Classes

M. M. Smirnov

ABSTRACT. I consider here Chern-Simons classes, their generalizations and related algebraic and analytic problems. I describe a new class of algebras whose elements contain Chern and generalized Chern-Simons classes. There is a Poisson bracket in these algebras, similar to the bracket in Kontsevich's non-commutative symplectic geometry [14]. I prove that the Poisson bracket gives rise to a graded Lie algebra containing differential forms representing Chern and Chern-Simons classes. This is a new result. I describe algebraic analogs of the dilogarithm and higher polylogarithms in the algebra corresponding to Chern-Simons classes. I study the properties of this bracket. It is possible to write the exterior differential and other operations in the algebra using this bracket. The bracket of any two Chern classes is zero and the bracket of a Chern class and a Chern-Simons class is d-closed. The construction developed here easily gives explicit formulas for known secondary classes and makes it possible to construct new ones. I develop an algebraic model for the action of the gauge group and describe how elements of algebra corresponding to the secondary characteristic classes change under this action (see theorem 3). It is possible give new explicit formulas for cocycles on a gauge group of a bundle and for the corresponding cocycles on the Lie algebra of the gauge group. I use formulas for secondary characteristic classes and an algebraic approach developed in [8].

There are several approaches to combinatorial formulas for characteristic classes. These approaches are due to Gelfand-Gabrielov-Losik, MacPherson, Patodi, Ranicki-Sullivan, Cheeger, Brylinski, and others. One of the main problems in the field is the problem of explicit description of secondary characteristic classes and difference cocycles (cf. Chern-Simons [4], Cheeger, Bott-Shulman-Stasheff [2], Youssin [20], and others). In some sense, this problem is solved here.

One of the other main problems related to combinatorial formulas for characteristic classes is the problem of writing topological invariants using local data given as fields of geometric objects. For Chern classes, curvature enables us to write these formulas using ordinary Chern-Weil theory.

In [8], Gelfand suggested the idea of using formal power series of geometric objects and constructing the universal field and variational bicomplex for them (see

---

1991 *Mathematics Subject Classification.* Primary 55R40, 58B30; Secondary 16W30, 81R50.
This work was partially supported by NSF grant DMS 92-13357.

also Gelfand-Kazhdan-Fuks [9]). Such a universal construction is given in chapter 3 for some particular cases.

Let us illustrate the problem of writing topological invariants using local data. Suppose we have an oriented graph consisting of points and arrows. To each vertex we associate a complex vector space E_i, and to each arrow we associate a linear mapping s_{ij}, as in the picture below. We suppose that the graph is parametrized by a manifold X; this means that $E_i(x)$ is a vector bundle and $s_{ij}(x)$ is a bundle map.

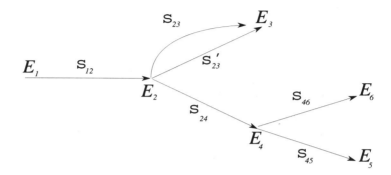

It is important to describe "Chern classes" and "secondary Chern classes" for such objects using local differential geometry. For a graph consisting of one vertex, the Chern-Weil theory gives ordinary Chern classes.

Consider first a graph with one vertex and a loop. The vertex corresponds to an n-dimensional vector bundle over a manifold X and the loop corresponds to an automorphism σ of the bundle. In topological K-theory, this data (the bundle and its automorphism) defines an element of $K_1(X)$. We can write in this case the usual Chern class for $K_1(X)$. For the trivial bundle, this class is given by

$$\mathrm{ch}_m(\sigma) = \frac{(-1)^{m-1}(m-1)!}{(2m-1)!}\mathrm{tr}(\sigma^{-1} \cdot d\sigma)^{2m-1}.$$

This class can also be defined as a secondary Chern class associated to the two connections $\nabla = d$ and $\nabla^\sigma = d + \sigma^{-1}d\sigma$. Locally, σ is a nondegenerate $n \times n$ matrix of functions of $x \in X$ defining the automorphism σ.

Let us now return to the problem of explicit calculation of Chern-Simons classes and illustrate it by a simple example.

It is well known that the Chern-Simons class constructed from two connections ω_0 and ω_1 has the form

$$\mathrm{ch}_2^1(\omega_0, \omega_1) = \frac{1}{2}\mathrm{tr}\left(\left(\omega_0 d\omega_0 + \frac{2}{3}\omega_0^3\right) - \left(\omega_1 d\omega_1 + \frac{2}{3}\omega_1^3\right) + d(\omega_0\omega_1)\right),$$

where the ω_i are matrices of 1-forms which define the connections. For higher secondary classes, the calculations give longer polynomials. For example,

$$\mathrm{ch}_3^1(\omega_0, \omega_1) = \frac{1}{2}\mathrm{tr}\left[\left(\frac{1}{5}a_0^5 + \frac{1}{3}a_0 b_0 b_0 + \frac{1}{2}a_0^3 b_0\right) - \left(\frac{1}{5}a_1^5 + \frac{1}{3}a_1 b_1 b_1 + \frac{1}{2}a_1^3 b_1\right)\right.$$
$$\left.\frac{1}{6}d\left((a_1 a_0 - a_0 a_1)(b_0 + b_1) - \left(a_0^2 + a_1^2 + \frac{1}{2}a_0 a_1\right)a_0 a_1\right)\right],$$

where $a_i = \omega_i$ and $b_i = d\omega_i$.

Thus, even in the simplest case for "normal" secondary characteristic classes (i.e., for a graph consisting of one point and zero arrows), formulas are lengthy and we have to handle the combinatorics of long noncommutative expressions. The algebraic language introduced in chapter 1 allows us to do this easily.

The construction of the algebra of Chern classes

We construct first a graded free associative algebra A generated by elements a_i of degree 1 and elements b_i of degree 2 ($i = 1, \ldots, N$), together with a differential d, such that $da_i = b_i$, $db_i = 0$. Then we consider the space V of cyclic words of A (i.e., a complex vector space with a basis consisting of equivalence classes of monomials, where two monomials obtained by cyclic permutation are considered to be the same up to a sign). Strictly speaking, V is the factor space of A by the subspace $\{PQ - (-1)^{|P||Q|}QP\}$ spanned by the graded commutators of all monomials in A.
That is,

$$V = A/\{PQ - (-1)^{|P||Q|}QP\}.$$

We also refer to monomials in A as *words* and to their equivalence classes as *cyclic words* (they form a basis of V).

The sign $\overset{\text{tr}}{=}$ distinguishes equality up to cyclic permutation (i.e., equality in V) from equality in the algebra A, which is denoted by $=$.

Let us now define a Poisson bracket on V. The Poisson bracket gives V the structure of a graded Lie algebra. A partial derivative $(\partial/\partial z)P$, or $\partial_z P$, where $z = a_i$ or b_i, is defined on monomials by the following rule: we take a monomial P and look at all appearances of z in it. For each letter z which appears in P we cyclically permute the word P so that this letter z becomes the first letter of the permuted word. We then delete this letter z. The sum of the resulting monomials will be a partial derivative.

For example, suppose A is generated by two letters a and b. Then

$$\partial_a(aaabab) = \not{a}aabab + (-1)^{1\cdot 7}\not{a}ababa + (-1)^{2\cdot 6}\not{a}babaa + (-1)^{5\cdot 3}\not{a}baaab$$
$$= aabab - ababa + babaa - baaab.$$
$$\partial_b(aaabab) = (-1)^{3\cdot 5}\not{b}abaaa + (-1)^{2\cdot 6}\not{b}aaaba = aaaba - abaaa.$$

The partial derivative is well-defined in the space of cyclic words V.

Now we may define the Poisson bracket of P, $Q \in V$ to be

$$\{P, Q\} \overset{\text{tr}}{=} \sum_{i=1}^{N}(\partial_{a_i}P \cdot \partial_{b_i}Q + (-1)^{|P||Q|}\partial_{a_i}Q \cdot \partial_{b_i}P).$$

In order to make sense of this formula. we take any of preimages of P and Q in A, also denoted by P and Q, and apply the operators ∂_{a_i} and ∂_{b_i} to these preimages. We multiply $\partial_{a_i}P$ and $\partial_{b_i}P$ in A and only then do we take the corresponding cyclic element in V. This is independent of our choices.

THEOREM 0. *The Poisson bracket is well defined on V and is linear in its arguments. It has the property*

$$\{P, \{Q, R\}\} + (-1)^{|P|(|Q|+|R|)}\{Q, \{R, P\}\} + (-1)^{|R||\{Q,P\}|}\{R, \{P, Q\}\} \overset{\text{tr}}{=} 0.$$

($\mathbf{Z}_2$-graded Jacobi identity).

Let us now define algebraic analogs of Chern and Chern-Simons classes. We fix an index i, so a_i and b_i are denoted by a and b. Choose a positive integer k. Consider the image in V of a polynomial $(1/k!)(a^2 + b)^k \in A$. We call this image the k-th Chern character associated to a (recall $b = da$), and denote it by $\mathrm{ch}_k(a)$:

$$\mathrm{ch}_k(a) \overset{\mathrm{tr}}{=} \frac{1}{k!}(a^2 + b)^k.$$

This corresponds to the differential form

$$\mathrm{tr}\,\frac{1}{k!}(\omega^2 + d\omega)^k.$$

Now we define algebraic analogs of the Chern-Simons classes ch_k^1. These are the cyclic images in V of the following polynomials in A:

$$\frac{1}{(k-1)!}a\Big(\frac{1}{k}b^{k-1} + \frac{1}{k+1}\Sigma_{1,k-2} + \frac{1}{k+2}\Sigma_{2,k-3} + \cdots + \frac{1}{2k-1}a^{2(k-1)}\Big),$$

where $\Sigma_{p,q}$ is the sum of all possible noncommutative monomials in a^2 and b with p appearances of a^2 and q appearances of b. For example, $\Sigma_{1,2} = a^2bb + ba^2b + bba^2$. We shall call the corresponding element of V the k−th secondary Chern character (Chern–Simons class) and denote it by $\mathrm{ch}_k^1(a)$:

$$\mathrm{ch}_k^1(a) \overset{\mathrm{tr}}{=} \frac{1}{(k-1)!}a\left(\frac{1}{k}b^{k-1} + \frac{1}{k+1}\Sigma_{1,k-2} + \frac{1}{k+2}\Sigma_{2,k-3} + \cdots + \frac{1}{2k-1}a^{2(k-1)}\right).$$

EXAMPLE 0. The second Chern character in V is

$$\mathrm{ch}_2(a) \overset{\mathrm{tr}}{=} \frac{1}{2}(a^2 + b)^2 \overset{\mathrm{tr}}{=} \frac{1}{2}(b^2 + 2a^2b).$$

$(a^4 \overset{\mathrm{tr}}{=} 0)$. This corresponds to the differential form

$$\frac{1}{2}\mathrm{tr}(\omega^2 + d\omega)^2.$$

The Chern-Simons class in V is

$$\mathrm{ch}_2^1(a) \overset{\mathrm{tr}}{=} \frac{1}{2}\left(ab + \frac{2}{3}a^3\right).$$

This corresponds to the Chern-Simons differential 3-form

$$\frac{1}{2}\mathrm{tr}\left(\omega d\omega + \frac{2}{3}\omega^3\right).$$

THEOREM 1.
1. *Chern characters are closed elements of the cyclic space V:*

$$d\,\mathrm{ch}_k(a) \overset{\mathrm{tr}}{=} 0.$$

2. *The differential in V of the Chern–Simons class defined by the above formula is the k-th Chern character:*

$$d\,\mathrm{ch}_k^1(a) \overset{\mathrm{tr}}{=} \mathrm{ch}_k(a).$$

We develop a similar theory for the algebra with generators depending on a real parameter t. Let A be a free associative algebra generated by elements $a(t)$, $\dot{a}(t)$ of degree 1 and elements $b(t)$, $\dot{b}(t)$ of degree 2, where $t \in [0,1]$. This algebra has a differential d such that

$$d\, a(t) = b(t), \quad d\dot{a}(t) = \dot{b}(t), \quad db(t) = 0, \quad d\dot{b}(t) = 0.$$

We shall multiply the elements of our algebra A by polynomials in t and by a differential 1-form dt. We may treat dt formally as an element of degree 1 that commutes with $b(t)$ and $\dot{b}(t)$ and anticommutes with $a(t)$ and $\dot{a}(t)$, also $(dt)(dt) = 0$. Polynomials in t commutes with elements of A. We define another differential $\delta = d_t$ that is formally not included in the algebraic structure of A. It maps elements $a(t)$ and $b(t)$ of A into a product of a formal element dt and $\dot{a}(t)$ and $\dot{b}(t)$ correspondingly.

$$d_t a(t) = dt\dot{a}(t) \quad \text{and} \quad d_t b(t) = dt\dot{b}(t).$$

Furthermore, there is a derivative in t for the expressions in a and b given by the formal rule $(d/dt)a = \dot{a}$ and $(d/dt)b = \dot{b}$. We are not going to apply d_t to expressions involving $\dot{a}$ and $\dot{b}$ (otherwise we would have to introduce letters with more dots, but we do not want to deal with this). We do not want to make our algebra very complicated at this point. It is just an abstraction of the algebra of matrices of 1 and 2 forms on a manifold X, when these forms depend on a real parameter t.

From this algebra A we construct as before the factor space

$$V = A/\{CB - (-1)^{|C||B|}BC\},$$

where $\{CB - (-1)^{|C||B|}BC\}$ is the subspace of A generated by commutators. We call V the *space of cyclic words* and we denote the equality in V by $\overset{\text{tr}}{=}$.

Consider a path $a(t)$, $0 \le t \le 1$, connecting a_0 and a_1 and consider

$$\operatorname{ch}_k^1(a(t)) \overset{\text{def}}{=} \int_0^1 [(d_t + d)a(t) + a(t)^2]^k.$$

We can now write an explicit formula for $\operatorname{ch}_k^1(a(t))$.

In addition, to $\Sigma_{p,q}$ as defined above, we also need to consider the expression $\Sigma_{\dot{n},p,q}$ which denotes the sum of all possible noncommutative monomials in $\dot{a}$, a^2 and b, where $\dot{a}$ appears n times, a^2 appears p times and b appears q times.

THEOREM 2. *The secondary characteristic class defined by a path $a(t)$ is given by*

$\operatorname{ch}_k^1(a(t))$

$$\overset{\text{tr}}{=} \frac{1}{(k-1)!}\Big(a\big(\frac{1}{k}b^{k-1} + \frac{1}{k+1}\Sigma_{1,k-2} + \frac{1}{k+2}\Sigma_{2,k-3} + \cdots + \frac{1}{2k-1}a^{2(k-1)}\big)\big)\Big)\Big|_1^0$$

$$+ \frac{1}{(k-1)!}\int_0^1 dt\, d\big(a\big(\frac{1}{k}\Sigma_{\dot{i},0,k-2} + \frac{1}{k+1}\Sigma_{\dot{i},1,k-3} + \cdots + \frac{1}{2k-1}\Sigma_{\dot{i},k-2,0}\big)\big).$$

The gauge group action on the algebra of Chern classes

Consider now the action of the gauge group on connections (or on elements a_i of the algebra A corresponding to connections).

The gauge transformation acts in a natural way:

$$g : a \to g^{-1}dg + g^{-1}ag.$$

Let us denote by c the expression $dg \cdot g^{-1}$.

THEOREM 3. *Under the gauge transformation on the space of cyclic words in the letters g, g^{-1}, dg, dg^{-1}, a_i, and b_i, the Chern–Simons class $\mathrm{ch}_k^1(a)$ transforms in the following way:*

$$\mathrm{ch}_k^1(a) \mapsto \tilde{\mathrm{ch}}_k^1 \stackrel{\mathrm{tr}}{=} \frac{1}{(k-1)!}(a+c) \sum_{\alpha+\beta+\gamma=k-1} (-1)^\gamma \frac{(k+\beta-1)!\gamma!}{(k+\beta+\gamma)!} \Sigma[(b)^\alpha, (a^2)^\beta, (u)^\gamma],$$

where $u = ca + ac + c^2$ and $\Sigma[(b)^\alpha, (a^2)^\beta, (u)^\gamma]$ is the sum of all possible words in which b, a^2, and u appear α times, β times, and γ times respectively.

Cocycles on the gauge group

Let M be a smooth real m-dimensional manifold. Let G be a connected Lie group, $\mathfrak{g}$ its Lie algebra. Let $E \to M$ be a principal G bundle with a base manifold M and let $G(E)$ be its gauge group. If we trivialize the bundle E over an open set $U \subset M$ then the trivialization gives us an isomorphism between the gauge group and the group of G valued functions on U.

We use ideas of Faddeev, Reiman and Semyonov-Tian-Shanskii [6] in the construction of cocycles on the gauge group of the bundle E. We construct new cocycles on the group and its Lie algebra.

Let us fix a connection $B(x)$, $x \in M$ on a bundle E. Locally we can look at B as a $\mathfrak{g}$ valued 1-form on M. Let us take $n+1$ connections ω_0, ..., ω_n which are gauge equivalent to B under the action of elements $g_0,\dots,g_n$ of the gauge group.

$$\omega_0(x) = g_0^{-1}dg_0 + g_0^{-1}Bg_0,$$

$$\dots$$

$$\omega_n(x) = g_n^{-1}dg_n + g_n^{-1}Bg_n.$$

Here d is a differential with respect to $x \in M$. Consider a product space $M \times \Delta_n$ where Δ_n is a standard n-dimensional simplex:

$$\Delta_n = \{(t_0, t_1, \dots, t_n)|t_0 + t_1 + \cdots + t_n = 1, t_0, t_1, \dots, t_n > 0\}.$$

On $M \times \Delta_n$ define a a connection

$$\omega(x,t) = t_0\omega_0(x) + \cdots + t_n\omega_n(x). \tag{0}$$

Let d be a differential with respect to $x \in M$ and δ be a differential with respect to $t \in \Delta_n$. Then $(d +)$ is the total differential on $M \times \Delta_n$. The curvature of $\omega(x,t)$ is

$$R_\omega = (d + \delta)\omega + \omega^2$$

We shall use here a standard definition of coboundary operator in group cohomology. Let G be a group and V be a left G module. Let $A^p(G, V)$ be the space of functions on $\underbrace{G \times \cdots \times G}_{p+1 \times}$ taking values in V. For $\alpha(h_0, \ldots, h_n) \in A^p(G, V)$ define the action of $g \in G$ as

$$(g\alpha)(h_0, \ldots, h_n) = g(\alpha(g^{-1}h_0, \ldots, g^{-1}h_n)).$$

Define a differential as

$$(\partial\alpha)(h_0, \ldots, h_n) \overset{\text{def}}{=} \sum (-1)^i \alpha(h_0, \ldots, \hat{h}_i, \ldots, h_n).$$

DEFINITION. Let $P(R_1, \ldots, R_N)$ be an invariant symmetric N form. We define cochains $a_n(g_0, \ldots, g_n)$ on the group G as

$$a_n(g_0, \ldots, g_n) = \int_{\Delta_n} P(R_\omega, \ldots, R_\omega)$$

Cochains $a_n(g_0, \ldots, g_n)$ are differential $(2N-n)$-forms on M depending on $g_0, \ldots, g_n$.

The important property of these cochains with values in differential forms is that

$$(\partial a_{n-1})(g_0, \ldots, g_n) = -d\, a_n(g_0, \ldots, g_n).$$

PROPOSITION 0. *If V is a closed submanifold of M of dimension $(2N - n)$ then*

$$a_n^V(g_0, \ldots, g_n) \overset{\text{def}}{=} \int_V a_n(g_0, \ldots, g_n)$$

is a cocycle on the group $G(E)$ (in the sense of cohomology of groups). If V is homologous to V' then cocycles a_n^V and $a_n^{V'}$ are homologous.

Cocycles on the Lie algebra of the gauge group

PROPOSITION 1. *The following cochains on the Lie algebra correspond to cochains*

$$a_n(1, g_1, \ldots, g_n)$$

on the gauge group (here $N =$degree of P).

$$\bar{a}_n(X_1, \ldots, X_n)$$

$$= \frac{N!}{(N-n)!}(-1)^{\frac{(n-1)}{2}} P\big(\underbrace{(X_1 \cdot B), \ldots, (X_n \cdot B)}_{N \ times}, \overbrace{R_B, \ldots, R_B}^{N-n \ times} \big).$$

Here vector fields $X_1, \ldots, X_n$ are the elements of the Lie algebra of the gauge group, $(X \cdot B)$ is the Lie derivative of the connection B along the vector field X and $R_B = dB + B^2$ is the curvature of connection B.

REMARK. We are working here with matrix groups. The Lie derivative of B is $(X \cdot B) = dX + XB - BX$. X is an element of the Lie algebra of the Gauge group, so it is a matrix of 0-forms, B is a connection, so it is a matrix of 1-forms. And BX and XB are just products of the matrices of 0 and 1-forms. So $(X \cdot B) = dX + XB - BX^1$ is a matrix of 1-forms.

---

[1]We assume that the connection B is transformed by the infinitesimal gauge transformations e^{-sX} as $e^{-sX}de^{sX} + e^{-sX}Be^{sX}$.

EXAMPLE 1. Elements of the group are 1, g_1, g_2. Take $P = (1/2!)\mathrm{tr}R^2$. The cocycle on the Lie algebra is

$$Const \cdot \mathrm{tr}\big((X_1 \cdot B)(X_2 \cdot B)\big).$$

Take $P = (1/3!)\mathrm{tr}R^3$. Take $P = (1/3!)\mathrm{tr}R^3$. The cocycle on the Lie algebra is

$$Const \cdot \mathrm{tr}\big((X_1 \cdot B)(X_2 \cdot B)(dB + B^2)\big).$$

DEFINITION. Let us consider k connections A_1, ..., A_k on the bundle over M and let us fix a connection B. We construct $n + 1$ connections that are gauge equivalent to B using transformations g_0, ..., g_n in the gauge group G. Altogether we have $k + n + 1$ connections A_1, ..., A_k, $g_0^{-1}dg_0 + g_0^{-1}Bg_0$, ..., $g_n^{-1}dg + g_n^{-1}Bg_n$. We construct on the group G a cochain $c_n(A_1, \ldots, A_k \mid g_0, \ldots, g_n)$. As before, we consider a connection ω on $M \times \Delta_{k+n}$, where

$$\Delta_{k+n} = \{\tau_1 + \cdots + \tau_k + t_0 + \cdots + t_n = 1 \mid \tau_i, t_p \geq 0\}$$

is a standard $k + n$ dimensional simplex. This connection ω linearly approximates our $k+n+1$ connections that are placed in the vertices of the simplex. We consider an invariant N form P and and define

$$c_n(A_1, \ldots, A_k \mid R_\omega, \ldots, R_\omega)$$

Cochains $c_n(A_1, \ldots, A_k \mid g_0, \ldots, g_n)$ are $(2N - n - k)$-forms on M depending on $g_0, \ldots, g_n$ and on k connections $A_1, \ldots, A_k$ that are "external fields".

THEOREM 4. *The cochain* $\bar{c}_n(A_1, \ldots, A_k \mid X_1, \ldots, X_n)$ *on the Lie algebra that corresponds to a cochain*

$$c_n(A_1, \ldots, A_k \mid 1, g_1, \ldots, g_n)$$

on the Lie group G can be written as

$$\int\limits_{\{\tau_0 + \ldots \tau_k = 1\}} \int\limits_{\{t_0 + \cdots + t_n \leq 1\}} \frac{1}{n!}\tau_0^n P\Big(d\tau_1(A_1 - B), \ldots, d\tau_k(A_k - B), dt_1(X_1 \cdot B), \ldots,$$

$$dt_n(X_n \cdot B), R_\tau, \ldots, R_\tau\Big),$$

where

$$R_\tau = \tau_0 dB + \tau_1 dA_1 + \ldots \tau_k dA_k + (\tau_0 B + \tau_1 A_1 + \cdots + \tau_k A_k)^2,$$

$(X_i \cdot B)$ *is the Lie derivative of the connection B along the vector field X_i and* $\tau_0 = 1 - \tau_1 - \cdots - \tau_k$.

EXAMPLE 2. Let

$$P(R_1, R_2, R_3, R_4) = \sum_{\sigma \in S_4} \mathrm{tr}(R_{\sigma(1)}R_{\sigma(2)}, R_{\sigma(3)}, R_{\sigma(4)})$$

Then

$$\bar{c}_1(A_1, A_2, B, X_1) = \frac{1}{2}\int\limits_{t_1=0}^{t_1=1} \int_{\Delta_2} \tau_0^2 d\tau_1(A_1 - B)d\tau_2(A_2 - B)dt_1(X_1 \cdot B)$$

$(\tau_0 dB + \tau_1 dA_1 + \tau_2 dA_2 (\tau_0 B + \tau_1 A_1 + \tau_2 A_2)^2)$.

In fact let

$y_1 = A_1 - B$,

$y_2 = A_2 - B$,

$y_3 = X_1 \cdot B = dX_1 + BX_1 + X_1 B$,

$y_4 = R_\tau = (1/20)dB + (1/60)dA_1 + (1/60)dA_2 + (1/30)B^2 + (1/180)A_1^2 +$

$\qquad (1/180)A_2^2 + (1/120)(A_1 B + BA_1) + (1/120)(A_2 B + BA_2) + (1/360)(A_2 A_1 +$ $A_1 A_2)$.

Then cocycle on the Lie algebra is

$\bar{c}_1(A_1, A_2, B, X_1)$

$\quad = 4\,\mathrm{tr}(y_1 y_3 y_4 y_2 + y_1 y_3 y_2 y_4 + y_1 y_4 y_3 y_2 - y_1 y_2 y_3 y_4 - y_1 y_4 y_2 y_3 - y_1 y_2 y_4 y_3)$

We used here

$$\int_{\Delta^2} \tau_0^3 = \frac{1}{20}, \quad \int_{\Delta^2} \tau_0^2 \tau_1 = \frac{1}{60}, \quad \int_{\Delta^2} \tau_0^2 \tau_1^2 = \frac{1}{180},$$

$$\int_{\Delta^2} \tau_0^4 = \frac{1}{30}, \quad \int_{\Delta^2} \tau_0^3 \tau_1 = \frac{1}{120}, \quad \int_{\Delta^2} \tau_0^2 \tau_1 \tau_2 = \frac{1}{360},$$

EXAMPLE 3.
$$\bar{c}_1(A, B, X_1) = 3\,\mathrm{tr}(y_1 y_2 y_3 + y_1 y_3 y_2),$$

where

$y_1 = A - B$,

$y_2 = X \cdot B = dX + XB - BX$,

$y_3 = \int_{\{\tau_0 + \tau_1 = 1\}} \tau_0 (\tau_0 dB + \tau_1 dA + \tau_0^2 B^2 + \tau_1 A^2 + \tau_0 \tau_1 (AB + BA))$

$\quad = (1/3)dB + (1/6)dA + (1/4)B^2 + (1/12)A^2 + (1/12)(AB + BA)$.

References

[1] R. Bott, *On the Chern-Weil homomorphism and the continuous cohomology of Lie groups*, Advances in Math. **11** (1973), 289–303.

[2] R. Bott, H. Shulman, and J. Stasheff, *On the de Rham theory of certain classifying spaces*, Advances in Math. **20** (1976), 43–56.

[3] J. Cheeger, *Spectral geometry of singular Riemannian spaces*, J. Differential Geom. **18** (1983), 575–657.

[4] S. S. Chern and J. Simons, *Characteristic forms and geometric invariants*, Ann. Math. **99** (1974), 48–69.

[5] J. Dupont, *The Dilogarithm as a characteristic class of a flat bundle*, J. Pure and Appl. Algebra **44** (1987), 137–164.

[6] L. D. Faddeev, A. G. Reiman, and M. A. Semyonov-Tian-Shanskii, *Quantum anomalies and cocycles on gauge groups*, Funct. Anal. Appl. **18** (1984), 64–72.

[7] L. D. Faddeev and S. L. Shatashvili, *Algebraic and Hamiltonian methods in the theory of non-Abelian anomalies*, Theor. Math. Phys **60** (1984), 206–217.

[8] I. M. Gelfand and M. M.Smirnov, *The Algebra of Chern–Simons Classes the Poisson Bracket and the Gauge Group action on it*, Lie Theory and Geometry in Honor of B. Kostant, Progress in Mathematics Vol. 123, J.-L. Brylinski, R. Brylinski, V. Guillemin, V. Kac eds., Birkhauser, Boston, 1994, pp. 261–288.

[9] I. M. Gelfand, D. B. Fuks, and D. A. Kazhdan, *The actions of infinite dimensional Lie algebras*, Funct. Anal. Appl. **6** (1972), 9–13.

[10] A. Gabrielov, I. M. Gelfand, and M. Losik, *Combinatorial calculation of characteristic class*, Funktsional Anal. i Prilozhen. (Funct. Anal. Appl.) **9** (1975), 12–28.

[11] I. M. Gelfand and R. MacPherson, *A combinatorial formula for the Pontrjagin classes*, Bull AMS **26** (1992), 304–309.

[12] I. M. Gelfand and B. L. Tsygan, *On the localization of topological invariants*, Comm. Math. Phys. **146** (1992), 73–90.

[13] F.Hirzebruch, *Topological Methods in Algebraic Geometry*, Springer, 1965.

[14] M. Kontsevich, *Formal (Non)-Commutative Symplectic Geometry*, The Gelfand Mathematical Seminars 1990–92, L.Corwin, I.Gelfand, and J. Lepowsky, eds., Birkhäuser, Boston, 1993, pp. 173–189.

[15] J. L. Loday, *Cyclic homology*, Springer, Berlin, 1992.

[16] B. H. Lian and G. J. Zuckerman, *New Perspectives on the BRST-Algebraic Structure of String Theory*.

[17] R. MacPherson, *The combinatorial formula of Garielov, Gelfand, and Losik for the first Pontrjagin class*, Séminaire Bourbaki No. 497, Lecture Notes in Math., **667**, Springer, Heidelberg, 1977.

[18] L. Positselskii, *Quadratic duality and curvature*, Funct. Anal. Appl. **27** (1993), 197–204.

[19] T. Tsujishita, *On variation bicomplex associated to differential equations*, Osaka J. Math. **19** (1982), 311–363.

[20] B. V. Youssin, *Sur les formes $S^{p,q}$ apparaissant dans le calcul combinatoire de la deuxième classe de Pontriaguine par la méthode de Gabrielov, Gel'fand et Losik*, C.R. Acad. Sci. Paris **292**, 641–644.

[21] E. Witten, *Quantum Field Theory and the Jones Polynomial*, Comm. Math. Phys. **121** (1989), 351–299.

[22] E. Witten, *On Quantum Gauge Theories in Two Dimensions*, Comm. Math. Phys. **141** (1991), 153–209.

MAX-PLANCK INSTITUT FÜR MATHEMATIK, BONN, GERMANY

Current address: Department of Mathematics, Princeton University, Princeton, NJ 08544, USA

E-mail address: `smirnov@math.princeton.edu`

Contemporary Mathematics
Volume **212**, 1998

Boundary Value Problems Treated with Methods of Clifford Analysis

Wolfgang Sprössig

ABSTRACT. A growing number of people is using Clifford analytic methods for the consideration of partial differential equations. In this paper we intent to explain applications to some types of nonlinear boundary value problems for special systems of partial differential equations.

1. Introduction

The study and application of methods of Clifford analysis and especially quaternionic analysis has recently pursued by a growing number of mathematicians and physicists. Starting with fundamental papers of R. Fueter [7], [8] in the early thirties and his follower W. Nef [19] in 1944 then Rose's work [20] on quaternion velocity potentials for axial-symmetric fluid flow were some of the first who applied analytic methods of (complex) quaternions representing solutions of partial differential equations. In this context should be mentioned the paper of Stein-Weiss [24] on generalized Cauchy-Riemann systems. Several authors ([4], [5], [13], [14], [15], [16], [23]) have been studied Maxwell equations and general field equations. Using special results of Clifford analysis (cf. [3]) in [10] firstly were investigated with quaternionic methods wide classes of boundary value problems of elliptic partial differential equations including a suitable numerical approach. Also other authors (cf. [21], [22], [25], [27]) dealt with boundary value problems in connection with hypercomplex analysis.

The aim of this contribution is to apply Clifford regular methods for the treatment of classes of non-linear boundary value problems.

2. Preliminaries and basic formulas

Let $\{e_0 = 1, e_1, \dots, e_n\}$ be a basis in $\mathbb{R}^{n+1}$. Consider the Clifford algebra $Cl_{0,n}$ with the basis $\{1; e_1, \dots, e_n; e_1 e_2, e_1 e_3, \dots, e_{n-1} e_n; \dots; e_1 e_2 \dots e_n\}$. Note that in our case $e_i^2 = -e_0$ for $i = 1, 2, \dots, n$. Clearly are fulfilled the relations $e_i e_j + e_j e_i =$

1991 *Mathematics Subject Classification.* Primary 35Q20, 30G35; Secondary 35J65, 78A30.

0. With the abbreviation $e_{i_1 \ldots i_k} := e_{i_1} \ldots e_{i_k}$ each element $u \in Cl_{0,n}$ has the form

$$u = u_0 + \sum_{k=1}^{n} \sum_{(i_1, \ldots, i_k)} u_{i_1 \ldots i_k} e_{i_1 \ldots i_k},$$

where $1 \leq i_1 \leq \cdots \leq i_k \leq n$. Furthermore we will define by the help of

$$\overline{e_{i_1 \ldots i_k}} := (-1)^{\frac{k(k+1)}{2}} e_{i_1 \ldots i_k}$$

the conjugation of u by

$$\bar{u} = \sum_{k=1}^{n} \sum_{(i_1, \ldots, i_k)} u_{i_1 \ldots i_k} \overline{e_{i_1 \ldots i_k}}.$$

Now we write for short $u_0 =: \operatorname{Re} u$ and $u - u_0 =: \operatorname{Im} u$. We assume $G \subset \mathbb{R}^n$ be a domain with a Liapunov boundary Γ and $x = x_0 + \underline{x}$ with $\underline{x} = \sum_{k=1}^{n} x_k e_k$. For each paravector $x \in \mathbb{R} + \mathbb{R}^n$ we have $x\bar{x} = x_0^2 + x_1^2 + \cdots + x_n^2$. Let $B(\Omega)$ be a Banach space of functions defined on Ω. Then the space $B(\Omega) \otimes Cl_{0,n}$ is denoted by $B(\Omega, Cl_{0,n})$, $\Omega \in \{G, \Gamma\}$.

Define the *Dirac operator*

$$D = \sum_{k=1}^{n} e_k \partial_i$$

in the vector space $\mathbb{R}^n$, where $\partial_i = \partial/\partial x_i$. Note that $D^2 = -\Delta$ and $\overline{D} = -D$. Functions which are defined in G and fulfil the equation

$$Du = 0 \qquad ((uD) = O)$$

are called *left (right) Clifford regular*. Functions which are both left and right Clifford regular are called *two-sided Clifford regular*. In this paper we will use only left Clifford regular functions which we will call more simple *Clifford regular*. We define now the so-called *Cauchy kernel* on $\mathbb{R}^n \setminus \{0\}$ by

$$e(x) = \frac{1}{\omega_n} \frac{-x}{|x|^n},$$

where ω_n denotes the surface of the unit ball in $\mathbb{R}^n$. It is well-known that $e(x)$ is two-sided Clifford regular.

Using the function $e(x)$ we introduce the following integral operators namely the so-called *Teodorescu transform*

$$(T_G u)(x) := - \int_G e(x - y) u(y) dy, \qquad x \in \mathbb{R}^n,$$

the *Cauchy-type operator*

$$(F_\Gamma u)(x) := \int_\Gamma e(x - y) n(y) u(y) d\Gamma_y, \qquad x \notin \Gamma,$$

and a multidimensional *singular integraloperator of Cauchy-type*

$$(S_\Gamma u)(x) := 2 \int_\Gamma e(x - y) n(y) u(y) d\Gamma_y, \qquad x \in \Gamma,$$

where $n(y) = \sum_{k=1}^{n} e_k n_k(y)$ is the unit vector of the outer normal at the point y. The integral which defines the operator S_Γ is to understand in the sense of Cauchy's principal value.

PROPOSITION 2.1. *Let be* Γ *a Lipschitz surface and let the singular integral* $(S_\Gamma u)(x)$ *exists in this point* $x \in \Gamma$. *Then the operator* S_Γ^β *which is defined by*

$$(S_\Gamma^\beta u)(x) := \frac{(2\beta(x) - \omega_n)}{\omega_n} u(x) + (S_\Gamma u)(x)$$

with $0 < \beta(x) < \omega_n$ *satisfies again the algebraic relation* $(S_\Gamma^\beta)^2 = I$. *Here* $\beta(x)$ *denotes the outer angle in the point* $x \in \Gamma$ *taken as part of the unit ball.*

PROPOSITION 2.2. *Introducing the operators* $P_\Gamma^\beta := 1/2(I + S_\Gamma^\beta)$ *and* $Q_\Gamma^\beta := 1/2(I - S_\Gamma^\beta)$, *then we have*
   (i) $((P_\Gamma^\beta)^2 u)(\xi) = (P_\Gamma^\beta u)(\xi)$, $((Q_\Gamma^\beta)^2 u)(\xi) = (Q_\Gamma^\beta u)(\xi)$.
   (ii) *The subspace* $\operatorname{im} P_\Gamma^\beta \cap C^{0,\alpha}(\Gamma, Cl_{0,n})$, $0 < \alpha < 1$ *describes the space of all* $Cl_{0,n}$*-valued functions which are left Clifford regular extendable into the domain* G.
   (iii) *The subspace* $\operatorname{im} Q_\Gamma^\beta \cap C^{0,\alpha}(\Gamma, Cl_{0,n})$ *describes the space of all* $Cl_{0,n}$*-valued functions which are left Clifford regular extendable into the domain* $R^n \setminus \overline{G}$ *and vanish at infinity.*

REMARK 2.3. In case of $n = 3$ the singular integral operator S_Γ was firstly studied by A. W. Bizadse [2] (1953–1955) on Liapunov surfaces. L. G. Magnaradse gave in 1957 an elementary proof for the algebraic relation $(S_\Gamma^2 u)(x) = u(x)$ in case of Hoelder continous functions. Recently the operator S_Γ is being studied on Lipschitz surfaces [18].

Very important are relations between the operators D, T_G and F_Γ as well as the behaviour of F_Γ near to the boundary Γ.

PROPOSITION 2.4. (Gürlebeck-Sprössig [11]). *Let*

$$u \in C^1(G, Cl_{0,n}) \cap C(\bar{G}, Cl_{0,n}).$$

Then we have
   (i) *Borel-Pompeiu formula:*

$$(F_\Gamma u)(x) + (T_G Du)(x) = \begin{cases} u(x), & x \in G \\ 0, & x \in \mathbb{R}^n \setminus \overline{G} \end{cases}$$

   (ii) *Right invertibility:*

$$(DT_G u)(x) = \begin{cases} u(x), & x \in G \\ 0, & x \in \mathbb{R}^n \setminus \overline{G} \end{cases}$$

   (iii) *Plemelj-Sokhotzki formulae: Let* $u \in C^{0,\alpha}(\Gamma, Cl_{0,n})$, $0 < \alpha < 1$, Γ *a Lipschitz surface and* $(S_\Gamma u)(\xi)$ *exists, then*

$$\lim_{\substack{x \to \xi \in \Gamma \\ x \in G}} (F_\Gamma u)(x) = (P_\Gamma^\beta u)(\xi)$$

$$\lim_{\substack{x \to \xi \in \Gamma \\ x \in \mathbb{R}^n \setminus \overline{G}}} (F_\Gamma u)(x) = -(Q_\Gamma^\beta u)(\xi).$$

REMARK 2.5. The operators F_Γ, S_Γ, P_Γ and Q_Γ can be extented to $L_p(\Gamma)$, $1 < p < \infty$.

3. On classes of boundary value problems

Let be $G \subset \mathbb{R}^3$ a domain with the Liapunov boundary Γ and $e_0 = 1$, e_1, e_2, $e_1 e_2 = e_3$ a basis in the algebra of real quaternions $\mathbb{H} = Cl_{0,2}$. An arbitrary element $u \in \mathbb{H}$ can be written as

$$u = \sum_{i=1}^{3} u_i e_i = u_0 + \underline{u}.$$

Furthermore let J be an operator defined by $Ju := u_0 - \underline{u}$. The Dirac operator D has in this algebra the form $D = \sum_{i=1}^{3} e_i \partial_i$.

Consider now a boundary value problem of the type:

(1) $$D(a + bJ)Du = f(u) \quad \text{in} \quad G$$

(2) $$u = g \quad \text{on} \quad \Gamma,$$

where $f : W_2^1(G, \mathbb{H}) \longmapsto L_2(G, \mathbb{H})$ is a non-linear operator which satisfies the conditions:

(i) $\|f(\cdot, u)\|_{L_2} \le M$, $M > 0$,
(ii) $\|f(\cdot, u_1) - f(\cdot, u_2)\|_{L_2} \le L \|u_1 - u_2\|_{W_2^1}$, $L > 0$.

The $\mathbb{H}$-valued function g on the right hand side belongs to $W_2^{3/2}(\Gamma, \mathbb{H})$.

EXAMPLE 3.1. (Kravchenko [17]). Choosing $a = -m/4(m-1) + 1$, $b = -m/4(m-1)$, where m denotes the Poisson number ($2 \le m \le 4$). Then the problem (1)–(2) with $f(\cdot, \underline{u}) = h(\cdot, \underline{u}) - \Theta^2 \rho \underline{u}$ and $\|h(\cdot, \underline{u}_1) - h(\cdot, \underline{u}_2)\|_{L_2} \le L \|\underline{u}_1 - \underline{u}_2\|_{W_2^1}$ is just the first boundary value problem of the equations of steady oscillation in elasticity. Hereby Θ denotes the oscillation frequency, $\underline{u}$ the displacement vector and ρ the density.

LEMMA 3.2. *The quaternionic Hilbert space $L_2(G, \mathbb{H})$ permits the orthogonal decomposition*

$$L_2(G, \mathbb{H}) = (c + dJ)\ker D \cap L_2(G, \mathbb{H}) \oplus_{a+bJ} DW_2^1(G, \mathbb{H})$$

relatively to the inner product

$$[u, v]_{a+bJ} := \int_G (c + dJ)\overline{u}(c + dJ)v \, dG,$$

where $c + dJ = (a + bJ)^{-1}$, $c = (1/2)[(a + b)^{-1} + (a - b)^{-1}]$ and $d = (1/2)[(a + b)^{-1} - (a - b)^{-1}]$

COROLLARY 3.3. *The orthoprojections onto the subspaces $(c + dJ)\ker D \cap L_2(G, \mathbb{H})$, $DW_2^1(G, \mathbb{H})$ are operators*

$$\mathcal{P}_{c+dJ} := (c + dJ)F_\Gamma (tr_\Gamma T_G(c + dJ)F_\Gamma)^{-1} tr_\Gamma T_G,$$
$$\mathcal{Q}_{c+dJ} := I - \mathcal{P}_{c+dJ}$$

PROOF. This follows analogously to Proposition 4.36 in [10].

THEOREM 3.4. *The boundary value problem* (1)–(2) *permits in the neighborhood* $B(r(g)) \subset W_2^1(G, H)$ *of* $r(g) := F_\Gamma g + T_G \mathcal{P}_{c+dJ} Dg$, *where* g *is a smooth continuation of* g *into the domain* G *a unique solution which can be achieved by the iteration procedure*

$$u_n - r(g) = T_G \mathcal{Q}_{c+dJ}(c + dJ) T_G f(u_{n-1}), \qquad tr_\Gamma u_n = g \quad, u_0 = r(g).$$

This procedure converges in $W_2^1(G, \mathbb{H})$ *if the following condition*

$$(3) \qquad L|G|^{1/3} \max_G |a + bJ| < \left(\frac{4\pi}{3} \right)^{1/3} \left(\frac{\lambda_1}{\lambda_1 + 1} \right)^{1/2},$$

is fulfilled, where λ_1 *is the first eigenvalue of Dirichlet's problem over the domain* G.

PROOF. Similarly to Theorem 3 in [11] we get from Proposition 2.4 and the property $F_\Gamma T_G = 0$ in G that the boundary value problem (1)–(2) is equivalent to the operator integral equation

$$u = F_\Gamma g + T_G \mathcal{P}_{c+dJ} D\tilde{g} + T_G \mathcal{Q}_{c+dJ} T_G f(u)$$

with $\tilde{g}$ denotes a smooth extension of g into the domain G. Furthermore Theorem 4.1.14 [10] yields us immediately

$$\lambda_1(G) = \frac{1}{\dfrac{\|T_G\|^2}{[\operatorname{im} \mathcal{Q}_{c+dJ}, \overset{\circ}{W_2^1}]} - 1};$$

here λ_1 is the first eigenvalue of Dirichlet's problem. Obviously the norms of $\mathcal{P}_{c+dJ}$ and $\mathcal{Q}_{c+dJ}$ in $L_2(G, \mathbb{H})$ equal one. It remains only to use the estimation in Theorem 2.3 in [12] namely

$$\frac{1}{\sqrt{\lambda_1}} \leq \|T_G\|_{L(C)} \leq \left(\frac{3}{4\pi} \right) |G|^{1/3}.$$

Then condition (3) is necessary in order to prove convergence by Banach's fixed-point principle.

These results can be transferred to more general situations. For this reason we introduce the matrix operators

$$\mathcal{D} = \begin{pmatrix} D & 0 \\ 0 & D \end{pmatrix}, \qquad \mathcal{M} = \begin{pmatrix} \alpha & \beta \\ \gamma & \delta \end{pmatrix},$$

where $\alpha, \beta, \gamma, \delta$ are multiplication operators of the type $\alpha = \alpha_1 + \alpha_2 J$, $a_i \in C_{\mathbb{R}}^\infty(G)$.
Let

$$U = \begin{pmatrix} u_1 \\ u_2 \end{pmatrix}, \qquad u_i = \sum_{k=0}^3 u_{ik} e_k, \qquad \Phi(U) = \begin{pmatrix} \phi_1 & \phi_2 \\ \phi_3 & \phi_4 \end{pmatrix}(U),$$

where ϕ_i, $i = 1, 2, 3, 4$ are differential operators of first order,

$$H = \begin{pmatrix} h_1 \\ h_2 \end{pmatrix}, \qquad g = \begin{pmatrix} g_1 \\ g_2 \end{pmatrix}.$$

Now we consider the non-linear boundary value problem

$$(4) \qquad \mathcal{D}\mathcal{M}\mathcal{D}U = \Phi(U) + H \quad \text{in} \quad G$$

$$(5) \qquad U = g \quad \text{on} \quad \Gamma$$

with $\det \mathcal{M} \neq 0$.

COROLLARY 3.5. *Let*

$$g \in W_2^{3/2}(\Gamma, \mathbb{H}) \quad and \quad \|\Phi(U^1) - \Phi(U^2)\|_{L_2} \leq L\|U^1 - U^2\|_{W_2^1}.$$

The boundary value problem (3)-(4) permits in the neighborhood of

$$R(g) := \mathcal{F}_\Gamma g + \mathcal{V}_n (tr_\Gamma \mathcal{V}_n)^{-1} \mathcal{Q}_\Gamma$$

with

$$\mathcal{F}_\Gamma = \begin{pmatrix} F_\Gamma & 0 \\ 0 & F_\Gamma \end{pmatrix}, \qquad \mathcal{Q}_\Gamma = \begin{pmatrix} Q_\Gamma & 0 \\ 0 & Q_\Gamma \end{pmatrix},$$

$$\mathcal{V}_n = \begin{pmatrix} V_n & 0 \\ 0 & V_n \end{pmatrix}, \qquad (V_n u)(x) = \frac{1}{4\pi} \int_\Gamma \frac{n(y)}{x - y} u(y) d\Gamma_y$$

a unique solution in $W_2^1(G, \mathbb{H})$ which can be obtained by the iteration procedure

$$U_n - R(g) = \mathcal{T}\mathcal{Q}\mathcal{M}^{-1}\mathcal{T}\mathcal{H}(U_{n-1})$$
$$U_0 = R(g) \qquad n = 1, 2, \ldots$$

where $\mathcal{H}(U) = \Phi(U) + H$.
 This procedure converges in $W_2^1(G, \mathbb{H})$ if

$$L|G|^{1/3}\|\mathcal{M}^{-1}\|_{C(G)} < \frac{1}{2}\left(\frac{4\pi}{3}\right)^{1/3}\left(\frac{\lambda_1}{\lambda_1 + 1}\right)^{1/2}.$$

PROOF. The proof is an immediately consequence of Theorem 3.4.

EXAMPLE 3.6. (Kravchenko [17]). The time harmonic equations in the moment theory in linear elasticity which are defined over a domain G read as follows

$$D(\mu + \gamma)D\underline{u} + \text{grad}(\lambda + \mu - \gamma)\text{div}\,\underline{u} + 2\Lambda\text{rot}\,\underline{w} = h_1$$
$$D(\nu + \Theta)D\underline{w} + \text{grad}(\epsilon - \nu - \Theta)\text{div}\,\underline{w} + 2\Lambda\text{rot}\,\underline{u} - 4\Lambda\underline{w} = h_2.$$

On the boundary Γ we have Dirichlet conditions e.g., $\underline{u} = g_1$ and $\underline{w} = g_2$.
 All occuring constants have a well-defined physical meaning and have for this reason to satisfy the conditions $\mu > 0$, $3\lambda + 2\mu > 0$, $\Lambda > 0$, $\varepsilon > 0$, $3\varepsilon + 2\nu > 0$, $\Theta > 0$. We should mention that $\underline{u}$ is the vector of the displacements and with $\underline{w}$ we denote the rotation vector.
 The entries of the above introduced matrix $\mathcal{M}$ we have to choose now in the following way

$$\beta := \gamma := 0, \quad \alpha := \alpha_1 + \alpha_2 J = \frac{\lambda + 3\mu + \gamma}{2} + \frac{\lambda + \mu - \gamma}{2}J$$
$$\delta := \delta_1 + \delta_2 J = \frac{\varepsilon + 3\nu + \Theta}{2} + \frac{\varepsilon + \nu - \Theta}{2}J.$$

Furthermore we have to set

$$\phi_1 = 0, \quad \phi_2 = \phi_3 = (-DJ + JD), \quad \phi_4 = -4$$
$$\mathcal{M}^{-1} = \left\{ \begin{vmatrix} \alpha_1 & \alpha_2 \\ \delta_1 & \delta_2 \end{vmatrix} + \begin{vmatrix} \alpha_1 & -\alpha_2 \\ \delta_1 & \delta_2 \end{vmatrix} J \right\}^{-1} \begin{pmatrix} \delta & 0 \\ 0 & \alpha \end{pmatrix}.$$

4. On a modified quaternionic operator calculus

In [16] is given the fundamental solution of the operator $D+a$, $a = a_0+a_1i \in \mathbb{C}$. This solution reads as follows

$$e_a(x) = -\frac{1}{4\pi|x|}e^{-ia|x|}\left(a + \frac{x}{|x|^2} + ia\frac{x}{|x|}\right)$$

For $a = a_0 + \underline{a} \in \mathbb{H}$ we get the fundamental solution of $(D + a)u = 0$ (cf.[16])

$$e_a(x) = e^{-\langle \underline{a},x\rangle}\operatorname{Re}_{\mathbb{C}}e_{a_0}(x), \qquad x \neq 0$$

where

$$(6) \qquad \operatorname{Re}_{\mathbb{C}}e_{a_0}(x) = -\frac{1}{4\pi|x|}\left[a_0\cos a_0|x| + \frac{x}{|x|^2}\cos a_0|x| + a_0\frac{x}{|x|}\sin a_0|x|\right].$$

Similarly to the real case (cf. [10]) one also could deduce formulae of Borel-Pompeiu type

$$F_au + T_aD_au = D_aT_au = \begin{cases} u & \text{in } G \\ 0 & \text{in } \mathbb{R}^3 \setminus G \end{cases}$$

and corresponding Plemelj-Sokhotzki's formulae. Basing on results obtained in [10] the following orthogonal decomposition of the quaternionic Hilbert space is available.

PROPOSITION 4.1. (Sprössig [23]). *Let G be a domain which is symmetrically to the origin. If we furnish $L_2(G,\mathbb{H})$ with the inner product*

$$(u, v) = \int_G \overline{u(x)}v(x)dx,$$

then one gets the orthogonal decomposition

$$L_2(G,\mathbb{H}) = \ker D_a \cap L_2(G,\mathbb{H}) \oplus D_{\overline{a}}\overset{\circ}{W_2^1}(G,\mathbb{H}).$$

The corresponding orthoprojections are given by

$$\mathcal{P}_a = F_a(tr_\Gamma V_a)^{-1}tr_\Gamma T_{\overline{a}} \quad \text{and} \quad \mathcal{Q} = I - \mathcal{P}_a.$$

Hereby V_a denotes the following single layer potential

$$(V_au)(x) = \frac{1}{4\pi}\int_\Gamma \frac{\cos(a_0|x-y|)e^{-\langle \underline{a},x-y\rangle}}{|x-y|}n(y)u(y)d\Gamma_y.$$

PROPOSITION 4.2. *For $u \in \operatorname{im}\mathcal{Q}_a$ is necessary and sufficient that $tr_\Gamma T_{\overline{a}}u = 0$.*

PROOF. 1. Let $u \in \operatorname{im}\mathcal{Q}$ then

$$tr_\Gamma T_{\overline{a}}u = tr_\Gamma T_{\overline{a}}\mathcal{Q}_au = tr_\Gamma T_{\overline{a}}D_{\overline{a}}w = tr_\Gamma w = 0,$$

where $w \in \overset{\circ}{W_2^1}(G,\mathbb{H})$.

2. Assume now that u satisfy the condition $tr_\Gamma T_{\overline{a}}u = 0$. From Proposition 4.1 we get that u admits the decomposition $u = u_1 + u_2$ where $u_1 \in \ker D_a$ and $u_2 \in \operatorname{im}\mathcal{Q}_a$. Denote with $A := -\Delta + 2\overline{a}D - \langle \underline{a}, D\rangle$. Obviously the function $T_{\overline{a}}u_1$ is a solution of the special Dirichlet problem

$$(A + |a|^2)v = 0 \quad \text{in } G \quad \text{and} \quad v = 0 \quad \text{on } \Gamma.$$

On the boundary it follows

$$0 = tr_\Gamma T_{\bar{a}} u_1 + tr_\Gamma T_a u_2 = tr_\Gamma T_{\bar{a}} u_1.$$

Because of $\text{sign}|a|^2 > 0$, we have $T_{\bar{a}} u_1 = 0$ in G and so $u_1 = 0$. Hence $u = u_2 \in \text{im} \mathcal{Q}_a$.

EXAMPLE 4.3. We consider the boundary value problem of the following partial differential equation of order $2n$: Let

$$\prod_{i=1}^{n} D_{a_i} D_{\bar{a}_i} u = f \quad \text{in } G, \qquad a_i \in \mathbb{H}, \ f \in L_2(G, \mathbb{H}), \ g_i \in W_2^{2n-(4i+1)/2}(\Gamma, \mathbb{H}).$$

On the boundary Γ we have the conditions

$$u = g_0, \qquad D_{a_1} D_{\bar{a}_1} u = g_1, \dots, D_{a_{n-1}} D_{\bar{a}_{n-1}} \cdots D_{a_1} D_{\bar{a}_1} u = g_{n-1}.$$

The unique solution $u \in W_2^{2n}(G, \mathbb{H})$ is then explicitly given by the formula

$$u = r_1(g) + T_{a_1} \mathcal{Q}_{a_1} T_{\bar{a}_1} r_2(g) + \cdots + \prod_{i=1}^{n-1} T_{a_i} \mathcal{Q}_{a_i} T_{\bar{a}_i} r_n(g_{n-1}) + \prod_{i=1}^{n} T_{a_i} \mathcal{Q}_{a_i} T_{\bar{a}_i} f,$$

where

$$r_k(g_{k-1}) := F_{\bar{a}_k} g_{k-1} + T_{\bar{a}_k} F_{a_k} (tr_\Gamma T_{\bar{a}_k} F_{a_k})^{-1} \mathcal{Q}_{\bar{a}_k} g_{k-1}.$$

Special case: $(-1)^n \Delta^n u = f$ in G and $\Delta^i u = g_i$, $i = 1, \dots, n-1$ on Γ permits the unique solution

$$u = r(g_0) + T\mathcal{Q}Tr(g_1) + \cdots + (T\mathcal{Q}T)^{n-1} r(g_{n-1}) + (T\mathcal{Q}T)^n f$$

with $r(g_k) := \left[F_\Gamma + V_0 (tr_\Gamma V_0)^{-1} \mathcal{Q}_\Gamma \right] g_k$.

REMARK 4.4. Under the assumption that componentwise $ab > 0$, where $a, b \in \mathbb{H}$ then by the aid of Hopf's maximum principle we obtain also a decomposition of the quaternionic Hilbert space in the direct sum:

$$L_2(G, \mathbb{H}) = \ker D_a \cap L_2(G, \mathbb{H}) + D_b \overset{\circ}{W_2^1}(G, \mathbb{H}).$$

Projections onto these subspaces are explicitly described by the formulae

$$\mathcal{P}_{ab} := F_a (tr_\Gamma T_b F_a)^{-1} tr_\Gamma T_b \quad \text{and} \quad \mathcal{Q}_{ab} := I - \mathcal{P}_{ab}.$$

In this way one gains representation formulae of problems of the type

$$(D + a)(D + b)u = f \quad \text{in} \quad G$$
$$u = g \quad \text{on} \quad \Gamma.$$

EXAMPLE 4.5. Let be $G \subset \mathbb{R}^3$ a bounded domain symmetric relatively to the origin. An electric charge is distributed with the density $\rho(x)$. We shall compute the electric field $\underline{E}(t, x) = E_0(t)\underline{E}_1(x)$ and the magnetic field $\underline{H}(t, x) = H_0(t)\underline{H}_1(x)$. Assume the dielectric "constant" ϵ, the permeability μ and the electric conductivity κ depend on the position x and the time t e.g., $\epsilon = \epsilon(t, x)$, $\mu = \mu(t, x)$ and $\kappa = \kappa(t, x)$.

Maxwell's equations reads now as follows:

$$\text{rot}\,\underline{E} = \partial_t(\mu\underline{H})$$
$$\text{rot}\,\underline{H} = \partial_t(\epsilon\underline{E}) + \kappa\underline{E}$$
$$\text{div}\,(\epsilon\underline{E}) = \rho$$
$$\text{div}(\mu\underline{H}) = 0.$$

Suppose now that $\underline{E}$ and $\underline{H}$ admit the factorizations

$$\underline{E} = E_0(t)\underline{E}_1(x)$$
$$\underline{H} = H_0(t)\underline{H}_1(x).$$

Now we put these expressions in Maxwell's equations and obtain

$$E_0\,\text{rot}\,\underline{E}_1 = -\partial_t(\mu H_0)\underline{H}_1$$
$$H_0\,\text{rot}\,\underline{H}_1 = (\partial_t(\epsilon E_0) + \kappa E_0)\underline{E}_1$$
$$E_0\,\text{grad}\,\epsilon \cdot \underline{E}_1 + \epsilon\,\text{div}\,\underline{E}_1) = \rho$$
$$\text{grad}\,\mu \cdot \underline{H}_1 + \mu\,\text{div}\,\underline{H}_1 = 0.$$

In this way we obtain with $D = \sum_1^3 \partial_i e_i$

$$\text{rot}\,\underline{E}_1 = -\frac{\partial_t \mu H_0}{E_0}\underline{H}_1$$
$$\text{rot}\,\underline{H}_1 = \frac{\partial_t(\epsilon E_0) + \kappa E_0}{H_0}\underline{E}_1$$
$$-\text{div}\,\underline{E}_1 = -\frac{\rho}{\epsilon E_0} + \left(\frac{D\epsilon}{\epsilon}\right) \cdot \underline{E}_1$$
$$-\text{div}\,\underline{H}_1 = \left(\frac{D\mu}{\mu}\right) \cdot \underline{H}_1.$$

Note that with "$\cdot$" is just denoted the scalar product, e.g., $\underline{u} \cdot \underline{v} = \sum_1^3 u_i v_i$. Writing for short

$$\underline{a} = \frac{D\mu}{\mu}, \qquad \underline{d} = \frac{D\epsilon}{\epsilon}, \qquad \rho' = -\frac{\rho}{\epsilon E_0},$$
$$b_0 = \frac{\partial_t(\epsilon E_0) + \kappa E_0}{H_0}, \qquad c_0 = -\frac{\partial_t(\mu H_0)}{E_0}$$

and such

(7) $$\qquad D\underline{E}_1 = c_0\underline{H}_1 + \underline{d} \cdot \underline{E}_1 + \rho'$$
(8) $$\qquad D\underline{H}_1 = \underline{a} \cdot \underline{H}_1 + b_0\underline{E}_1.$$

We get the matrix equation

$$\begin{pmatrix} D & 0 \\ 0 & D \end{pmatrix} \begin{pmatrix} \underline{E}_1 \\ \underline{H}_1 \end{pmatrix} = \begin{pmatrix} \underline{d}\cdot & c_0 \\ b_0 & \underline{a}\cdot \end{pmatrix} \begin{pmatrix} \underline{E}_1 \\ \underline{H}_1 \end{pmatrix} + \begin{pmatrix} \rho' \\ 0 \end{pmatrix}.$$

Assume now that $\underline{d}$ is linearly dependent on $\underline{E}_1$ and $\underline{a}$ is linearly dependent on $\underline{H}_1$, then one has $\underline{d} \cdot \underline{E}_1 = -d\underline{E}_1$ and also $\underline{a} \cdot \underline{H}_1 = -a\underline{H}_1$. From (1)–(2) we obtain the matrix equation

$$\begin{pmatrix} D - \underline{d} & 0 \\ 0 & D - \underline{a} \end{pmatrix} \begin{pmatrix} \underline{E}_1 \\ \underline{H}_1 \end{pmatrix} = \begin{pmatrix} 0 & c_0 \\ b_0 & 0 \end{pmatrix} \begin{pmatrix} \underline{E}_1 \\ \underline{H}_1 \end{pmatrix} + \begin{pmatrix} \rho' \\ 0 \end{pmatrix}.$$

In case of $\underline{a}$ and $\underline{d}$ do not depend of the position, e.g., $\underline{a} = \underline{a}(t)$ and also $\underline{d} = \underline{d}(t)$ then we have only to use the generalized Borel-Pompeiu formula

$$u(x) = (F_\lambda u)(x) + (T_\lambda(D + \lambda))u(x)$$

which is valid for $u \in W_2^1$. $\lambda = \underline{\lambda}$ belongs to the real quaternions. Obviously $\lambda = \sum_1^3 \lambda_i e_i$. Furthermore the fundamental solution of the operator $D + \lambda$ reads as follows

$$e_\lambda(x) = \left(-\frac{x}{4\pi |x|^3} \right) e^{-\lambda \cdot x}$$

with $x = \underline{x} = \sum_1^3 x_i e_i$.

The corresponding operators reads

$$(T_\lambda u)(x) = - \int_G e_\lambda(x - y)u(y)dG_y$$

$$(F_\lambda u)(x) = \int_\Gamma e_\lambda(x - y)n(y)u(y)d\Gamma_y$$

with $\Gamma = \partial G$. By multiplication from the left with the matrix operator

$$\begin{pmatrix} T_{-\underline{d}} & 0 \\ 0 & T_{-\underline{a}} \end{pmatrix}$$

yields

$$\begin{pmatrix} \underline{E}_1 \\ \underline{H}_1 \end{pmatrix} = \begin{pmatrix} 0 & T_{-\underline{d}}c_0 \\ T_{-\underline{a}}b_0 & 0 \end{pmatrix} \begin{pmatrix} \underline{E}_1 \\ \underline{H}_1 \end{pmatrix} + \begin{pmatrix} T_{-\underline{d}}\rho' \\ 0 \end{pmatrix} + \begin{pmatrix} \Phi_{-\underline{d}} \\ \Phi_{-\underline{a}} \end{pmatrix}$$

where the functions Φ_α in ker D_α. This leads now to the integral representations

$$\underline{E}_1 = T_{-\underline{d}}c_0\underline{H}_1 + T_{-\underline{d}}\rho' + \Phi_{-\underline{d}}$$
$$\underline{H}_1 = T_{-\underline{a}}b_0\underline{E}_1 + \Phi_{-\underline{a}}.$$

Separating these equations we find

$$\underline{E}_1 = T_{-\underline{d}}c_0 T_{-\underline{a}}b_0\underline{E}_1 + T_{-\underline{d}}c_0\Phi_{-\underline{a}} + T_{-\underline{d}}\rho' + \Phi_{-\underline{d}}.$$

REMARK 4.6. Multiplying equations (1)–(2) separately by the integral operator $T := T_0$ then one get integral equations without restrictions for the coefficients. Nevertheless seems the use of the kernels $e_{-a}(x - y)$ and $e_{-d}(x - y)$ gives more informations on the behaviour of the solutions as the use of $e(x - y)$.

We will consider Dirichlet's boundary value problem of the Yukawa equation. It is well-known from several authors (cf. [16], [27], [1]) that the fundamental solution of $(-D + a_0)u = 0$ can be given by

$$E_{ia_0} = e(x) \left((1 + a_0(x))e^{-a_0|x|} - i\frac{a_0 e^{-a_0|x|}}{2\pi|x|} \right), \qquad a_0 \in \mathbb{R}.$$

In two dimensions there is a complete theory worked out by R. J. Duffin in 1956 [6]. U. Wimmer [26] could generalize Proposition 4.1 to the Hilbert space of complex quaternions $L_2(G, \mathbb{C}\mathbb{H})$. Using this decomposition one get a representation of the solution of the boundary value problem

$$-\Delta u + |a|^2 u = f \qquad \text{in} \quad G$$
$$u = g \qquad \text{on} \quad \Gamma.$$

We have the formula

$$u = \operatorname{Re}_{\mathbb{C}} \left(F_{ai}g + T_{ai}F_{-\bar{a}i}(tr_\Gamma T_{ai}F_{-\bar{a}i})^{-1}Q_{ai}g + T_{ai}Q_{ai(-ai)}T_{-\bar{a}i}f \right).$$

REMARK 4.7. In the habilitation paper of K. Gürlebeck [9] (see also [10]) has been developed a corresponding discrete operator calculus. In this way all representation formulas have the additional advantage to allow numerical implementations.

REMARK 4.8. Note here that boundary value problems of Yukawa's equation arise in natural way in the treatment of parabolic equations with the method of semi-discretization. The corresponding coefficient a_0 is just the inverse meshwidth of the time-discretization.

5. Harmonic approximation

We consider in a bounded domain G the following non-linear boundary value problem:

$$Du = f(u) \qquad \text{in} \quad G$$
$$u = g \qquad \text{on} \quad \Gamma$$

The operator $f : W_2^1(G, \mathbb{H}) \longmapsto W_2^2(G, \mathbb{H})$ satisfy the condition $\|f(u)\|_{W_2^1(G,\mathbb{H})} \leq M$ and $g \in W_2^{3/2}(\Gamma, H)$. Suppose now that we already know that a unique solution $u^* \in W_2^1(G, \mathbb{H})$ is existing.

THEOREM 5.1. *The solution u^* can be approximated by the harmonic quaternionic valued function u_h in the domain G_0*

$$u_h = F_{\Gamma_0}g + T_{G_0}F_{\Gamma_0}f(\tilde{g}),$$

where $\tilde{g}$ is a smooth continuation of the function g into the domain G up to a parallel surface Γ_1 and G_0 is just the "annulus" domain between both surfaces Γ_0 and Γ_1. Furthermore the following error estimate is valid:

$$\|u_h - u^*\|_{C(G,\mathbb{H})} \leq C|G_0|^{1/3}.$$

The constant C is bounded by $M(\frac{3}{4\pi})^{1/3}(|G|^{1/3} + \|\mathcal{P}_\Gamma\|)$ with $\Gamma = \Gamma_0 + \Gamma_1$.

6. On a class of strong non-linear boundary value problems

We consider boundary value problems of the type

(9) $$D(\phi(|Du|)Du = f \quad \text{in} \quad G$$

(10) $$u = g \quad \text{on} \quad \Gamma$$

with $\phi > 0$, $\phi \in C_\mathbb{R}^\infty(\mathbb{R}^+)$.

We try to find solutions with the following ansatz:

$$u = Dv, \qquad v \in \overset{\circ}{W}_2^1(G, \mathbb{H}).$$

After substitution we get the new problem

$$D(\phi(|\Delta v|)\Delta v) = f \quad \text{in} \quad G.$$

Next we look for solutions of the linear auxiliary problem

$$\Delta v = \psi \quad \text{in} \quad G$$
$$v = 0 \quad \text{on} \quad \Gamma$$

where $\psi \in W_2^1(G, \mathbb{H})$. By the help of our explanations in [10] we get immediately for the solution of this auxiliary problem the representation

$$v = T_G \mathcal{Q} T_G \psi.$$

Inserting this in (7)–(8) we achieve

$$D(\phi(|\psi|)\psi) = f \iff \phi(|\psi|)\psi = T_G f + F_\Gamma \eta$$

and from this

(11) $$\phi(|\psi|)|\psi| = |T_G f + F_\Gamma \eta|$$

where $\eta = \phi(|\psi|)\psi$. Assume now that there exists a function χ such that $|\psi| = \chi(|T_G f + F_\Gamma \eta|)$. Eventually we gain from (9) the subsequent representation of ψ

(12) $$\psi = \frac{T_G f + F_\Gamma \eta}{\phi(\chi(|T_G f + F_\Gamma \eta|))} = (T_G f + F_\Gamma \eta)m(f, \eta).$$

m is defined by the latter equation. Finally we get for u

$$u = \mathcal{Q} T_G (T_G f + F_\Gamma \eta)m(f, \eta)$$

We have only to prove that u fulfils the boundary condition (8). In order to verify this we start with the very simple decomposition of the expression $\mathcal{Q} T_G z$. It holds

$$\mathcal{Q} T_G z = T \mathcal{Q} z + (T_G \mathcal{P} z - \mathcal{P} T_G)z.$$

For the first term $tr_\Gamma T_G \mathcal{Q} z = 0$. The second part is harmonic as it is easy to see. From this we find

$$u = \Phi + (T_G \mathcal{P} - \mathcal{P} T_G)m(f, \eta)(T_G f + F_\Gamma \eta) + T_G \mathcal{Q} m(f, \eta)(T_G f + F_\Gamma \eta).$$

With

$$F_\Gamma g := \Phi - \mathcal{P} T_G m(f, \eta)(T_G f + F_\Gamma \eta)$$

we obtain now

$$u = F_\Gamma g + T_G \mathcal{P} m(f,\eta)(T_G f + F_\Gamma \eta) + T_g \mathcal{Q} m(f,\eta)(T_G f + F_\Gamma \eta)$$

This leads to the necessary condition

(13) $$Q_\Gamma g = tr_\Gamma T_G m(f,\eta)(T_G f + F_\Gamma \eta)$$

Special cases: In case of $\phi \equiv 1$ we have for

$$\eta = F_\Gamma (tr_\Gamma V_0)^{-1}(Q_\Gamma g - tr_\Gamma T_G^2 f).$$

Under the assumption $\eta|_\Gamma = 0$ it follows in another case

$$Q_\Gamma g = tr_\Gamma T_G \left(\frac{T_G f}{\phi(\chi(T_G f))} \right).$$

Let us finish our explanations with a simple but nevertheless important example.

EXAMPLE 6.1. Let be $\phi(t) = t^k$, $t > 0$ and $\phi > 0$. Then we get immediately

$$|\psi|^k \psi = T_G f + F_\Gamma |\psi|^k \psi = T_G f + F_\Gamma \eta.$$

Hence

$$|\psi|^{k+1} = |T_G f + F_\Gamma \eta|$$
$$\psi = \frac{T_G f + F_\Gamma \eta}{|T_G f + F_\Gamma \eta|^{k/(k+1)}}.$$

Obviously we have then $m(f,\eta) = |T_G f + F_\Gamma \eta|^{k/(k+1)}$. With $\eta = 0$ on the boundary Γ there holds

$$Q_\Gamma g = tr_\Gamma T_G \left(\frac{T_G f}{|T_G f|^{k/(k+1)}} \right).$$

If $g = 0$ on Γ then condition (10) reads as follows

$$\frac{T_G f}{|T_G f|^{k/(k+1)}} \in \text{im}\mathcal{Q}.$$

References

[1] S. Bernstein, *Fundamental solution of Dirac type operators*, Banach Center Publications, Banach Center Symposium: Generalizations of Complex Analysis, May 30–July 1,1994, Warsaw, 1995.

[2] A. W. Bizadse, *On two-dimensional integrals of Cauchy-type*, AN Grus. SSR **XVI** (1955), 177–184.

[3] F. Brackx, R. Delanghe, and F. Sommen, *Clifford analysis*, Pitman, Boston-London-Melbourne, 1982.

[4] G. Casanova, *L'algbre vectorielle*, Presses Universitaires de France, 1976.

[5] C. A. Deavours, *The quaternionic calculus*, Amer. Math. Month. **80** (1973), 995–1008.

[6] R. J. Duffin, *Basic properties of discrete analytic functions*, Duke Math. J. **31** (1956), 33–363.

[7] R. Fueter, *Die Funktionentheorie der Differentialgleichungen $\Delta u = 0$ und $\Delta\Delta u = 0$ mit vier reellen Variablen*, Comm. Math. Helv. **7** (1935), 307–330.

[8] R. Fueter, *Reguläre Funktionen einer Quaternionenvariablen*, Math. Inst. d. Universität Zürich, 1949.

[9] K. Gürlebeck, *Grundlagen einer diskreten räumlich verallgemeinerten Funktionentheorie und ihrer Anwendungen* Thesis, TU Karl-Marx-Stadt, 1988.

[10] K. Gürlebeck and W. Sprössig, *Quaternionic Analysis and Elliptic Boundary Value Problems*, Birkhäuser-Verlag, Basel, 1990.

[11] K. Gürlebeck and W.Sprössig, *Clifford Analysis and Elliptic Boundary Value Problems*, Clifford algebras and Spinor structures R. Ablamowicz and P. Lounesto (eds.), Kluver Academ. Publ., 1995, pp. 325–334.

[12] K. Gürlebeck and W.Sprössig, *A unified Approach to Estimation of Lower Bounds for the first Eigenvalue of several Elliptic Boundary Value Problems*, Math.Nachr. **131** (1987), 183–199.

[13] D. Hestenes, *New Foundations for Classical Mechanics*, Reidel Publ. Co., Dortrecht-Boston, 1985.

[14] C. Imaeda, *A new formulation of classical electrodynamics*, Nuovo Cimento **32 A** (1976), 138–162.

[15] B. Jancewicz, *Multivectors and Clifford Algebra in Electrodynamics*, World Scientific Publ., Singapore, 1988.

[16] V. V. Kravchenko, *On the relation between holomorphic biquaternionic functions and time-harmonic electromagnetic fields*, Deposited in UKREINTEI 29.12.92 under Nr. 2073-Uk-92 (in Russian).

[17] V. V. Kravchenko, *On a hypercomplex factorization of some equations of mathematical physics*, Preprint #1, Dept. of Math., Instituto Superior Tecnico, Lisboa, Portugal (1993), 8 pp.

[18] A. McIntosh, *Clifford algebras and the higher dimensional Cauchy integral*, Approximation theory and Function spaces, Banach Center Publications **22** (1989), pp. 253–267.

[19] W. Nef, *Die Funktionentheorie der partiellen Differentialgleichungen zweiter Ordnung (Hyperkomplexe Funktionentheorie)*, Bull. Soc. Fribourgoise Sc. Nat. **37** (1944), 348–37.

[20] A. Rose, *On the use of a complex (quaternion) velocity potential in three dimensions*, Comment. Math. Helv. **24** (1950), 135–148.

[21] J. Ryan, *Dirac oerators, Schrödinger-type operators in C^n and Huygens' principle*, J. Funct. Anal. **87** (1990), 295–318.

[22] M. V. Shapiro and N. L. Vasilevski, *Quaternionic ψ-hyperholomorphic functions, singular integral operators and boundary value problems. I, II*, Complex Variables. Theory and Applications, 1994, pp. 17–46 and 67–96.

[23] W. Sprössig, *Quaternionic Operators and Dirichlet Problems*, in preparation.

[24] E. M.Stein and G. Weiss, *Generalization of the Cauchy-Riemann equations and representations of the rotation group*, Amer. J. Math. **90** (1968), 163–196.

[25] I. Stern, *Boundary value problems for generalized Cauchy-Riemann systems in the space*, Boundary value and initial value problems in complex analysis R.Khönau and W. Tutschke (eds.), Pitman Res. Not. in Math. 256, 1991, pp. 159–183.

[26] U. Wimmer, *Orthogonal Decomposition of the space $L_{2,\mathbb{CH}}(G)$*, Proc.of the 19-th Summer School "Applications of Mathematics in Engineering August 24–September 02, 1993, Varna, 203–210.

[27] Z. Xu, *A function theory for the operator $D - \lambda$*, Complex Variables, Theory and Appl. **16** (1991), 27–42.

UNIVERSITY OF MINING AND TECHNOLOGY FREIBERG, FACULTY OF MATHEMATICS AND INFORMATICS, BERNHARD-VON-COTTA STRASSE 2, D-09596 FREIBERG, GERMANY
E-mail address: `sproessig@mathe.tu-freiberg.de`

Contemporary Mathematics
Volume **212**, 1998

Analytic Models of the Quantum Harmonic Oscillator

Franciszek Hugon Szafraniec

ABSTRACT. After some discussion of the creation and annihilation operators we provide with two, additional to existing so far, models in spaces of entire functions. The first, in the Bargmann-Segal space provides us with a family of creation and annihilation operators coming from the operator of multiplication by the independent variable and acting in its domain. The other is in spaces introduced in [**2**] and realizes a kind of "homotopy" between the very classical oscillator in $\mathcal{L}^2(\mathbb{R})$ and that in the Bargmann-Segal space.

1.

The harmonic oscillator of quantum mechanics[1] in the simplest case is usually described algebraically by two "operators" a^+ and a^- satisfying the commutation relation

$$(*) \qquad\qquad a^- a^+ - a^+ a^- = 1.$$

The "operators" a^+ and a^- are called the creation and the annihilation ones respectively. The way to make these "operators" just operators is to consider them as (necessarily unbounded) Hilbert space ones taking care of their domains. It is physically and historically justified to start with the "operator" H (the Hamiltonian) defined as

$$H = N + \frac{1}{2}, \quad N = a^+ a^-$$

and consider it, still roughly, as a selfadjoint operator in a separable Hilbert space with the pure point spectrum greater or equal $1/2$. Thus, because N becomes positive, it is reasonable to assume, still formally, that a^+ and a^- are adjoint each to the other and, because N is diagonal with respect to the basis of the eigenvectors of H, that they are (properly directed) weighted shift operators. That this is a fortunate choise convices us the uniqueness result of [**9**] saying that this or a direct

---

1991 *Mathematics Subject Classification.* Primary: 47B20. Secondary: 47B38, 81S05.

Key words and phrases. quantum harmonic oscillator; creation, annihilation and number operators; the Hamiltonian; weighted shift operators; unbounded subnormal operators; the Bargmann-Segal space; the Hermite polynomials; the Charlier polynomials.

[1]We would like to recomend here, by the way, a beautiful overview [**3**] of the story of the quantum harmonic oscillator written by mathematicians.

sum of copies of this, up to a unitary equvalence, is the only possibility. Now it is
time to be more precise.

2.

Let $\mathcal{H}$ be a complex Hilbert space. Thus the commutation relation (*) can be
restated as

(**) $S^*Sf - SS^*f = f$

for f in a suitable dense subset $\mathcal{D}$ of $\mathcal{H}$. In order to make the right hand side any
sense it would be convenient to assume first that $\mathcal{D}$ is invariant for both S and
S^*. Assuming that $\mathcal{D}$ is a core[2] of S would give a chance the operator S to be
determined[3] by (**). Now to find S and $\mathcal{D}$ we may, according to what has been
said in Section 1, proceed as follows: choose any orthonormal basis $\{e_n\}_{n=0}^\infty$ in $\mathcal{H}$
such that $\{e_n\}_{n=0}^\infty \subset \mathcal{D}(S)$ and assume that S is a weighted shift with respect to
this basis, that is

$$Se_n = \lambda_n e_{n+1}, \quad n = 0, 1, 2, \ldots,$$

and, consequently, that

$$S^*e_n = \lambda_{n-1} e_{n-1}, \quad n = 1, 2, \ldots, \quad S^*e_0 = 0.$$

Then we get from (**) $\lambda_n = \sqrt{n+1}$, $n = 0, 1, \ldots$ Now we come to the definitions.

Let now the Hilbert space $\mathcal{H}$ be separable and let $e = \{e_n\}_{n=0}^\infty$ be an orthogonal
basis in it and set $\mathcal{D}_e = \lin\{e_n\}_{n=0}^\infty$. Then S is called a *creation operator* with
respect to the basis e if

$$\mathcal{D}(S) = \mathcal{D}_e, \quad Se_n = \sqrt{n+1}\,e_{n+1}, \quad n = 0, 1, \ldots$$

Thus the *annihilation operator* $S^\times$ with respect to e is defined by

$$\mathcal{D}(S^\times) = \mathcal{D}_e, \quad S^\times e_n = \sqrt{n}\,e_{n-1}, \quad S^\times e_0 = 0.$$

The operators S and $S^\times$ are *formally adjoint* each to the other, that is

$$\langle Sf, g \rangle = \langle f, S^\times g \rangle, \quad f, g \in \mathcal{D}_e.$$

Set

$$\mathcal{D}_{\max} = \left\{ f \in \mathcal{H}; \sum_{n=0}^\infty n\,|\langle f, e_n \rangle|^2 < +\infty \right\}$$

and define

$$\mathcal{D}(S_{\max}) = \mathcal{D}_{\max}, \quad S_{\max}f = \sum_{n=0}^\infty \sqrt{n+1}\langle f, e_n \rangle e_{n+1}, \quad f \in \mathcal{D}_{\max},$$

$$\mathcal{D}(S_{\max}^\times) = \mathcal{D}_{\max}, \quad S_{\max}^\times f = \sum_{n=1}^\infty \sqrt{n}\langle f, e_n \rangle e_{n-1}, \quad f \in \mathcal{D}_{\max}.$$

Then it can be get by standard argumentation that

(1) $S^- = S_{\max}, \quad (S^\times)^- = S_{\max}^\times, \quad S^* = S_{\max}^\times$ and $(S^\times)^* = S_{\max}.$

---

[2]$\mathcal{D}$ is a *core* of an operator A if $(A|_\mathcal{D})^- = A^-$.

[3]Notice that $\mathcal{D}$ can not be taken equal to the domain of the closure S^- of S, otherwise, due
to a result of [4], S would be bounded.

Notice that the maximal set $\mathcal{D}$ for which (**) can be considered is

$$(2) \quad \mathcal{D}_{ccr} = \mathcal{D}(S^* S_{\max}) \cap \mathcal{D}(S_{\max} S^*) = \left\{ f \in \mathcal{H}; \sum_{n=0}^{\infty} n^2 \left| \langle f, e_n \rangle \right|^2 < +\infty \right\}.$$

Since $\mathcal{D}_e \subset \mathcal{D}_{ccr} \subset \mathcal{D}_{\max}$ and $\mathcal{D}_e$ is a core of $S_{\max} = S^-$, $\mathcal{D}_{ccr}$ is a core of $S_{\max}$ as well.

Call a closed densely defined operator S the *maximal creation operator* if $\mathcal{D}(S^* S) = \mathcal{D}(SS^*)$ and (**) holds on this subspace. Then S^* is called the *maximal annihilation operator*. Thus we come to the following conclusion:

(a) for any orthonormal basis $e = \{e_n\}_{n=0}^{\infty}$ the corresponding creation and annihilation operators satisfy (**) on $\mathcal{D}_e$, the operators $S_{\max}$ and $S_{\max}^{\times}$ are, according to (1), the maximal creation and annihilation operators with $\mathcal{D}(S^* S_{\max}) = \mathcal{D}(S_{\max} S^*) = \mathcal{D}_{ccr}$.

The converse, which can be also read as the uniqueness result, is in [9] and also [5], pp. 68–70 and can be restated here as

(b) for any maximal creation operator S such that the null space $\mathcal{N}(SS^*)$ is one dimensional[4] there is a basis $e = \{e_n\}_{n=0}^{\infty}$ in $\mathcal{D}(S)$ such that $S|_{\text{lin}\{e_n\}_{n=0}^{\infty}}$ is the creation operator with respect to e and $S^*|_{\text{lin}\{e_n\}_{n=0}^{\infty}}$ is the annihilation operator with respect to e.

In the domain of a maximal creation operator (in fact in any dense subspace) one may find, according to (a), plenty of creation operators. The following observation provides us with some of them which are related to the operator in question.

(c) if $\mathcal{D}$ is any dense subspace of $\mathcal{D}_{ccr}$ which is invariant for both $S_{\max}$ and S^*, and A is any symmetric operator with $\mathcal{D} \subset \mathcal{D}(A)$ such that $S_{\max} A f = A S_{\max} f$ for $f \in \mathcal{D}$ then the operator

$$S_{\max}|_{\mathcal{D}} + A|_{\mathcal{D}}$$

satisfies (**) on $\mathcal{D}$.

Indeed, because $(S_{\max}|_{\mathcal{D}} + A|_{\mathcal{D}})^* f = S^* f + A f$ for $f \in \mathcal{D}$ the conclusion of (c) can be check straightforwardly.

One may take, for instance $A = \lambda I$, $\lambda \in \mathbb{R}$, and any $\mathcal{D}$ invariant for $S_{\max}$ and S^* to get plenty of creation operators in the domain of a maximal creation operator. One of our aims is to illustrate this.

3.

The oldest model of the quantum harmonic oscillator couple, the creation and the annihilition operator, is

$$(3) \quad S = \frac{1}{\sqrt{2}} \left(x - \frac{\mathrm{d}}{\mathrm{d}x} \right), \quad S^{\times} = \frac{1}{\sqrt{2}} \left(x + \frac{\mathrm{d}}{\mathrm{d}x} \right)$$

considered in $\mathcal{L}^2(\mathbb{R})$ with $\mathcal{D}(S) = \mathcal{D}(S^{\times}) = \text{lin}\{h_n\}_{n=0}^{\infty}$ where h_n is the n-th Hermite function

$$h_n = 2^{-n/2}(n!)^{-1/2}\pi^{-1/4}\,\mathrm{e}^{-x^2/2}\,H_n$$

---

[4]If $\dim \mathcal{N}(SS^*) > 1$, there is a family of bases $e_\alpha = \{e_{\alpha,n}\}_{n=0}^{\infty}$, $\alpha \in A$, in $\mathcal{D}(S)$ such that for any α, $e_\alpha = \{e_{\alpha,0}\}_{\alpha \in A}$ is a basis in $\mathcal{N}(SS^*)$ and such that $S|_{\text{lin}\{e_{\alpha,n}\}_{n=0}^{\infty}}$ is the creation operator with respect to e_α and $S^*|_{\text{lin}\{e_{\alpha,n}\}_{n=0}^{\infty}}$ is the annihilation operator with respect to e_α.

with H_n, the n-Hermite polynomial, defined as

(4)
$$H_n(x) = (-1)^n \, \mathrm{e}^{x^2} \, \frac{\mathrm{d}^n}{\mathrm{d}x^n} \, \mathrm{e}^{-x^2}.$$

Another model is in ℓ^2 [8]. Define the Charlier functions $c_n^{(a)}$, $\quad n = 0, 1, \dots$ in discrete variable x as

$$c_n^{(a)}(x) = a^{-n/2}(n!)^{-1/2} C_n^{(a)}(x) \mathrm{e}^{-a/2} a^{x/2}(x!)^{-1/2}, \quad \text{for } x = 0, 1, \dots,$$

where $\{C_n^{(a)}\}_{n=0}^\infty$, $a > 0$ are the Charlier polynomials determined by

$$\mathrm{e}^{-az}(1+z)^x = \sum_{n=0}^\infty C_n^{(a)}(x) \frac{z^n}{n!}.$$

Now define the operators S_a and $S_a^\times$, $a > 0$, by $\mathcal{D}(S_a) = \mathcal{D}(S_a^\times) = \mathrm{lin}\{c_n^{(a)}\}_{n=0}^\infty$, and $S_a f = g$, $S_a^\times f = h$ where g and h are given by

$$g(x) = \sqrt{x} f(x-1) - \sqrt{a} f(x), \quad x = 1, 2, \dots \quad \text{and} \quad g(0) = -\sqrt{a} f(0)$$

and

$$h(x) = \sqrt{x+1} f(x+1) - \sqrt{a} f(x), \quad x = 0, 1, \dots$$

It turns out [8] that S_a and $S_a^\times$ are the creation and the annihilation operators with respect the orthonormal basis $\{c_a^{(a)}\}_{n=0}^\infty$; they both are finite difference operators of the first order. Not going into details here we would rather bring this ℓ^2 situation of [8] over to the analytic case of $\mathcal{A}^2(\exp(-|z|^2) \, \mathrm{d}x\mathrm{d}y)$ described in Section which follows.

4.

An analytic model of the quantum oscillator, known for long time [1], is in $\mathcal{A}^2(\exp(-|z|^2 \, \mathrm{d}x\mathrm{d}y)$, called the Bargmann-Segal space, which is composed of all entire functions in $\mathcal{L}^2(\exp(-|z|^2 \, \mathrm{d}x\mathrm{d}y)$ and it is, in fact, a reproducing kernel Hilbert space with the kernel $(z, w) \mapsto \exp(z\bar{w})$. The standard orthonormal basis in the space $\mathcal{A}^2(\exp(-|z|^2)) \, \mathrm{d}x\mathrm{d}y)$ is $\{e_n\}_{n=0}^\infty$ defined as

$$e_n = \frac{z^n}{\sqrt{n!}}, \quad z \in \mathbb{C}, \quad n = 0, 1, \dots$$

Set $\mathcal{D}_0 = \mathrm{lin}\{e_n\}_{n=0}^\infty$. Then the operators S and $S^\times$ defined as

(5)
$$Sf(z) = zf(z), z \in \mathbb{C}, \quad S^\times f = \frac{\mathrm{d}}{\mathrm{d}z} f, \quad f \in \mathcal{D}(S) = \mathcal{D}(S^\times) = \mathcal{D}_0$$

are the creation and the annihilation operators. The unitary equivalence between $\mathcal{L}^2(\mathbb{R})$ and $\mathcal{A}^2(\exp(-|z|^2 \, \mathrm{d}x\mathrm{d}y)$ and its inverse can be implemented by integral transforms whose kernels comes from the generating function of the Hermite polynomials.

Notice that in the natural extension $\mathcal{L}^2(\exp(-|z|^2 \, \mathrm{d}x\mathrm{d}y)$ of $\mathcal{A}^2(\exp(-|z|^2) \, \mathrm{d}x\mathrm{d}y)$ the creation operator S, which is the operator of multiplication by the independent variable, extends to the operator which acts in the same way in the larger space. Because the latter operator is normal, the creation operator is (the most spectacular and also the best behaving) example of an unbounded *subnormal* operator; look at [6, 7] for some details concerning the theory. The annihilation operator, is the projection of the operator of multiplication by $\bar{z}$ in $\mathcal{L}^2(\exp(-|z|^2 \, \mathrm{d}x\mathrm{d}y)$ to the Bargmann-Segal space.

Now for $a \geq 0$ set $e^{(a)} = \{e_n^{(a)}\}_{n=0}^{\infty}$ with

$$e_n^{(a)}(z) = \frac{1}{\sqrt{n!}} e^{-a/2 - \sqrt{a}z} (\sqrt{a} + z)^n, \quad z \in \mathbb{C}.$$

It is clear that $e^{(a)}$ is an orthonormal basis in $\mathcal{A}^2(\exp(-|z|^2 \, dx dy)$. Set $\mathcal{D}_a = \lim\{e_n^{(a)}\}_{n=0}^{\infty}$. No $e_n^{(a)}$ is in $\mathcal{D}_0$ or, in other words,

(6) $$\mathcal{D}_a \cap \mathcal{D}_0 = \{0\} \quad \text{for any } a > 0.$$

Define the operators S_a and $S_a^{\times}$ by

$$S_a f(z) = (\sqrt{a} + z) f(z), \quad z \in \mathbb{C},$$

$$S_a^{\times} f = \left(\sqrt{a} + \frac{d}{dz} \right) f, \quad f \in \mathcal{D}(S_a) = \mathcal{D}(S_a^{\times}) = \mathcal{D}_a.$$

They are the creation and the annihilation operators with respect to the basis $e^{(a)}$.

Notice that if we consider the maximal creation operator $S_{a,\max}$ corresponding to S_a in the way described in Section 2, we get (it comes from the ℓ^2 argument used in [8]) that

$$\mathcal{D}(S_{a,\max}) = \mathcal{D}(S_{\max})$$

where $S_{\max}$ is the maximal creation operator corresponding to S defined in (5). This means that $\mathcal{D}(S_{a,\max})$ is independent of a and the operators $S_{a,\max}$, $S_{a,\max}^{\times}$, $S_{\max}$ and $S_{\max}^{\times}$ have the same domain (equal to $\mathcal{D}_{\max}$) while the domains of S_a and S are different. In addition to (6), it can be checked easily that

$$\mathcal{D}_a \subset \mathcal{D}_{\max} \text{ for any } a > 0.$$

Moreover $\mathcal{D}_0$ and any $\mathcal{D}_a$ is a core of $S_{\max}$ as well as of $S_{a,\max}$ ([8] again). All these considerations lead us to the following

THEOREM 1. *For any $a > 0$*

$$S_{a,\max} = S_{\max} + \sqrt{a} I|_{\mathcal{D}_{\max}}$$

and

$$S_{a,\max}^{\times} = S_{\max}^{\times} + \sqrt{a} I|_{\mathcal{D}_{\max}}.$$

Illustrating observation (c) Theorem 1 provides us with a family of bases in the maximal domain of the operator of multiplication by the independent variable in $\mathcal{A}^2(\exp(-|z|^2) \, dx dy)$ which give rise to creation operators coming from that. The symmetric perturbation appearing there is here just $\sqrt{a} I|_{\mathcal{D}_{\max}}$.

COROLLARY 1. *The following conclusions hold true:*
1° *for any $f \in \mathcal{D}_{\max}$, $S_{a,\max} f \to S_{\max} f$ and $S_{a,\max}^{\times} f \to S_{\max}^{\times} f$ as $a \to 0+$;*
2° *for any $a, b > 0$*

$$S_{a,\max} - S_{b,\max} = (\sqrt{a} - \sqrt{b}) I|_{\mathcal{D}_{\max}}$$

and

$$S_{a,\max}^{\times} - S_{b,\max}^{\times} = (\sqrt{a} - \sqrt{b}) I|_{\mathcal{D}_{\max}};$$

3° *$\mathcal{D}_0$ is invariant for for any $S_{a,\max}$ and for any $a, b > 0$ the operators $S_{a,\max}$ and $S_{b,\max}$ commute on $\mathcal{D}_0$ and so do $S_{a,\max}^{\times}$ and $S_{b,\max}^{\times}$.*

5.

The analytic model we intend to present here is based on orthogonality of Hermite polynomials discovered in [**2**]. They are orthogonal with respect to a measure (in fact, to a one parameter family of measures) which, in contrast to the Bargmann-Segal situation, is not rotationally invariant. This implies that the model for the creation operator is different from multiplication by the independent variable. It turns out that it is, nevertheless, close to both this defined by (3) as well as to that by (5). Let us go into details.

The Hermite polynomials, defined as in (4), are now considered as those in a complex variable. Let $0 < A < 1$. Then

$$\int_{\mathbb{R}^2} H_m(x + iy) H_n(x - iy) \exp\left[-(1 - A)x^2 - \left(\frac{1}{A} - 1\right)y^2\right] dx dy = b_n(A) \delta_{m,n}$$

where

$$b_n(A) = \frac{\pi\sqrt{A}}{1 - A}\left(2\frac{1 + A}{1 - A}\right)^n n!.$$

Introducing the Hilbert space $\mathcal{X}_A$ of entire functions f such that

$$\int_{\mathbb{R}^2} |f(x + iy)|^2 \exp\left[Ax^2 - \frac{1}{A}y^2\right] dx dy < \infty$$

and defining

$$h_n^A(z) = b_n(A)^{-1/2} e^{-z^2/2} H_n(z), \quad z \in \mathbb{C}$$

it was shown in [**2**] that $\{h_n^A\}_{n=0}^\infty$ is an orthonormal basis in $\mathcal{X}_A$. From the algebraic relation $H_{n+1} = 2zH_n - H_n'$ we get directly

$$\sqrt{n + 1}\, h_{n+1}^A = \sqrt{\frac{1 - A}{2(1 + A)}}\, [zh_n^A - (h_n^A)'].$$

$2nH_{n-1} = H_n'$ implies

$$\sqrt{n}h_n^A = \sqrt{\frac{1 + A}{2(1 - A)}}[zh_n^A + (h_n^A)'].$$

Thus we arrived at the following

THEOREM 2. *The operators S_A and $S_A^\times$ defined as*

$$S_A f(z) = \sqrt{\frac{1 - A}{2(1 + A)}}\, [zf(z) - f'(z)],$$

$$S_A^\times f(z) = \sqrt{\frac{1 + A}{2(1 - A)}}\, [zf(z) + f'(z)], \quad z \in \mathbb{C}, \quad f \in \mathcal{D}_A$$

are the creation and the annihilation operator in $\mathcal{X}_A$.

It is clear that

$$0 < A < B < 1 \text{ implies } \mathcal{X}_B \subset \mathcal{X}_A.$$

On the other hand, as shown in [**2**],

$$\text{any } \mathcal{X}_A \subset \mathcal{L}^2(\mathbb{R})$$

if one thinks of this inclusion as taking restrictions of functions in $\mathcal{X}_A$ to the real axis. All these inclusions are continous. So let R_A be the injection of $\mathcal{X}_A$ into $\mathcal{L}^2(\mathbb{R})$.

Denote by S_0 and $S_0^\times$ the creation and the annihilation operators in $\mathcal{L}^2(\mathbb{R})$ as by (3). Then $\mathcal{D}_0$ stands for their domain, $S_{0,\max}$ and $S_{0,\max}^\times$ do for their maximal operators and $\mathcal{D}_{0,\max}$ for their common domain. Similarly, denote[5] by S_1 and $S_1^\times$ the creation and the annihilation operators in $\mathcal{A}^2(\exp(-|z|^2)\,\mathrm{d}x\mathrm{d}y)$ defined by (5) and by $\mathcal{D}_1$ their domain.

Notice that, in fact, $\mathcal{D}_A$ is independent of A and that

$$R_A \mathcal{D}_A = \mathcal{D}_0.$$

On the other hand, since $R_A S_{A,\max}|_{\mathcal{D}_A} \subset S_{0,\max}|_{\mathcal{D}_0} R_A$ and, after passing to the closures (continuity of R_A), since

$$R_A S_{A,\max} \subset S_{0,\max} R_A,$$

we get

$$R_A \mathcal{D}_{A,\max} \subset \mathcal{D}_{0,\max}.$$

Thus Theorem 2 implies immediately

COROLLARY 2. For any $f \in \mathcal{D}_{A,\max}$,

$$R_A S_{A,\max} f \to S_{0,\max} R_A f, \quad R_A S_{A,\max}^\times f \to S_{0,\max}^\times R_A f$$

as $A \to 0+$, the convergence being in the $\mathcal{L}^2(\mathbb{R})$-norm.

Notice that, because for any $f \in \mathcal{A}^2(\exp(-|z|^2)\,\mathrm{d}x\mathrm{d}y)$,

$$\int_{\mathbb{R}^2} |f(\sqrt{1-A}(x+iy))\,e^{-z^2/2}|^2 \exp\left[Ax^2 - \frac{1}{A}y^2\right] \mathrm{d}x\mathrm{d}y$$

$$= \frac{1}{1-A} \int_{\mathbb{R}^2} |f(x+iy)|^2 \exp\left[-x^2 - \frac{1}{A}y^2\right] \mathrm{d}x\mathrm{d}y$$

$$\leq \frac{1}{1-A} \int_{\mathbb{R}^2} |f(x+iy)|^2 \exp(-x^2 - y^2)\,\mathrm{d}x\mathrm{d}y,$$

the function $g(z) = f(\sqrt{1-A}z)\,e^{-z^2/2}$, $z \in \mathbb{C}$ is in $\mathcal{X}_A$. Thus the operator $T_A f = g$ maps $\mathcal{A}^2(\exp(-|z|^2)\,\mathrm{d}x\mathrm{d}y)$ into $\mathcal{X}_A$ and is bounded with the norm $\|T_A\| \leq \sqrt{1/(1-A)}$. It is injective and $T_A^{-1} g = h$, where $h(z) = g\left(z/\sqrt{1-A}\right) e^{z^2/2(1-A)}$, $z \in \mathbb{C}$. Moreover, notice that

(7) $$T_A(\mathcal{D}_1) = \mathcal{D}_A.$$

Thus we come, after some calculation, to

$$S_A T_A f = \sqrt{\frac{2}{1+A}} T_A \left(zf(z) - \frac{1}{2}\sqrt{1-A}f'(z)\right)$$

and similarily

$$S_A^\times T_A f = \sqrt{\frac{1+A}{2}} T_A f'.$$

This, due to (7), leads us to the following

---

[5]The parameter a appearing in Section 4 can be equal to 1 but never to $\mathbf{1}$. So S_1 is not the same as S_1.

THEOREM 3. *For any $f \in \mathcal{D}_1$*

$$T_A^{-1} S_A T_A f \to S_1 f$$

and

$$T_A^{-1} S_A^\times T_A f \to S_1^\times f$$

as $A \to 1-$, the convergence being in the $\mathcal{A}^2(\exp(-|z|^2 \, \mathrm{d}x \mathrm{d}y)$-norm.

Corollary 2 and Theorem 3 show that the spaces $\mathcal{X}_A$, $0 < A < 1$ realize a kind of "homotopy" for the quantum harmonic oscillator in $\mathcal{L}^2(\mathbb{R})$ and that in the space $\mathcal{A}^2(\exp(-|z|^2) \, \mathrm{d}x \mathrm{d}y)$.

References

1. V. Bargmann, *On a Hilbert space of analytic functions and an associated integral transform*, Comm. Pure App. Math. **14** (1961), 187–214.
2. S. L. L. van Eijndhoven, J. L. H. Meyers, *New orthogonality relations for the Hermite polynomials and related Hilbert spaces*, J. Math. Ann. Appl. **146** (1990), 89–98.
3. W. Mlak, M. Słociński, *Quantum phase and circular operators*, Univ. Iagell. Acta Math. **24** (1992), 133–144.
4. Y. Okazaki, *Boundedness of closed linear operator T satisfying $\mathcal{R}(T) \subset \mathcal{D}(T)$*, Proc. Japan Acad. **62** (1986), 294–296.
5. C. R. Putnam, *Commutation properties of Hilbert space operators and related topics*, Springer-Verlag, Berlin, Heidelberg, New York, 1967.
6. J. Stochel, F. H. Szafraniec, *On normal extensions of unbounded operators. II*, Acta Sci. Math. (Szeged) **53** (1989), 153–177.
7. J. Stochel, F. H. Szafraniec, *On normal extensions of unbounded operators. III. Spectral properties*, Publ. RIMS, Kyoto Univ. **25** (1989), 105–139.
8. F. H. Szafraniec, *Yet another face of the creation operator*, Operator Theory Adv. Appl. **80** (1995), 266–275.
9. H. G. Tillmann, *Zur Eindeutigkeit der Lösungen der quantummechanischen Vertauschungsrelationen*, Acta Sci. Math. (Szeged) **24** (1963), 258–270.

INSTYTUT MATEMATYKI UJ, UL.REYMONTA 4, PL-30059 KRAKÓW, POLAND
E-mail address: fhszafra@im.uj.edu.pl

Contemporary Mathematics
Volume **212**, 1998

Interesting Relations in Fock Space

Alexander Turbiner

ABSTRACT. A certain non-linear relations between the generators of (q-deformed) Heisenberg algebra are found. Some of these relations are invariant under quantization and q-deformation.

In this Talk I want to present different relations appearing in the Fock space generated by the q-deformed Heisenberg algebra. In a certain particular case, some of those relations can be summarized by the theorem:

THEOREM. *The following differential identities hold for any natural number n and $\delta \in \mathbb{C}$*

$$(1) \quad \left(x^2 \frac{d}{dx} - nx \right)^{n+1} = x^{2n+2} \frac{d^{n+1}}{dx^{n+1}} \, , \quad \left(\frac{d^2}{dx^2} x - n \frac{d}{dx} \right)^{n+1} = \frac{d^{2n+2}}{dx^{2n+2}} x^{n+1}$$

$$(2) \quad \left(x \frac{d}{dx} x \right)^n = x^n \frac{d^n}{dx^n} x^n \, , \quad \left(\frac{d}{dx} x \frac{d}{dx} \right)^n = \frac{d^n}{dx^n} x^n \frac{d^n}{dx^n} .$$

$$(3) \quad x \frac{d}{dx} \left(x \frac{d}{dx} - 1 \right) \ldots \left(x \frac{d}{dx} - n \right) = x^{n+1} \frac{d^{n+1}}{dx^{n+1}}$$

$$(4) \quad \prod_{k=0}^{n} \left[x \left(1 - e^{-\delta \frac{d}{dx}} \right) - k\delta \right] = x^{(n+1)} \left(1 - e^{-\delta \frac{d}{dx}} \right)^{n+1}$$

where $x^{(n+1)} = x(x - \delta)(x - 2\delta) \ldots (x - n\delta).$

The proof can be carried out straightforwardly by induction. Later we will show that these relations can be easily modified to quite general Fock spaces.

Take some 3-dimensional complex algebra with an identity operator and with two elements a and b, obeying the relation

$$(5) \quad ab - qba = p,$$

1991 *Mathematics Subject Classification.* Primary 47B48, 17B15.

where p, q are any complex numbers. The algebra with the identity operator generated by the elements a, b and obeying (1) is usually called the q-deformed Heisenberg algebra h_q (See, for example, papers [1–4] and references and a discussion therein). The parameter q is called the parameter of quantum deformation. If $p = 0$, then (5) describes the non-commutative (quantum) plane (See, for example, [5–6]), while if $q = 1$, then h_q becomes the ordinary (classical) Heisenberg algebra and the parameter p plays a role of the Planck constant (See, for example, [7]). The operator linear space of all holomorphic functions in a, b with vacuum

$$a|0> = 0$$

is called *Fock* space. Then the following theorem holds:

THEOREM 1 [8]. *For any $p, q \in \mathbb{C}$ in (5) the following identities are true*

(6.1) $$(aba)^n = a^n b^n a^n, \quad n = 1, 2, 3, \ldots$$

and, if $q \neq 0$, also

(6.2) $$(bab)^n = b^n a^n b^n, \quad n = 1, 2, 3, \ldots$$

The proof can be carried out by induction using the following easy lemma

LEMMA 1. *For any $p, q \in \mathbb{C}$ in (5),*

(7.1) $$ab^n - q^n b^n a = p\{n\}b^{n-1},$$

(7.2) $$a^n b - q^n b a^n = p\{n\}a^{n-1},$$

where $n = 1, 2, 3 \ldots$ and $\{n\} = (1 - q^n)/(1 - q)$ is the so-called q-number (see, for example, [9]).

Now let us proceed to the proof of (6.1). For $n = 1$ the relation (6.1) is fulfilled trivially. Assume that (6.1) holds for some n and check that it holds for $n + 1$. It is easy to write the chain of equalities :

$$(aba)^{n+1} = (aba)^n aba = a^n b^n a^n aba = a^n (b^n a)(a^n b)a$$

$$= a^n \left(\frac{ab^n}{q^n} - \frac{p}{q^n}\{n\}b^{n-1} \right) (q^n ba^n + p\{n\}a^{n-1})a = a^{n+1} b^{n+1} a^{n+1}, \quad (\diamond)$$

where the relations (7.1) and (7.2) were used in this chain. In an analogous way, one can prove (6.2). □

Theorem 1 leads to two corollaries. Both of them can be easily verified:

COROLLARY 1. *For any $p, q \in \mathbb{C}$ in (5) and natural numbers n, k,*

(8) $$(\underbrace{ababa \cdots aba}_{2k+1})^n = \underbrace{a^n b^n a^n \cdots b^n a^n}_{2k+1}, \quad n, k = 1, 2, 3, \ldots$$

COROLLARY 2. *Let* $T_k^{(n)} = \underbrace{a^n b^n a^n \dots b^n a^n}_{2k+1}$. *Then, for any* $p, q \in \mathbb{C}$ *in* (5), *the following relation holds*

(9.1) $$[T_k^{(n)}, T_k^{(m)}] = 0 ,$$

as well as the more general relation:

(9.2) $$[T_k^{(n_1)} T_k^{(n_2)} \dots T_k^{(n_i)}, T_k^{(m_1)} T_k^{(m_2)} \dots T_k^{(m_j)}] = 0 ,$$

where $[\alpha, \beta] \equiv \alpha\beta - \beta\alpha$ *is the standard commutator and* $\langle n \rangle, \langle m \rangle$ *are sets of any non-negative, integer numbers.*

It is evident that formulas (8), (9.1) and (9.2) remain correct under the replacement $a \rightleftharpoons b$, if $q \neq 0$. It is worth noting that the algebra h_q under an appropriate choice of the parameters p, q has a natural representation

$$a = x , \quad b = D \qquad (\star)$$

where the operator $Df(x) = [f(x) - f(qx)]/[x(1 - q)]$ is dilationally-invariant shift operator and usually it is called the Jackson symbol (See, for example, [9]). The relation (6.1) then becomes

$$(xDx)^n = x^n D^n x^n ,$$

while relation (6.2)

$$(DxD)^n = D^n x^n D^n .$$

Since at $q \rightarrow 1$, the operator $D \rightarrow \frac{d}{dx}$, the above relations become differential identities (2).

THEOREM 2 [10]. *For any* $p, q \in \mathbb{C}$ *in* (5) *and natural* n, m *the following identities hold:*

(10.1) $$[a^n b^n, a^m b^m] = 0 ,$$

(10.2) $$[a^n b^n, b^m a^m] = 0 ,$$

(10.3) $$[b^n a^n, b^m a^m] = 0 .$$

The proof is based on the following easy lemma

LEMMA 2. *For any* $p, q \in \mathbb{C}$ *in* (5) *and natural number* n

(11.1) $$a^n b^n = P(ab) ,$$

(11.2) $$b^n a^n = Q(ab) ,$$

where P, Q *are some polynomials in one variable of order not higher than* n.

Let us introduce a notation $t^{(n)} = a^n b^n$ or $b^n a^n$, in which the order of the multipliers is not essential. The statement of the Theorem 2 can now be written as $[t^{(n)}, t^{(m)}] = 0$ and the following is true

COROLLARY 3. *The commutator*

(12) $$[t^{(n_1)}t^{(n_2)}\ldots t^{(n_i)},\ t^{(m_1)}t^{(m_2)}\ldots t^{(m_j)}]\ =\ 0$$

holds for any $p, q \in \mathbb{C}$ *in* (5) *and any sets* $\langle n \rangle, \langle m \rangle$ *of non-negative, integer numbers.*

One can make sense of (6.1) and (6.2), (8), (9.1) and (9.2), (10.1)–(10.3), (12) as follows: In the algebra of polynomials in a, b there exist relations invariant under a variation of the parameters p, q in (5). Also formulas (6.1) and (6.2), (8) can be interpreted as formulas of a certain special ordering other than the standard lexicographical one.

A natural question can be raised: Is the existence of the relations (6.1) and (6.2), (8), (9.1) and (9.2), (10.1)–(10.3), (12) connected unambiguously to the algebra h_q, or are there more general algebra(s) leading to those relations? A certain answer is given by the following theorem

THEOREM 3 [10]. *If two elements* a, b *of a Banach algebra with unit are related by*

(13) $$ba = f(ab)\ ,$$

where f *is a holomorphic function in a neighbourhood of* $Spec\,\{ab\} \cup Spec\,\{ba\}$, *then relations* (6.1), (8), (10.3) *hold. If, in addition, the function* f *is single-sheeted, then* (10.2) *also holds.*

The proof is essentially based on the fact that, if the function f in (13) is holomorphic, then for *any* holomorphic F

(14) $$bF(ab)\ =\ F(ba)b\ ,$$

which guarantees the correctness of the statement (11.2) of Lemma 2, although Q is no longer polynomial. This immediately proves (10.3). The relation (6.1) can be proven by induction and an analogue of the logical chain ($\Diamond$) is

$$(aba)^{n+1}\ =\ (aba)^n aba\ =\ a^n b^n a^n aba\ =\ a^n (b^n a^n) aba$$

$$=\ [b^n a^n = Q(ab),\ \text{see Lemma 2}]$$

$$=\ a^n Q(ab)(ab)a\ =\ a^n (ab)Q(ab)aa^{n+1}b^{n+1}a^{n+1}\ .$$

An extra condition that f is single-sheeted implies that $ab = f^{-1}(ba)$, which immediately leads to the statement (10.2). □

It is evident that the replacement $a \rightleftharpoons b$ in Theorem 3 leads to the fulfilment of the equalities (6.2), (10.1) and (10.2) as well.

There exists another type of relations in Fock space stemming from a fact that some algebras are contained in the Fock space and these algebras can possess finite-dimensional representations. One can prove the theorem

THEOREM 4. *For any* $p, q \in \mathbb{C}$ *in* (5) *and* $n = 1, 2, 3, \ldots$ *the following identities are true*

(15.1) $$(b^2 a - \{n\}b)^{n+1} = q^{n(n+1)} b^{2n+2} a^{n+1}\ ,$$

and, if $q \neq 0$, *also*

(15.2) $$(ba^2 - \{n\}a)^{n+1} = q^{n(n+1)} b^{n+1} a^{2n+2}\ ,$$

where $\{n\} = (1 - q^n)/(1 - q)$ *is q-number.*

In order to prove this theorem, we need, at first, to state the following observation

LEMMA 3 [11, 12, 13]. *For any* $p, q \in \mathbb{C}$ *in* (5) *and* $\alpha \in \mathbb{C}$, *three elements of the Fock space*

$$j^+ = b^2 a - \{\alpha\} b,$$

(16)
$$j^0 = ba - \frac{\{\alpha\}\{\alpha + 1\}}{\{2\alpha + 2\}},$$

$$j^- = a,$$

(modified by some multiplicative factors) obey q-deformed commutation relations

$$q\tilde{j}^0\tilde{j}^- - \tilde{j}^-\tilde{j}^0 = -\tilde{j}^-,$$

(17)
$$q^2\tilde{j}^+\tilde{j}^- - \tilde{j}^-\tilde{j}^+ = -(q+1)\tilde{j}^0,$$

$$\tilde{j}^0\tilde{j}^+ - q\tilde{j}^+\tilde{j}^0 = \tilde{j}^+,$$

forming the algebra $s\ell_{2q}$. *If* $q \to 1$, *these commutation relations become the standard* $s\ell_2$ *ones. If* $\alpha = n$ *is a non-negative integer, the generators* (16) *form the finite-dimensional representation corresponding the (operator) finite-dimensional representation space*

(18)
$$V_n = \langle 1, b, b^2, \ldots, b^n \rangle$$

in the Fock space.

Validity of this Lemma can be checked by direct calculation. The proof of the Theorem 4 is based on an evident fact [14] that a positive-root (negative-root) generator in finite-dimensional representation taken in the power of the dimension of the finite-dimensional representation annihilates the space of the finite-dimensional representation and, correspondingly, acts in its complement. The operator $(j^+)^{n+1}$ at $\alpha = n$ annihilates (18) and hence it must be proportional (from the right) to a^{n+1}. Since j^+ is graded with the grading equals to $(+1)$, $(j^+)^{n+1}$ must be proportional (from the left) to b^{2n+2}. What is left in this consideration is a value of possible multiplicative factor appearing in r.h.s. (15.1). This factor is equal to $q^{n(n+1)}$ and can be found by direct calculation. Equation (15.2) can be proved analogously, with only minor modifications.

THEOREM 5 [8]. *For any* $p, q \in \mathbb{C}$ *in* (5) *and* $n = 1, 2, 3, \ldots$ *the following identity holds :*

(19)
$$ba(ba - \{1\}) \ldots (ba - \{n\}) = q^{n(n+1)/2} b^{n+1} a^{n+1}$$

The proof can be carried out by induction. Now let us consider how the identity (19) looks for different representations of the algebra (5).

Taking the representation $(\star)$ for the algebra (5) and plugging it into (19), we arrive at an identity for dilatationally-invariant shift operator D :

(20)
$$xD(xD - \{1\}) \ldots (xD - \{n\}) = q^{n(n+1)/2} x^{n+1} D^{n+1}$$

If $q \to 1$ the representation $(\star)$ degenerates into

$$a = \frac{d}{dx} \, , \quad b = x \qquad (\star\star)$$

and (19) coincides to the identity (3). Recently it was found a more general representation of the algebra (5) then $(\star\star)$ at $q = 1$ characterized by a free parameter $\delta \in C$ [15]:

(21.1)
$$a = \frac{(e^{\delta(d/dx)} - 1)}{\delta} \, , \quad b = xe^{-\delta(d/dx)} \, ,$$

or, in other form,

(21.2)
$$a = \mathcal{D}_+ \, , \quad b = x(1 - \delta\mathcal{D}_-) \, ,$$

where $\mathcal{D}_{\pm}f(x) = [f(x \pm \delta) - f(x)]/[\pm\delta]$ are (translationally-invariant) finite-difference operators. After substitution of (21.1) in (19) for $q = 1$ and a simple transformation the differential identity (4) appears. This identity takes a slightly different form if the realization (21.2) is used :

(22)
$$xD_-(xD_- - 1) \dots (xD_- - n) = \delta^{n+1} x^{(n+1)} D_-^{n+1}$$

(cf. (20)).

Now take an algebra generated by three generators

(23)
$$ab - pba = F(N), \quad qNa - aN = -a, \quad Nb - qbN = b,$$

(cf. (5)), where $p, q \in C$, F is a holomorphic function in a neighbourhood of $Spec\{N\} \cup \{0\}$.

THEOREM 6. *For $p, q \neq 0 \in \mathbf{C}$ in (23) the identities (6.1), (6.2) and (10.1)–(10.3) hold, and also*
(24)
$$ba \left(ba - F\left(\frac{N-1}{q}\right)\right) \cdots \left(ba - \sum_{k=1}^{n} p^{k-1} F\left(\frac{N - \{k\}}{q^k}\right)\right) = p^{\frac{n(n+1)}{2}} b^{n+1} a^{n+1}.$$

It is worth mentioning that for certain cases the formula (25) was known: $p = 1$ [8], $p = q$ [16].

The proof is carried out by induction and is based on a certain generalization of Lemma 2:

LEMMA 4. *For any $p, q \neq 0 \in \mathbf{C}$ in (23) and n a natural number*

(25.1)
$$a^n b^n = \sum_{k=0}^{n} \alpha_k(N)(ab)^k,$$

(25.2)
$$b^n a^n = \sum_{k=0}^{n} \beta_k(N)(ab)^k,$$

where α, β are calculable functions.

We also use the simple observation that in (23)

$$af(N) = f(qN + 1)a, \quad f(N)b = bf(qN + 1),$$
$$[f(N), a^n b^n] = [f(N), b^n a^n] = 0, \quad n = 1, 2, \dots,$$

where $f(N)$ is a holomorphic function in a neighborhood of $Spec\{N\} \cup \{0\}$.

It is worth noting that the algebra (5) or (13) can be interpreted as a deformation of the Heisenberg algebra. In turn, the algebra (23) contains as special cases:

(i) sl_2 ($q = 1$, $F = 2N$),

(ii) the q-deformated algebra sl_{2q} ($p = q^2$, $F = (q + 1)N$, see (17)),

(iii) the quantum group $U_q(sl_2)$ ($p, q = 1$, $F = (\sin \tilde{q}N)/(\tilde{q} - \tilde{q}^{-1})$) and even,

(iv) the Heisenberg algebra ($q = 1$, $F = 1$) as a sub-algebra, and

(v) the q-deformed Heisenberg algebra ($F = 1$) as a sub-algebra.

All the above demonstrates the general nature of the invariant identities (6.1), (6.2) and (10.1)–(10.3).

References

[1]   E. P. Wigner, Phys. Rev. **77** (1950), 711.

[2]   M. Chaichian and P. Kulish, *Preprint CERN TH-5969/90* (1990).

[3]   P. Kulish and V. Damaskinsky, J. Phys. **A23** (1990), L415.

[4]   Y. I. Manin, *Bonn preprint MP/91-60* (1991).

[5]   J. Wess and B. Zumino, Nucl. Phys. (Proceedings Suppl.) **B18** (1990), 302.

[6]   B. Zumino, Mod. Phys. Lett. **A6** (1991), 1225.

[7]   L. D. Landau and E. M. Lifshitz, *Quantum Mechanics*, Fizmatgiz, Moscow, 1974 (in Russian).

[8]   N. Fleury and A. Turbiner, *Preprint CRN 94/08 and IFUNAM FT 94-41(1994)*, J. Math. Phys. **35** (1994), 6144.

[9]   G. Gasper and M. Rahman, *Basic Hypergeometric Series*, Cambridge University Press, Cambridge, 1990.

[10] A. Turbiner, *Invariant identities in the Heisenberg algebra*, Funktsional'nyi Analiz i eqo Prilozhenia **29** (1995), 88–91 (in Russian); Soviet Math.—Functional Analysis and its Application (English Translation).

[11] O. Ogievetsky and A. Turbiner, $sl(2, \mathbb{R})_q$ *and quasi-exactly-solvable problems*, Preprint CERN-TH: 6212/91 (1991) (unpublished).

[12] A. Turbiner, *Lie algebras and linear operators with invariant subspace*, Lie Algebras, Cohomologies and New Findings in Quantum Mechanics, *Contemporary Mathematics*, v. 160, N. Kamran and P. Olver (eds.), AMS, Providence, RI, 1994, pp. 263–310.

[13] A. Turbiner, *Quasi-exactly-solvable differential equations*, CRC Handbook of Lie Group Analysis of Differential Equations, vol. 3: New Trends in Theoretical Developments and Computational Methods, Chapter 12, N. Ibragimov (ed.), CRC Press, 1995, pp. 331–366.

[14] A. Turbiner and G. Post, J. Phys. **A27** (1994), L9.

[15] Yu. F. Smirnov, A. V. Turbiner, *Lie-algebraic discretization of differential equations*, Mod. Phys. Lett. **A10** (1995), 1795–1802.

[16] W.-S. Chung, *Private communication* (June 1995).

INSTITUTE FOR THEORETICAL AND EXPERIMENTAL PHYSICS, MOSCOW 117259, RUSSIA

Current address: Instituto de Ciencias Nucleares, UNAM, A.P. 70-543, 04510 México D.F

E-mail address: turbiner@nuclecu.unam.mx or turbiner@axcrna.cern.ch

Contemporary Mathematics
Volume **212**, 1998

Quantization: Some Problems, Tools, and Applications

André Unterberger

ABSTRACT. This mostly expository paper deals with various topics associated with quantization theory or (generalized pseudodifferential analysis. The relationship between pseudodifferential analtsis and families of (possibly discretely parametrized) coherent states is first considered; next, the relativistic Klein–Gordon analysis is shown to bring to light a generalization of hypergeometric functions. Finally, composition formulas are described in connection with the quantization of the two main families of coadjoint orbits of $SL(2, \mathbb{R})$.

1. Introduction

The quantization process plays a major role in quantum mechanics, in the theory of representations and in partial differential equations. It can be roughly described in terms of a linear map Op that associates operators on some fixed Hilbert space $\mathcal{H}$ to functions defined on some given symplectic manifold X (f is then called the *symbol* of $Op(f)$); besides, with certain automorphisms ϕ of X one should be able to associate unitary transformations $\pi(\phi)$ of $\mathcal{H}$, in such a way that, looking at one-parameter groups (ϕ_t) and their associated one-parameter groups $(\pi(\phi_t))$ of unitary transformations, the generators of these two objects (a certain self- adjoint operator on $\mathcal{H}$ and a function f on X whose hamiltonian vector field generates (ϕ_t)) should correspond under the map Op. In quantum mechanics, the rule Op is the one that associates observables (operators) to classical observables, which are just functions living on a certain classical phase space: even though this is somewhat misleading since it refers to the extended phase space rather than the genuine one, let us recall that it is in this way (substituting $(h/2i\pi)(\partial/\partial t)$ for p_0 and $(h/2i\pi)(\partial/\partial x_j)$ for $\mathbf{p}_j$) that one gets the Schrödinger equation or the Klein-Gordon equation from the equations $p_0 = |\mathbf{p}|^2/2m$ or $p_0^2 = m^2c^4 + c^2|\mathbf{p}|^2$ which normalize the energy-momentum covector of a free particle in terms of its mass in non-relativistic or in relativistic mechanics. In the theory of representations, X usually appears as an orbit in the coadjoint representation of some Lie group G: the main problem is to build $\mathcal{H}$ from X, which (very roughly) usually involves some method of (real or complex) polarization. Quantization enters partial differential

1991 *Mathematics Subject Classification.* Primary 58F05, 35S99; Secondary 11F99, 33C05, 43A90.

equations as a box of tools, in that it permits to build *pseudodifferential analyses*: in the scheme above, only the case when $X = \mathbb{R}^n \times \mathbb{R}^n$ and $\mathcal{H} = L^2(\mathbb{R}^n)$, Op being either the *standard* rule or *Weyl's* rule, has been practised and applied, up to now, to any considerable extent.

It is not our intention to deal with the quantization problem in general (i.e., the one in which X is a general symplectic manifold and one asks that $\pi(\phi)$ should be defined for fairly general symplectic transformations ϕ of X): in this case, at best some approximate answer (one may plug in some free parameter and ask for some asymptotic theory) can be given. Some work in this direction has been done recently by Fedosov [9]; also, one should recall that the problem might not even have been raised in any meaningful way had it not been for Egorov's result [7] on the possibility to quantize, in an approximate way, non-linear canonical transformations of $\mathbb{R}^n \times \mathbb{R}^n$. Rather, our aim is to adhere to a much more rigid theory, in which a (finite-dimensional) Lie group G of symplectic transformations of X, together with a representation π of G, is given. Even then, as will be seen on an array of examples, a variety of methods (developed over a course of fifteen years or so) is available, which we shall allude to rather than expound; instead, we shall illustrate these with incursions into three domains: the theory of coherent states, a study of generalized harmonic oscillators, and the discussion of composition formulas in symbolic calculi. The first and third of these topics are quite popular at present, which is why we think this is the proper time to expound our views on these questions, especially since they run off the main trends; the second topic involves a surprising application of some symbolic calculus to a generalization of the hypergeometric equation.

2. Coherent states

Before the terminology of *families of coherent states* had become fashionable, especially among physicists, some mathematically harder results had appeared: we wish here to call attention to these, at the same time pointing to the role of coherent states in pseudodifferential analysis.

Let $\mathcal{H}$ be a Hilbert space, let Ω be a (σ-finite) measure space, and let (ϕ_ω) be a weakly measurable family of vectors in $\mathcal{H}$ parametrized by $\omega \in \Omega$: we shall say that (ϕ_ω) is a family of coherent states in H if there exists a constant $C > 0$ such that, for every $u \in \mathcal{H}$, the inequality

$$(1) \qquad\qquad C^{-1}\|u\|^2 \leq \int_\Omega |(u, \phi_\omega)|^2 \, d\omega \leq C\|u\|^2$$

holds. For instance, if π is a square-integrable unitary representation of some Lie group G in $\mathcal{H}$, the family $(\phi_\gamma)_{\gamma \in G}$, with $\phi_\gamma = \pi(\gamma)\phi$ and an arbitrary $\phi \in \mathcal{H}$, will do, as Schur's Lemma shows: whether π is square-integrable or not, the idea to represent $u \in \mathcal{H}$ by the function $\gamma \mapsto (u, \pi(\gamma)\phi)$ occurs time and again, for instance, in Gelfand, Graev, Pyatetskii-Shapiro [10]. As a special example, choosing any $\phi \in \mathcal{S}(\mathbb{R}^n)$ with $\|\phi\|_{L^2} = 1$ and, using the (projective) Heisenberg representation of $\mathbb{R}^n \times \mathbb{R}^n$ in $L^2(\mathbb{R}^n)$, defining

$$(2) \qquad\qquad \phi_{x+i\xi}(t) = \phi(t - x) \exp(2i\pi < t - \frac{x}{2}, \xi >)$$

for $z = x + i\xi \in \mathbb{R}^n \times \mathbb{R}^n$, one has $\|u\|^2 = \int |(u, \phi_z)|^2 \, dx \, d\xi$ for all $u \in L^2(\mathbb{R}^n)$. Things do not work in the same way when π is not square-integrable, since the integral $\int |(u, \phi_\gamma)|^2 \, d\gamma$ then diverges, which calls for replacing G by some subset (or, in an equivalent way, a certain homogeneous space G/H by some subset in the case when ϕ is H-invariant under π). Often, this is done quite easily, giving up some group structure but not all, still preserving a sufficient amount of linear structure so as to integrate out, using the usual (commutative) Plancherel formula, half the variables: of course, it then yields only rather special families of coherent states, but it provides some useful examples, in particular in the representations that show up in special relativity, and has been tackled with, in recent years, by physicists.

On the other hand, we wish to recall here an important instance in which a non-trivial construction arose: also, what is important in our view is not only to *define* coherent states, but also to show how they can be used in pseudodifferential analysis, or in partial differential equations. Let us refer to Perelomov's book[14] for a popular survey on coherent states in general.

Back to $\mathcal{H} = L^2(\mathbb{R}^n)$, let us assume that one has given a sequence (ω_j), where each ω_j is a pair $(w_j, \| \ \|_j)$, w_j a point in $\mathbb{R}^{2n}$ and $\| \ \|_j$ a norm on $\mathbb{R}^{2n}$, the image of the canonical norm under some symplectic transformation: writing $\|T\|_j^2 = (\theta_j^{-1}T, T)$ for $T \in \mathbb{R}^{2n}$, with θ_j a symplectic positive-definite symmetric matrix, one may characterize θ_j, written in block-form as $\theta_j = \begin{pmatrix} A & B \\ B' & C \end{pmatrix}$, by its upper row (A, B); it satisfies the property that $A^{-1}(I - iB)$ is symmetric and $A > 0$ (cf. [18], Definition 4.3). One may then associate with θ_j the non-standard Gaussian function $\psi_{\theta_j} \in L^2(\mathbb{R}^n)$, defined as

$$(3) \qquad \psi_{\theta_j}(t) = (\det A)^{-1/4}\, 2^{n/4}\, e^{-\pi \langle A^{-1}(I-iB)t, t \rangle}.$$

Actually, considering the harmonic oscillator L_{θ_j} with Weyl symbol $T \mapsto \pi \|T\|_j^2 = \pi(\theta_j^{-1}T, T)$, ψ_{θ_j} is just the normalized ground state of L_{θ_j}, a whole orthonormal basis of eigenfunctions being $(\psi_{\theta_j}^\alpha)_{\alpha \in \mathbb{N}^n}$, a set of Hermite functions (everything is standard if $\theta_j = I$, otherwise what we get is the set of transforms, under some metaplectic transformation, of the collection of standard Hermite functions). Using the Heisenberg translations again, let us set, if $w_j = (x_j, \xi_j) \in \mathbb{R}^n \times \mathbb{R}^n$,

$$(4) \qquad \phi_{\omega_j}^\alpha(t) = \psi_{\theta_j}^\alpha(t - x_j) \exp\left(2i\pi \left\langle t - \frac{x_j}{2}, \xi_j \right\rangle\right).$$

We now want to give conditions relative to the sequence $(\omega_j) = (w_j, \theta_j)$ under which, for some non-negative integer ν, the family $(\phi_{\omega_j}^\alpha)_{|\alpha| \leq \nu}$ will become a family of coherent states in $L^2(\mathbb{R}^n)$.

THEOREM 2.1. *Set* $\|T\|_{j,k}^2 = 2 \inf\{\|T_1\|_j^2 + \|T_2\|_k^2 : T_1 + T_2 = T\}$ *for every* $T \in \mathbb{R}^{2n}$*: call* $\det_{j,k}$ *the discriminant of this quadratic form, which is* ≤ 1 *since the norms* $\| \ \|_j$ *are symplectic. Assume that the following four properties hold:*

(i) *for N sufficiently large, the kernel*

$$(j, k) \mapsto \det_{j,k}^{1/4}(1 + \|w_j - w_k\|_{j,k}^2)^{-N}$$

is the kernel of a bounded operator on $l^2 = L^2(\mathbb{N})$*;*

(ii) *for some $C_1 > 0$ and for all $T \in \mathbb{R}^{2n}$, there exists j with $\|T - w_j\|_j \leq C_1$;*

(iii) *for some $C_2 > C_1$, there exists an integer $m > 0$ such that, for every j, the number of k with $\|w_j - w_k\|_{j,k} \leq 2^{1/2}C_2$ is $\leq m$;*

(iv) *for some $C_3 > 0$, and the same C_2 as before, all j, k and all $T \in \mathbb{R}^{2n}$, the condition $\|w_j - w_k\|_{j,k} \leq 2^{1/2}C_2$ implies $\|T\|_j \leq C_3\|T\|_k$.*

Then, *for some $\nu \in \mathbb{N}$, the family $(\phi^\alpha_{\omega_j})_{|\alpha| \leq \nu}$ is a family of coherent states on $L^2(\mathbb{R}^n)$.*

PROOF. This was actually proved in [18] under slightly more general assumptions (a general measure space Ω rather than a sequence (ω_j) was used there). However, it requires some thumbing through the paper to reassemble the data, for which we shall now help the reader. Choose a non-negative $f \in C^\infty(\bar{\mathbb{R}}^+)$ with support in $[0, C_2^2]$, equal to 1 on $[0, C_1^2]$ and set

$$(5) \qquad a_j(T) = \frac{f(\|T - w_j\|_j^2)}{\sum_k f(\|T - w_k\|_k^2)}$$

for all j. Then the family (ω_j) together with the partition of unity $1 = \sum a_j$ satisfies the assumptions of Definition 8.1 in [18]. Indeed, according to Lemma 7.1 and Definition 5.2 there, what has to be found is, for all n, a bound not depending on j for the L^2-norm of any derivative $(\nabla_{X_1} \cdots \nabla_{X_r} a_j)(T)$ $(r \leq N)$ where all vectors X_s satisfy $\|X_s\|_j \leq 1$: now, if T lies in the support of the function g_j, $g_j(T) = f(\|T - w_j\|_j^2)$, it can only lie in the support of g_k if $\|w_j - w_k\|_{j,k} \leq 2^{1/2}C_2$. Thus Theorem 8.1 in [18] includes our present Theorem 2.1 as a special case.

We should mention that a complete proof of Theorem 8.1 depends on pseudodifferential analysis in an essential way (and on some possibly unexpected application of Hadamard's three-circle theorem, as the proof relies also on that of Theorem 4.1 in [17]); in the other way round, families of coherent states can help analyze pseudodifferential operators. Indeed, using the simplest family (2), one can characterize operators A with Weyl symbols f having bounded derivatives of all given orders as those operators which, for every given N, satisfy for some $C > 0$ the estimate

$$(6) \qquad |(A\phi_z, \phi_{z'})| \leq C(1 + \|z - z'\|^2)^{-N}$$

for all $z, z' \in \mathbb{C}^n \times \mathbb{C}^n$. This was shown in [17], section 1, to permit a very short treatment of classical pseudodifferential analysis on $\mathbb{R}^n$. More important, this type of characterization has turned out to be valid (with suitable families of coherent states) in a whole array of symbolic calculi, the *Fuchs calculi*. These are pseudodifferential analyses on symmetric cones Λ, defined in some appropriate way, taking into account both the intrinsic Riemannian (symmetric space) structure on Λ and its embedding into some space $\mathbb{R}^n$: the case when Λ is the solid light-cone was developed in [20]; before that, the half-line case had been treated[19], and the general case has been treated in some joint work, due to appear, with H. Upmeier.

Going back to the Weyl calculus of pseudodifferential operators, different treatments were given in 1979 by Hörmander [11] and by the author [17], of which the second one fits with our present subject. If one goes back to the situation described in Theorem 2.1, one may define a symbol f to be *of weight* 1 if it is C^∞ and satisfies the following estimates: given $T \in \mathbb{R}^{2n}$, any j with $\|T - w_j\|_j \leq C_1$ and any set $(X_1, \ldots, X_r)$ of vectors in $\mathbb{R}^{2n}$ with $\|X_s\|_j \leq 1$ for $s = 1, \ldots, r$, one should have $|(\nabla_{X_1} \cdots \nabla_{X_r} f)(T)| \leq C$ for C depending only on r (by (iv), this does not depend on the choice of j). From the results in [17] and in [18] it follows that (under the assumptions of Theorem 2.1) any symbol of weight 1 gives rise to a bounded operator $A = Op(f)$ on $L^2(\mathbb{R}^n)$, satisfying the estimates

$$(7) \qquad |(A\phi_{\omega_j}^\alpha, \phi_{\omega_k}^\beta)| \le C\det_{j,k}^{1/4}(1 + \|w_j - w_k\|_{j,k}^2)^{-N}$$

for all α, β with $|\alpha| \le \nu$, $|\beta| \le \nu$, and all N, with a constant C depending on N and ν but not on j, k. The converse was proved to be true by F. Bruyant [4], in a slightly different situation and under some extra assumption: the situation would be that in which some symplectic norm $\| \quad \|_T$ is assigned to each point $T \in \mathbb{R}^{2n}$ in a smooth way, and the extra assumption is that one should have, for some pair C_1, N_1,

$$(8) \qquad C_1^{-1}(1 + d(T_1, T_2))^{1/N_1} \le 1 + \|T_1 - T_2\|_{T_1, T_2} \le C_1(1 + d(T_1, T_2))^{N_1}$$

where d stands for the geodesic distance associated with the given Riemannian structure.

A weaker form of Theorem 2.1 appeared in 1978 in [16]; independently, Cordoba and Fefferman [6] gave a result in the same direction: however, their result (connected to the specific Riemannian structure on $\mathbb{R}^{2n}$ associated with classical symbol theory, i.e., the so-denoted $S_{1,0}$-class) involved some error term (of the order of magnitude of $\|u\|_{-1/2}^2$) on the left-hand side of (1).

Thus families of coherent states and pseudodifferential analysis should benefit from each other. On the other hand, in the case when

$$(9) \qquad \|u\|^2 = \int_\Omega |(u, \phi_\omega)|^2 \, d\omega$$

for some family (ϕ_ω) of coherent states on $\mathcal{H}$, one might advocate the formula

$$(10) \qquad (Au, v) = \int f(\omega)(u, \phi_\omega)(\phi_\omega, v) \, d\omega$$

as defining a quantization map from functions on Ω to operators on $\mathcal{H}$. Of obvious significance is the case when one has $\pi(\gamma)\phi_\omega = \phi_{\gamma \cdot \omega}$ for some representation π of a group G in $\mathcal{H}$ together with some action of G on Ω: (10) was introduced by Wick in some special case and by Berezin ([1], [2]) in general. However, experience shows that one never gets anything near a decent pseudodifferential analysis in this way, which is due to the fact that operators A defined like this are much too nice in order for the correspondence to be invertible: this does not mean that these calculi do not bring any interesting information, especially from the point of view of harmonic analysis, only that one should turn elsewhere for genuine new pseudodifferential analyses.

3. The chronogeometric oscillator

On $L^2(\mathbb{R}^n)$, the harmonic oscillator $L = \pi \sum (x_j^2 + D_j^2)$, with $D_j = (1/2i\pi) \cdot (\partial/\partial x_j)$, lies in the enveloping algebra of the Heisenberg representation (of which the operators x_j, D_j, $1 \le j \le n$ are a set of infinitesimal generators), but is itself one of the infinitesimal generators of the metaplectic representation. Operators that lie among the infinitesimal generators of some explicit representation hold few mysteries, which is why we shall turn to generalizing the first point of view.

Consider a free spinless particle of mass 1, of positive frequency: by this we mean a quantum object characterized by some wave fuction $\tilde{u}(t, \mathbf{x})$ on $\mathbb{R} \times \mathbb{R}^n$, satisfying the Klein-Gordon equation $\Box \tilde{u} = -4\pi^2 c^2 \tilde{u}$ (in which we have dispensed with Planck's constant) with $\Box = c^{-2}(\partial/\partial t^2) - \Delta$, together with the condition relative to the support of the Fourier transform of $\tilde{u}$: all this may be summed up by the equation

$$(11) \qquad (2i\pi)^{-1}(\partial \tilde{u}/\partial t) = c^2 \left(1 - \frac{\Delta}{4\pi^2 c^2}\right)^{1/2} \tilde{u}$$

in which the square root is taken in the spectral-theoretic sense. It is possible, through (11), to identify $\tilde{u}$ to its restriction u to $\mathbb{R}^n$: then, it turns out that the Hilbert space $H_c^{1/2}(\mathbb{R}^n)$, characterized by

$$(12) \qquad \|u\|^2 = \int |\hat{u}(\mathbf{p})|^2 \left(1 + \frac{\mathbf{p}^2}{c^2}\right)^{1/2} d\mathbf{p} \ < \infty,$$

where $\hat{u}(\mathbf{p}) = \int e^{-2i\pi\langle x, \mathbf{p}\rangle} dx$, is acted upon unitarily as follows. Consider the so-called orthochronous Poincaré group $\mathcal{P}$, which is the group of affine transformations of the space-time $\mathbb{R}^{n+1}$ generated by arbitrary translations together with Lorentz transformations (i.e., those linear transformations preserving the quadratic form $c^2 dt^2 - |dx|^2$ *that do not reverse time*: letting $\mathcal{P}$ act on $\tilde{u}$ in the obvious way (by composition), one gets the *Bargmann-Wigner representation*, which we shall presently describe in its $H_c^{1/2}(\mathbb{R}^n)$-realization. With $\langle D \rangle := (1 - (\Delta/4\pi^2 c^2))^{1/2}$, it is easier, and just what is needed for our purpose, to describe the infinitesimal operators of the representation: those corresponding respectively to time-translation, space-translations, rotations and the so-called relativistic boosts are

$$(13) \qquad \langle D \rangle, \quad D_j, \quad R_{jk} = x_j D_k - x_k D_j \quad \text{and} \quad B_j = x_j \langle D \rangle.$$

Observe that when $c \to \infty$, $\langle D \rangle$ reduces to 1, and the representation collapses, infinitesimally or globally, to the Heisenberg representation with rotation operators added (physicists would call this a representation of the Galilean group): this is an example of group contraction, another process quite frequent in quantization theory.

Does there exist a symbolic calculus of operators on $H_c^{1/2}(\mathbb{R}^n)$, covariant under the Bargmann-Wigner representation and some action of $\mathcal{P}$ on some phase space? Yes, moreover it is well-behaved as a pseudodifferential analysis: again, characterizations of classes of operators in terms of their action on sets of coherent states can be given (cf. our monograph [21]), but our present aim is not to describe it. What we are interested in is the operator L with

$$(14) \qquad \pi^{-1} L = \sum (B_j^2 + D_j^2) - c^{-2} \sum_{j<k} R_{jk}^2$$

which may be considered as a relativistic analogue of the harmonic oscillator. When made fully explicit, it is the case $h = 1$ of the operator L^n defined by
(15)

$$-4\pi L^n = h^2 \sum \frac{\partial^2}{\partial x_j^2} - 4\pi^2 |x|^2 + c^{-2} \left[h^2 \left(\sum x_j \frac{\partial}{\partial x_j}\right)^2 + (n-1)h \sum x_j \frac{\partial}{\partial x_j} \right].$$

This operator has some remarkable properties, some of which we now want to describe: to prevent misunderstanding, let us note that even though h accompanies $\partial/\partial x_j$ in the same way as a Planck's constant, its role is somewhat subtler than that of a rescaling constant. One may note that, with $\mu = (n-1)/(2h) - n/2$, L^n commutes with the operator $(1 + |x|^2/c^2)^{-\mu} \mathcal{F}_h$ where

$$(16) \qquad (\mathcal{F}_h u)(x) = h^{-n/2} \int e^{-(2i\pi/h)\langle x,y\rangle} u(y)\, dy.$$

Also, it can be proved that L^n, initially defined on $\mathcal{S}(\mathbb{R}^n)$, is essentially self-adjoint on the space $L^2(\mathbb{R}^n; (1 + |x|^2/c^2)^\mu\, dx)$.

Under the correspondence

$$(17) \qquad f(x) = u\left(\frac{|x|^2}{c^2}\right) = u(s),$$

the radial part of L^n reduces to

$$(18) \qquad L^n_{rad} = \frac{h^2}{\pi c^2} L(\lambda, \mu, \nu)$$

with $\lambda = (n-2)/2$, $\nu = \pi c^2/h$ and μ as above, where $L(\lambda, \mu, \nu)$ is the ordinary differential operator on $(0, \infty)$ defined as

$$(19) \qquad L(\lambda, \mu, \nu) = -s(1+s)\frac{d^2}{ds^2} - [\lambda + 1 + (\lambda + \mu + 2)s]\frac{d}{ds} + \nu^2 s.$$

Actually, the whole operator L^n, not only its radial part, can be solved in terms of the operators $L(\lambda, \mu, \nu)$: indeed, if P_m is any harmonic polynomial on $\mathbb{R}^n$, homogeneous of degree m, and if we set

$$(20) \qquad f(x) = P_m\left(\frac{x}{c}\right) v\left(\frac{|x|^2}{c^2}\right),$$

one may check that f will satisfy the eigenvalue equation

$$(21) \qquad L^n f = \frac{h^2}{\pi c^2} \rho f$$

provided that (with μ still as above)

$$(22) \qquad L\left(\frac{n-2}{2} + m, \mu, \pi c^2\right) v = \left[\rho + \frac{m}{4}(m + n + 2\mu)\right] v.$$

In this way, a complete set of eigenstates of L^n can be described in terms of eigenstates of the operators $L(\lambda, \mu, \nu)$.

In [22] we made a careful study of the equation $L(\lambda, \mu, \nu)u = \rho u$ (with $\nu > 0$, $\lambda > -1$ and $\mu > -1$), which we called the *chronogeometric equation*: observe that it has just one term more (the term $\nu^2 su$) than the general hypergeometric equation, which turns the point at infinity into a non-regular (non-Fuchs type) singular point. For this reason, we found it quite exciting that the solutions of this equation, though not in any explicit (series or integral) form, should satisfy several identities, of which we shall only quote one: in this identity, one has $\mu = -1/2$, i.e. $h = 1$, but other identities in [22] involve other values of μ as well.

PROPOSITION 3.1. *Assume* $\lambda > -1/2$. *Let* u *be an eigenfunction of the operator* $L(\lambda, -1/2, \nu)$, *analytic on* $[0, +\infty[$ *and such that* $u(0) = 1$. *Set* $\psi(s) = u(s^2 - 1)$, $s \geq 1$. *For all* $x \geq 0$, $y \geq 0$, *one has*

$$\psi(\cosh x)\,\psi(\cosh y) = \frac{\Gamma(\lambda+1)}{\pi^{1/2}\Gamma(\lambda+1/2)} \int_0^\pi \frac{\Gamma(\lambda+1/2)}{\nu^{\lambda-1/2}} j_{\lambda-1/2,\nu}(\sinh x \, \sinh y \, \sin\theta)$$

$$\cdot \, \psi(\cosh x \, \cosh y + \sinh x \, \sinh y \, \cos\theta)$$

$$\cdot \, \sin^{2\lambda}\theta \, d\theta,$$

where $j_{\lambda-1/2,\nu}$ *is linked to the Bessel function by*

$$j_{\lambda-1/2,\nu}(t) = t^{-\lambda+1/2}\, J_{\lambda-1/2}(2\nu t).$$

PROOF. This is Theorem 5.1 in [22].

REMARK. When $\nu = 0$, one has

$$\frac{\Gamma(\lambda+1/2)}{\nu^{\lambda-1/2}}\, j_{\lambda-1/2,\nu}(t) = 1$$

for all t, so that the identity reduces to an identity known for Gegenbauer functions: if, moreover, $\lambda = (n-2)/2$, this is none other than the usual spherical function identity for the symmetric space $SO_0(1,n)/SO(n)$.

In the case when $n = 1$ and $h = 1$, L transforms, under the change of variable $x = c\sinh t$, into

(23) $$L = (-4\pi c^2)^{-1}\left[\frac{d^2}{dt^2} - 4\pi^2 c^4 \sinh^2 t\right],$$

the so-called *modified Mathieu operator*. Using the Klein-Gordon symbolic calculus of operators on $H_c^{1/2}(\mathbb{R})$ alluded to in the beginning of this section, one may associate operators $Op_{KG}(f)$ to symbols $f = f(x, \mathbf{p})$. In particular, the symbol of L is the function $\pi r - 1/(16\pi c^2)$ with $r = x^2 + \mathbf{p}^2 + c^{-2}x^2\mathbf{p}^2$. It now turns out that operators which are functions of L in the spectral-theoretic sense are just those whose Klein-Gordon symbols are functions of r (when $c = \infty$, this reduces to a well-known fact connecting the harmonic oscillator to the Weyl calculus). This is what led to our study of the chronogeometric equation. In particular, let $(\phi_k)_{k\geq 0}$ be a complete sequence of eigenfunctions of L corresponding to an increasing sequence of eigenvalues, and let $\Phi_k(x^2 + \mathbf{p}^2 + c^{-2}x^2\mathbf{p}^2)$ be the Klein-Gordon symbol of the rank-one operator $u \mapsto (u, \phi_k)\,\phi_k$ (where the scalar product is taken in $H_c^{1/2}(\mathbb{R})$): with $(\mathcal{G}\phi_k)(x) = (1 + x^2/c^2)^{1/2}\hat{\phi}_k(x)$ (by what was said right before (16), $\mathcal{G}\phi_k$ is actually a constant times ϕ_k), one then has

(24) $$\Phi_k(r) = (-1)^k \left(-\frac{1}{\pi}\frac{d}{dr}\right)^{1/2}[(\mathcal{G}\phi_k)(r^{1/2})]^2.$$

This (non-trivial) identity yields interesting identities for chronogeometric functions, since, as shown in [22], Theorem 3.4, ϕ_{2k} (resp. ϕ_{2k+1}) can be expressed in terms of an eigenfunction of $L(-1/2, -1/2, \nu)$ (resp. $L(1/2, -1/2, \nu)$), whereas Φ_k can be expressed in terms of an eigenfunction of $L(0, 0, 2\nu)$.

Thus, the operator L^n can be regarded as a relativistic generalization of the harmonic oscillator (which is the $c = \infty$ case): its spectral decomposition calls for the introduction of eigenfunctions of an interesting new ordinary differential operator; at least in the one-dimensional case, the analogy with the non-relativistic case extends to the stability of the class of chronogeometric functions under the taking of *Wigner functions*, i.e. symbols of rank-one operators associated with given elements of $\mathcal{H} = H_c^{1/2}(\mathbb{R})$. Only this demands that a genuinely relativistic calculus (the Klein-Gordon calculus) should be used at the same time.

It seems to us an important question (whose solution, however, will require considerable work) whether it is possible to associate in some natural way symbolic calculi (i.e. quantization processes) to species of elementary particles, characterized by free field equations: what is clear at present (cf. [21], section 3) is that if one applies, starting from the (free) Schrödinger equation rather than the Klein-Gordon equation, the procedure that led to the Klein-Gordon calculus, it is the Weyl calculus that one gets as a result.

4. Algebras of symbols; contractions

Every decent symbolic calculus Op gives rise to a composition formula, i.e. a bilinear map # that expresses a product $Op(f)Op(g)$ as $Op(f\#g)$ (this requires of course that operators of the type $Op(f)$, for some classes of symbols f, should constitute algebras: this is seldom a trivial fact and it is not true, for instance, in the Berezin calculus). In the Weyl calculus, the composition formula can be described as follows: with $T' = (q', p') \in \mathbb{R}^n \times \mathbb{R}^n$ and $T'' = (q'', p'')$, set

$$
(25) \qquad i\pi L = (4i\pi)^{-1} \sum \left[-\frac{\partial^2}{\partial q_j' \, \partial p_j''} + \frac{\partial^2}{\partial q_j'' \, \partial p_j'} \right].
$$

Then, say for $f, g \in \mathcal{S}(\mathbb{R}^{2n})$, one has

$$
(26) \qquad (f\#g)(T) = [e^{i\pi L}(f(T+T')g(T+T''))] \quad (T' = T'' = 0).
$$

This identity can be expressed in a closed integral form; one can also, expanding $e^{i\pi L}$ as a series (or, if one prefers, $e^{i\pi hL}$ as a power series in the formal constant h), get the usual asymptotic expansion of $f\#g$ as a (non-convergent) sum of products of derivatives of f and g.

It was quite natural that one should, at first, believe (cf. [3]) that a similar rule would hold for other quantization methods as well, some appropriate parameter taking the role of h: actually, in the Fuchs calculi or in the Klein-Gordon calculus, expansions of such a kind do work (in a non-formal sense: but it requires much work to prove). We want to show, on two examples taken from quantizations associated with the group $G = SL(2, \mathbb{R})$, that phenomena of a much more varied nature may occur.

Recall (cf. e.g. Knapp [12]) that, given any integer $k \geq 2$, one can define the unitary representation π_k taken from the holomorphic discrete series of G as follows: call $\mathcal{D}_k^+$ the space of all functions f holomorphic on the Poincaré upper half-plane satisfying the square-integrability condition

$$
(27) \qquad \|f\|_k^2 := \int_\Pi |f(z)|^2 (\operatorname{Im} z)^k \, d\mu(z) < \infty
$$

with $d\mu(z) = (\operatorname{Im} z)^{-2}\, d(\operatorname{Re} z)\, d(\operatorname{Im} z)$; then, for $\gamma = \begin{pmatrix} a & b \\ c & d \end{pmatrix} \in G$, set

$$(28) \qquad (\pi_k(\gamma)f)(z) = (-cz + a)^{-k} f\left(\frac{dz - b}{-cz + a}\right).$$

Actually, allowing a projective representation only (this means that one is satisfied with $\pi(\gamma)\pi(\gamma') = \mu\pi(\gamma\gamma')$ with some μ, $|\mu| = 1$ depending on $\gamma, ,\gamma'$), one may extend the definition (28) above to that of π_λ with λ real > 0: only for $\lambda > 1$ does definition (27) extend, and for $\lambda = 1$ one gets the Hardy space. To permit a comparison with [23], let us note that our present λ would have been denoted $\lambda - 1$ there. Since Π is a symmetric space, one may consider, for every $z \in \Pi$, the geodesic symmetry S_z around z, which may be regarded as an element of $PSL(2,\mathbb{R})$: one can then associate with z, through (28), the unitary transform σ_z of $\mathcal{D}_\lambda^+$ defined by $\sigma_z = \mu\pi_\lambda(S_z)$, where the phase factor μ ($|\mu| = 1$) is the unique one that turns σ_z into a positive symmetric operator. In [23] and some preceding paper, (both joint work with J. Unterberger) we called *passive symbol* of a trace-class operator A on $\mathcal{D}_\lambda^+$ the function $\operatorname{Symb}(A)$ on Π defined as $(\operatorname{Symb}(A))(z) = 2\operatorname{Tr}(A\sigma_z)$: of course, it is just as well to define the map Symb from operators to symbols as to define a map Op in the reverse direction (which can, and has to be done anyway [23]). Let us turn to a description of the composition rule $\#$ for passive symbols.

At $i \in \Pi$, let (r, θ) be the pair of (geodesic) normal coordinates (i.e. z lies on the half-line, in the sense of hyperbolic geometry, obtained by turning the half-line from i to $i\infty$ by the angle θ, and $r = d(i,z)$): this defines a chart $\Psi : \mathbb{R}^2 \to \Pi$ through $\Psi(q,p) = z$ with

$$(29) \qquad q = \sinh r \cos\theta, \qquad p = \sinh r \sin\theta;$$

also, a chart $\Psi_z : \mathbb{R}^2 \to \Pi$ can be associated with any $z = x + iy \in \Pi$, setting

$$(30) \qquad \Psi_z(q,p) = x + y\,\Psi(q,p).$$

The following theorem was then proved in [23], théorème 5.1.

> Let $f, g \in C_0^\infty(\Pi)$ and assume that λ is large enough. Then

$$(31) \qquad (f\#g)(z) = [E(-i\pi L)((f \circ \Psi_z) \otimes (g \circ \Psi_z))] \quad (q' = q'' = p' = p'' = 0),$$

where $i\pi L$ is just the case $n = 1$ of (25), and the function $E(\zeta)$ is the extension, for $\zeta \notin -\mathbb{R}^+$, of the holomorphic function defined for $\operatorname{Re}\zeta > 0$ by

$$(32) \qquad E(\zeta) = 4\pi \int_0^\infty J_{\lambda-1}(4\pi t) e^{-t^{-1}\zeta}\, dt.$$

Observe the striking similarity of (31) with (26): only the charts $\Psi_T(T') = T + T'$ used in (26) have to be replaced by charts taking the hyperbolic geometry into account. But E is not the exponential function any more and it can be seen to have an essential singularity at $\lambda = \infty$: also, whether λ is an integer or not, $E(\zeta)$ is many-valued around $\zeta = 0$. Thus $f\#g$ cannot be fully described by any asymptotic expansion as a formal power series in λ^{-1}: neither does, for fixed λ, *any* asymptotic expansion of $f\#g$ as a sum of products of derivatives of f and g appear as meaningful.

The misconception that such an asymptotic expansion should be possible, at least in some formal sense, must of course be traced to an ill-founded assumed similarity with the Weyl case: however, the phenomenon alluded to can be related, in this case, to the contraction of Heisenberg's relation $[p, q] = h/2i\pi$, as h goes to zero, towards the relation defining a commutative algebra. Now, as λ goes to ∞, both the representation π_λ and the associated calculus just described contract to the Fuchs calculus on a half-line ([19], section 10), another model of which (only in the one-dimensional case!) can be seen to be the Klein-Gordon calculus together with the Bargmann-Wigner representation: the latter one can be made to depend on c in some explicit way, so that, letting c go to ∞, one gets a new contraction process, this time towards the Heisenberg representation together with the Weyl calculus. Summing up, we see that a *tower of three* contractions, not just *one* contraction, makes the (infinitesimal or global) representation π_λ collapse completely: each step may then be followed up through a corresponding contraction of the associated symbolic calculi.

We now switch to another situation in which we shall actually find a *convergent series* for $f \# g$! The discrepancy between the two situations is of the same nature as that between continuous and discrete spectra: in our opinion, spectral theory, much more than the asymptotic point of view, has a role to play in quantization theory. This time, we take $\mathcal{H} = L^2(\mathbb{R})$ together with the representation $\pi_{i\lambda}$ ($\lambda \in \mathbb{R}$) of $G = SL(2, \mathbb{R})$ given by ([12])

$$(33) \qquad (\pi_{i\lambda}(\gamma)u)(s) = |-cs + a|^{-1-i\lambda} \, u\left(\frac{ds - b}{-cs + a}\right).$$

Since $\pi_{i\lambda}$ is taken from the principal series, it is natural that the correct phase space should be the one-sheeted hyperboloid in the coadjoint representation of G, another model of which is the space Π_i, the complementary of the diagonal in $(\mathbb{R} \cup \{\infty\}) \times (\mathbb{R} \cup \{\infty\})$: Π_i is acted upon by G under the rule

$$(34) \qquad \gamma.(s, t) = \left(\frac{as + b}{cs + d}, \frac{at + b}{ct + d}\right);$$

it has an invariant measure $d\mu_i(s, t) = (s - t)^{-2} \, ds \, dt$ and admits an invariant differential operator

$$(35) \qquad \Box = (s - t)^2 \frac{\partial^2}{\partial s \, \partial t}.$$

Assuming $\lambda \neq 0$, let θ be the unitary transform of $L^2(\mathbb{R})$ characterized by

$$(36) \qquad \mathcal{F}(\theta u)(x) = |x|^{-i\lambda} \, \hat{u}(x).$$

Imitating this time the *standard*, not the Weyl, calculus of pseudodifferential operators, one is led to associating with every function $f \in L^2(\Pi_i)$ the Hilbert-Schmidt operator $Op(f)$ characterized by

$$(37) \qquad (Op(f)u)(s) = c_{-\lambda} \int f(s, t)|s - t|^{-1-i\lambda}(\theta u)(t) \, dt,$$

where $c_{-\lambda}$ is defined so that the inverse Fourier transform of the function $x \mapsto |x|^{-i\lambda}$ coincides in the complementary of zero with the function $s \mapsto c_\lambda |s|^{-1+i\lambda}$. Molchanov [13] was the first to introduce it as a generalization of the Berezin calculus; it was rediscovered as a generalization of the standard calculus in [24], where the calculus was pushed further: in particular, there is a global formula for $f \# g$ in this calculus, namely

$$(38) \qquad (f \# g)(s,t) = c_\lambda \, c_{-\lambda} \int f(s,x) \, g(y,t) |[s,y,x,t]|^{-1-i\lambda} \, d\mu_i(x,y)$$

where $[s,y,x,t]$ stands for the cross-ratio of the four given points, which seems to be of limited value.

However (cf. forthcoming paper by J. Unterberger and the present author), much more can be obtained if one takes into account the decomposition of $L^2(\Pi_i)$ under the quasi-regular action of G or, what amounts to the same, the spectral decomposition of $\square$. As shown by Strichartz [15] or Faraut [8], the spectrum of $\square$ has a continuous part $[1/4, \infty[$ together with a discrete one consisting of all numbers $-n(n+1)$, $n \in \mathbb{N}$. The discretely embedded subspace E_n corresponding to such an eigenvalue can be decomposed into irreducible terms under the quasi-regular action of G in $L^2(\Pi_i)$ (i.e. the one defined by $\pi(\gamma)f = f \circ \gamma^{-1}$) as a sum $E_n = E_n^+ \oplus E_n^-$: we shall now describe E_n^+, since it is for $f, g \in E^+$ (or E^-), the Hilbert sum of the E_n^+'s (or E_n^-'s) that we can give the formula for $f \# g$ which we have in view.

Given $n \in \mathbb{N}$ and $z \in \Pi$, set

$$(39) \qquad g_z^{n+1}(s,t) = \left[\frac{s-t}{(s-\bar{z})(t-\bar{z})} \right]^{n+1} :$$

then $g_z^{n+1} \in L^2(\Pi_i)$, and E_n^+ is by definition the closed subspace of $L^2(\Pi_i)$ generated by the functions g_z^{n+1}, $z \in \Pi$. Setting $\alpha_n = 2^{-2n} \binom{2n}{n}$ and

$$(40) \qquad (T_n f)(z) = \alpha_n^{-1} \int_{\Pi_i} f(s,t) \, \bar{g}_z^{n+1}(s,t) \, d\mu_i(s,t)$$

for every $f \in L^2(\Pi_i)$ and $z \in \Pi$, one can see, more precisely, that the operator $((2n+1)\alpha_n/4\pi)^{1/2} \, T_n$ is an isometry from E_n^+ onto the space $\mathcal{D}_{2n+2}^+$ defined in (27): it acts as an intertwining operator between the quasi-regular action of G in E_n^+ and the representation π_{2n+2} (defined in (28)) of G in $\mathcal{D}_{2n+2}^+$. In this way, one may use T_n in a systematic way to realize E_n^+ as a space of holomorphic functions on Π.

Next, consider the basis $\{\varepsilon_0, \varepsilon_1, \varepsilon_2\}$ of the Lie algebra of G given by

$$(41) \qquad \varepsilon_0 = \begin{pmatrix} 0 & \frac{1}{2} \\ -\frac{1}{2} & 0 \end{pmatrix}, \quad \varepsilon_1 = \begin{pmatrix} 0 & -\frac{1}{2} \\ -\frac{1}{2} & 0 \end{pmatrix}, \quad \varepsilon_2 = \begin{pmatrix} \frac{1}{2} & 0 \\ 0 & -\frac{1}{2} \end{pmatrix}.$$

Under $\pi_{i\lambda}$, the associated infinitesimal operators are certain self-adjoint realizations of the formal differential operators

$$(42) \qquad \begin{aligned} e_0 &= (-4i\pi)^{-1} \left[(s^2+1)\frac{d}{ds} + (i\lambda+1)s \right], \\ e_1 &= (-4i\pi)^{-1} \left[(s^2-1)\frac{d}{ds} + (i\lambda+1)s \right], \\ e_2 &= (2i\pi)^{-1} \left[s\frac{d}{ds} + \frac{i\lambda+1}{2} \right]. \end{aligned}$$

For each $\zeta = (\zeta_0, \zeta_1, \zeta_2) \in \mathbb{C}^3$ such that $\zeta_0^2 - \zeta_1^2 - \zeta_2^2 = 0$, $\zeta_0 + \zeta_1 = 1$ and Im $\zeta_2 > 0$, one can see that the operator $L_\zeta = \sum \zeta_j e_j$ is invertible as an operator in the space of C^∞ vectors of the representation $\pi_{i\lambda}$.

The following fact does not have any analogue in pseudodifferential analysis on $\mathbb{R}^n$: *the linear space consisting of all operators $Op(f)$, $f \in E^+$, is exactly the closure, in the Hilbert-Schmidt norm, of the algebra generated by the inverses of the operators L_ζ just defined.*

In particular, given $f \in E_m^+$ and $g \in E_n^+$, one can express their composition as a convergent series (in $E^+ \subset L^2(\Pi_i)$)

$$(43) \qquad f \# g = \sum_{j \geq 0} h_j, \qquad h_j \in E_{m+n+1+j}^+.$$

Transferring all symbols in this formula to functions on Π as explained above, one can make the various terms explicit as

$$(44) \qquad T_{m+n+1+j} h_j = \Phi_j(m, n, \lambda) F_j^{2m+2, 2n+2}(T_m f, T_n g)$$

where, given two holomorphic functions ϕ and ψ on the Poincaré half-plane, $F_j^{k_1, k_2}(\phi, \psi)$ is defined as

$$(45) \quad F_j^{k_1, k_2}(\phi, \psi) = \sum_{l=0}^{j} (-1)^l \binom{k_1 + j - 1}{l} \binom{k_2 + j - 1}{j - l} \left(\frac{\partial}{\partial z}\right)^{j-l} \phi \cdot \left(\frac{\partial}{\partial z}\right)^l \psi$$

and where the coefficients are given by

$$\Phi_j(m, n, \lambda) = 2^{2j} \frac{(2m+1)!(2n+1)!(m+n+1+j)!}{m!(2n+j+1)!(2m+2n+2+2j)!} \sum_{l=0}^{j} (-1)^{j-l} \binom{j}{l}$$

$$\cdot \frac{(m+l)!(2m+2n+2+j+l)!}{(2m+l+1)!(m+n+1+l)!} \frac{\Gamma(m+l+1-i\lambda)\Gamma(n+1-i\lambda)}{\Gamma(m+n+l+2-i\lambda)\Gamma(-i\lambda)}.$$

The bilinear expression (45) is just the same as the one introduced (without reference to any symbolic calculus or composition law) by Cohen in [5]: it plays a role in the theory of modular forms (cf. also Zagier [25]). Indeed Cohen showed that if ϕ and ψ are (weakly) modular forms of weights k_1 k_2, i.e. if the identities $\pi_{k_1}(\gamma)\phi = \phi$ and $\pi_{k_2}(\gamma)\psi = \psi$ hold for every γ belonging to some arithmetic subgroup Γ of $SL(2, \mathbb{R})$, then, for all $j \in \mathbb{N}$, $F_j^{k_1, k_2}(\phi, \psi)$ is a modular form of weight $k_1 + k_2 + 2j$ with respect to Γ. In the present context, this property appears as a consequence of the *covariance* of the symbolic calculus (i.e. the property that $\pi_{i\lambda}(\gamma) Op(f) \pi_{i\lambda}(\gamma^{-1}) = Op(f \circ \gamma^{-1})$ for every symbol f and every $\gamma \in G$), in view of the fact, mentioned above, that T_m intertwines π_{2m+2} with the quasi-regular action of G in E_m^+.

References

[1]   F. A. Berezin, *Quantization*, Math.USSR Izvestija **38** (1974), 1109–1165.

[2]   F. A. Berezin, *Quantization in complex symmetric spaces*, Math. USSR Izvestija **39** (1975), 341–379.

[3]   F. A. Berezin, *A connection between the co-and contravariant symbols of operators on classical complex symmetric spaces*, Soviet Math.Dokl. **19** (1978), 786–789.

[4]   F. Bruyant, *Estimations pour la composition d'un grand nombre d'opérateurs pseudodifférentiels et applications*, Thèse de 3ème cycle, Univ. of Reims (1979).

[5]   H. Cohen, *Sums involving the Values at negative Integers of L-Functions of Quadratic Characters*, Math.Ann. **217** (1975), 271–295.

[6]   A. Cordoba and C. Fefferman, *Wave packets and Fourier integral operators*, Comm. Part. Diff. Equ. **3** ((1978), 979–1006.

[7]   Yu. Egorov, *On canonical transformations of pseudodifferential operators*, Uspehi Mat. Nauk. **25** (1969), 235–236.

[8]   J. Faraut, *Distributions sphériques sur les espaces hyperboliques*, J. Math. Pures et Appl. **58** (1979), 369–444.

[9]   B. V. Fedosov, *A simple geometrical construction of deformation quantization*, J. Diff. Geom. (to appear).

[10] I. M. Gelfand, M. I. Graev, and I. I. Pyatetskii-Shapiro,, *Representation Theory and Automorphic Functions*, W. B. Saunders Comp., Philadelphia, 1969.

[11] L. Hörmander, *The Weyl calculus of pseudodifferential operators*, Comm. Pure Appl. Math. **23:3** (1979), 359–443.

[12] A. W. Knapp, *Representation theory of semi-simple Lie groups*, Princeton Univ. Press, Princeton, 1986.

[13] V. F. Molchanov, *Quantization on the imaginary Lobachevskii plane*, Funksional'nyi Analiz, Ego Prilozheniya **14** (1980), 73–74.

[14] A. Perelomov, *Generalized coherent states and their applications*, Springer Verlag, 1986.

[15] R. S. Strichartz, *Harmonic analysis on hyperboloids*, J. Funct. Anal **12** (1973), 341–383.

[16] A. Unterberger, *Décompositions spectrales approchées associées à une famille d'oscillateurs harmoniques*, C. R. Acad. Sci Paris **287** (1978), 783–785.

[17] A. Unterberger, , *Oscillateur harmonique et opérateurs pseudodifférentiels*, Ann. Inst. Fourier **29** (1979), 201–221.

[18] A. Unterberger, *Les opérateurs métadifférentiels*, Lecture Notes in Physics **126** (1980), 205–241.

[19] A. Unterberger, *The calculus of pseudodifferential operators of Fuchs type*, Comm. Part. Diff. Equ. **9** (1984), 1179–1236.

[20] A. Unterberger, *Analyse harmonique et analyse pseudodifférentielle du cône de lumière*, Astérisque #156, Soc. Math. de France, Paris (1987).

[21] A. Unterberger, *Quantification relativiste*, Mém. Soc. Math. de France #44–45, Paris (1991).

[22] A. Unterberger, *Relativity, spherical functions and the hypergeometric equation*, Ann. Inst. Henri Poincaré **62:2** (1995), 103–144.

[23] A. Unterberger and J. Unterberger, *Quantification et analyse pseudodifférentielle*, Ann. Ecole Norm. Sup. **21** (1988), 133–158.

[24] A. Unterberger and J. Unterberger, *Representations of $SL(2, \mathbb{R})$ and symbolic calculi*, Int. Equ. Oper. Theory **18** (1994), 303–334.

[25] D. Zagier, *Introduction to modular forms*, From Number Theory to Physics, Springer-Verlag, 1992, pp. 238–291.

DEPARTMENT OF MATHEMATICS, CNRS URA 1870, UNIVERSITY OF REIMS, BP 1039, 51687 REIMS CEDEX 2, FRANCE

E-mail address: `andre.unterberger@univ-reims.fr`

Selected Titles in This Series